Natural UV Radiation in Enhancing Survival Value and Quality of Plants

Swati Sen Mandi

Natural UV Radiation in Enhancing Survival Value and Quality of Plants

Springer

Swati Sen Mandi
Division of Plant Biology
Bose Institute
Kolkata, West Bengal
India

ISBN 978-81-322-3830-0 ISBN 978-81-322-2767-0 (eBook)
DOI 10.1007/978-81-322-2767-0

Softcover reprint of the hardcover 1st edition 2017

Printed on acid-free paper

This Springer imprint is published by Springer Nature
The registered company is Springer (India) Pvt. Ltd.
The registered company address is: 7th Floor, Vijaya Building, 17 Barakhamba Road, New Delhi 110 001, India

This book is dedicated to scientists and researchers with a mind to explore the effect of UV radiation as a beneficial environmental factor allowing survival of life through the ages of enhanced UV fluence on earth in post-ozone hole era.

Preface

Among the various environmental factors, one that constitutes the most damaging component, adversely affecting life forms on earth, is UV radiation – the adversity increasing as a consequence of enhanced UV fluence through the depleted stratospheric ozone layer that became evident as "ozone hole" in the Antarctic region during the later part of the last century. Realization that UV fluence on earth was gradually increasing (as a consequence of stratospheric ozone thinning) raised an alarm and stimulated urgent research on effect of UV on life forms particularly on plants that, despite being exposed maximally to the brunt of UV radiation due to their stationary habit, still fulfill the role of primary producers that directly or indirectly support sustenance of life on earth.

It is thus important to develop understanding on UV as a component of earth's atmosphere and its effect on cell molecular events particularly operative in plants, for conferring UV acclimation. While early studies (conducted in greenhouse/growth chambers) report on (exaggerated) damaging effects of artificial UV lamps (in absence of interaction with other natural spectral components of solar radiation), it has been established that realistic assessment of the effect of UV is possible only from studies in open air. Such studies establish that plants, possessing unique feature of cellular plasticity, are capable of developing UV acclimation process including production of personalized internal UV screen composed of compounds, viz., flavonoids, that are specifically synthesized in plants under weather-related variation in UV fluence. Thus, while UV at high intensity inflicts macromolecular damage, low-dose UV (resultant of passage through cloud cover) induces signal transduction-based epigenetic (temporal UV adaptive)/genetic (persistent UV adapted) traits through upregulation in synthesis of UV-screening compounds. Such cell protection strategy functions prophylactically by restricting UV entry. UV acclimation may also be affected retaliatorily by DNA repair affected by upregulation of DNA repair enzymes for protection from whatever UV radiation escapes internal UV screening and enters into internal tissue.

Massive thinning of the UV-screening ozone layer has brought into focus issues of enhanced "greenhouse gases" that, formed under enhancement of destructive anthropogenic activity, has given rise to a 2-pronged destructive effect, viz., "global warming" in the troposphere and concomitant cooling effect in the stratosphere, that enhances ozone-depleting conditions.

All such interrelated disruptive forces threaten to disrupt ecological balance thereby threatening to jeopardize sustenance of life on earth. This has drawn the attention of both scientists and common people. The gravity of the issue calls for undertaking appropriate measures, by all nations under the leadership of the United Nations, to reduce stratospheric ozone destruction by managing/restricting emission of "greenhouse gases" into the atmosphere (delineated in the Kyoto Protocol/Montreal Protocol). Although regulations of the Montreal Protocol have been effective in reducing ozone loss to an appreciable extent, staggering ozone destructive effects of some greenhouse gases, viz., N2O, and some species of CFCs – that exhibit long life in the atmosphere – still raise concern.

The most commonly studied UV-screening compounds that, due to their structural specialization, exhibit dual role are flavonoids, being specifically synthesized in plants, functioning as the key player in development of UV acclimation strategy, manifested as morphological/biochemical trait variation at the epigenetic/genetic level. In addition to providing UV adaptive traits in plants, flavonoids also bring about important value addition when used as dietary component and also as herbal medicine for humans/animals.

The present book titled *Natural UV Radiation in Enhancing Survival Value and Quality of Plants* focuses on UV-induced damaging effect on life forms and the countering cell molecular mechanisms (prophylactic/retaliatory) in plants. This book is intended to provide a comprehensive understanding on the prospects of survival of life forms under enhancing UV fluence in the post-ozone hole era. Experimental data used here to develop an understanding and related concept development in this regard include reports in the literature as well as case studies conducted in the laboratory of the author. Technology-based development of commercial deliverables from these studies through use of breeding/biotechnological strategies has also been discussed.

Kolkata, India Swati Sen Mandi

Acknowledgment

Preparation of this manuscript has also been an extensive task. In connection with the preparation of this manuscript, I have received untiring help from my present research group. I wish to acknowledge my heartfelt appreciation to them:

1. Dr. Subhabrata Ghosh who has provided untiring help and support to me for surfing the Internet in the processes of my knowledge development and the preparation of the manuscript together with figures and tables
2. Dr. Waikhom Sayanika Devi who has helped me in preparing the bibliography as well as looked into the minor details in preparation of the manuscript

The author expresses heartfelt gratitude to Dr. Kalyan Kumar Mandi, former principal, Scottish Church College, Kolkata, who has been particularly helpful in concept development through thought sharing in the context of writing this book.

Grateful acknowledgement is also due to Dr. E. A. Siddiq, honorary professor, Institute of Biotechnology, Acharya N. G. Ranga Agricultural University, Rajendra nagar, Hyderabad, for his inspiring support in writing this book.

Contents

Abbreviations

2,4-D	2,4-Dichlorophenoxyacetic acid
4CL	4-Coumarate CoA ligase
6-4PP	6-4 Pyrimidine-pyrimidone
8-OH-G	8-Hydroxy guanosine
ACCase	Acetyl CoA carboxylase
ADH	Alcohol dehydrogenase
APX	Ascorbate peroxidase
ATM	Ataxia telangiectasia-mutated
ATP	Adenosine triphosphate
ATR	ATM-Rad3-related
BER	Base excision repair
BLAST	Basic local alignment search tool
bya	Billion years ago
CAB	Chlorophyll a/b binding protein
CBCRs	Cyanobacteriochromes
CDK	Cyclin-dependent kinase
CDM	Clean Development Mechanism
CFC	Chlorofluorocarbon
CHI	Chalcone isomerase
ChIP	Chromatin immunoprecipitation
Chla	Chlorophyll a
Chlb	Chlorophyll b
CHS	Chalcone synthase
COP1	Constitutively photomorphogenic 1
CPD	Cyclobutane pyrimidine dimer
CRY	Cryptochromes
DFR	Dihydrofolate reductase
DGDG	Digalactosyldiacylglycerol
DNA	Deoxyribonucleic acid
DSB	Double-strand break
DU	Dobson units
dUTP	Deoxyuridine triphosphate
DXR	Deoxyxylulose phosphate reductoisomerase
ELIP1 and ELIP2	Early light-inducible protein 1 and 2
FAD	Flavin adenine dinucleotide
FADH	Reduced flavin adenine dinucleotide

fl-cDNAs	Full-length complementary deoxyribonucleic acid
FS	Flavonol synthase
GHG	Greenhouse gas
GTP	Guanosine triphosphate
HaCat	Human keratinocytes
HCA	Hydroxycinnamic acid
HCFC	Hydrochlorofluorocarbonss
HPLC	High-performance liquid chromatography
HPLC-MS/MS	HPLC coupled with tandem mass spectrometry
HPLC-ED	HPLC coupled with electrochemical detection
HY5	Elongated hypocotyl 5
IAA	Indole-3-acetic acid
IFS	Isoflavone synthase
IR	Infrared
MAA	Mycosporine-like amino acids
MAPK	Mitogen-activated protein kinase
MEP	Methylerythritol 4-phosphate
MGDG	Monogalactosyldiacylglycerol
MiRNA	Micro ribonucleic acid
MMR	Mismatch repair
MP	Montreal Protocol
MRNA	Messenger ribonucleic acid
MSRA	Methionine-*S*-sulfoxide reductase
MSRE	Methylation-sensitive restriction enzyme
MTP	Mitochondrial transmembrane potential
mya	Million years ago
MYB12 and MYB111	MYB domain protein 12 and 111
NADPH	Nicotinamide adenine dinucleotide phosphate
NDPI	Normalized Difference Pigment Reflectance Index
NER	Nucleotide excision repair
NO	Nitric oxide
ODS	Ozone-depleting substances
OECD	Organisation for Economic Co-operation and Development
PAL	Phenylalanine ammonia lyase
PAR	Photosynthetically active radiation
PCD	Programmed cell death
PHOT	Phototropins
PHY	Phytochromes
POD	Peroxidase
PR1	Pathogenesis-related protein
PSC	Polar stratospheric clouds
PSI	Photosystem I
PSII	Photosystem II
QRI	Quality Response Index
RAPD	Random amplified polymorphic DNA
RCC	Regulator of chromatin condensation

RE	Restriction enzyme
ROS	Reactive oxygen species
Rubisco	Ribulose-1,5-bisphosphate carboxylase
RUP1 and RUP2	Repressor of UV-B photomorphogenesis 1 and 2
SAR	Structure–activity relationship
SC	Stratum corneum
SLW	Specific leaf weight
SNP	Single nucleotide polymorphism
SOD	Superoxide dismutase
SSB	Single-strand break
TD	Tibial dyschondroplasia
TLC	Thin-layer chromatography
TOMS	Total Ozone Mapping Spectrometer
tt4	Transparent testa-4
UNFCCC or FCCC	United Nations Framework Convention on Climate Change
USEPA	United States Environmental Protection Agency
UV-A	Ultraviolet A
UV-B	Ultraviolet B
UV-C	Ultraviolet C
UVI	Ultraviolet Index
UVR	Ultraviolet radiation
UVr	UV resistant
UVR3	UV repair defective 3
UVR8	UV resistance locus 8
UVs	UV sensitive
WHO	World Health Organization
WL	White light
WT	Wild type
XP	Xeroderma pigmentosum

About the Author

Prof. Swati Sen Mandi, basically an experimental biologist specialized in plant molecular biology, since her PhD at Cambridge University, UK, has been devoting all her research career to understand how a plant system responds to abiotic stresses of all kinds. From studies on cell molecular events on (rice) seeds stored under ambient storage in tropical regions and also on tea leaf quality at high/low altitude, she has made commendable progress in the understanding on the positive aspects of UV radiation particularly with respect to DNA repair and formation of secondary metabolites towards developing an acclimation strategy. She has presented experimental data to demonstrate higher accumulation of flavonoids related to antioxidant and aroma potential in plants at high altitudes, viz. in Darjeeling tea (leaf). Her interest in addressing problems of societal relevance through research on medicinal plants collected from high/low altitude (growing under high/low UV fluence) has been useful in her continuing efforts for setting up a centre for research on medicinal plants of Eastern and North-Eastern India. Significantly, her excellent publications in reputed scientific journals and her experience as visiting scientist in different internationally renowned laboratories together with the international (Kenya) patent right she earned for her UV radiation-related work on tea and the national (India) patent right she earned for development of DNA marker-assisted selection of rice varieties with high seed vigour and viability traits bear ample testimony to her scientific standing. Her findings in the recent years on plant cell molecular mechanisms that constitute acclimation strategy-associated trait development in plants have been documented in this textbook entitled *Natural UV Radiation in Enhancing Survival Value and Quality of Plants*.

1 Natural Ultraviolet Radiation

1.1 Introduction

Rays from the sun consist of a wide range of electromagnetic waves of varying wavelength emitted as visible light of 390–700 nm to the extent of 40 %, invisible radiation viz. Ultraviolet (UV) radiation of 100–390 nm to the extent of 10 % and infrared (IR) radiation of 700 nm–1 mm to the extent of 50 %. The invisible radiations viz. UV and IR flank the visible light radiation on the two sides of solar spectrum (Fig. 1.1).

1.1.1 Ultraviolet Radiation

The fraction of ultraviolet radiation (UVR) of sun's rays comprises three wavebands, viz., UV-A (315–390 nm), UV-B (280–315 nm), and UV-C (100–280 nm), each with energy of impact corresponding to the respective wavelengths. Entry of these UV wavebands into the earth's atmosphere is differentially restricted by an "ozone layer" that spans the lower region of the stratosphere that constitutes one of the atmospheric layers enveloping the earth (Fig. 1.2).

Due to such restricted penetration of UV radiation into earth's atmosphere, through the stratospheric ozone layer, UV exhibits limited presence (only 0.5 % of total solar UV). Of the UV radiation penetrating in earth's atmosphere, UV-A comprises 90 % and UV-B comprises 5–10 %; UV-C is totally absent in the earth's atmosphere. Under such condition, a state of equilibrium was established through evolutionary development of UV acclimation strategy in life forms, particularly plants that have to endure UV exposure due to stationary lifestyle. Conditions in subsequent years, (under conditions of enhancing ozone depletion and associated enhancement in UV fluence) became a cause for enhanced cell molecular damages that threaten persistence of life forms on earth (Bornman et al. 2015). This has stimulated urgent research toward understanding UV damage and damage countering mechanism in life forms; such studies have also revealed the simultaneous development of quality traits as a manifestation of upregulation of UV protective compounds in plants.

UV-C radiation (100–280 nm) being low in wavelength is of very high energy of impact. Fortunately, this wavelength is totally restricted from entering into the earth's atmosphere through the stratospheric ozone layer (Fig. 1.2). UV-C thus is not naturally present in the earth's atmosphere.

UV-B radiation (280–315 nm) with its high energy of impact (corresponding to its short wavelength), despite its low penetration through the stratospheric ozone layer, constitutes the most damaging fraction of UV radiation present in the earth's atmosphere. Under such environ-

S. Sen Mandi, *Natural UV Radiation in Enhancing Survival Value and Quality of Plants*,
DOI 10.1007/978-81-322-2767-0_1

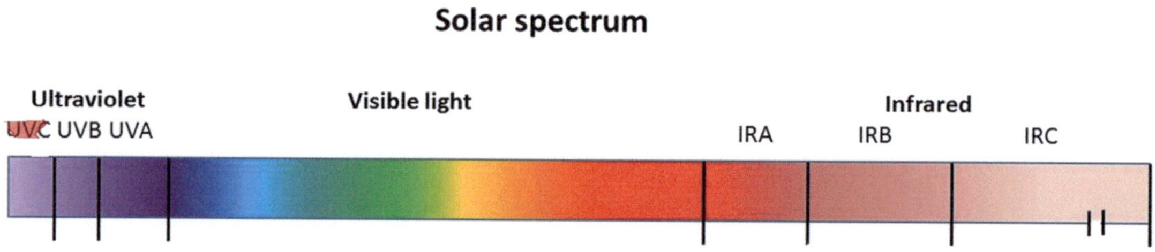

Fig. 1.1 Electromagnetic radiations of solar spectrum (Taken from web)

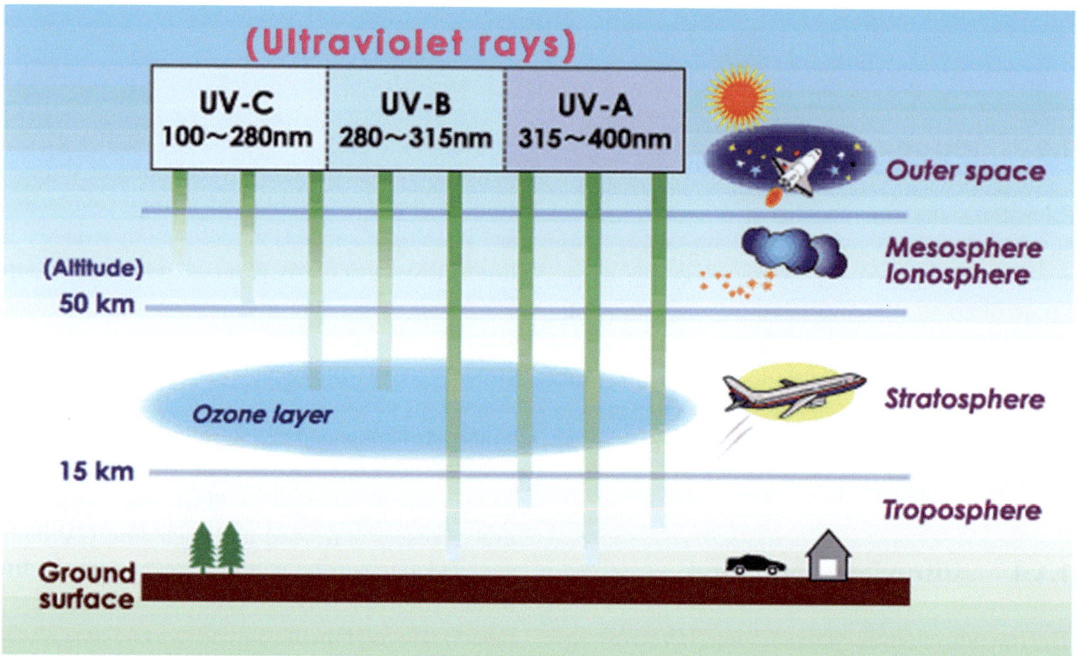

Fig. 1.2 Extent of permeability of different UV rays (UV-A, UV-B, and UB-C) through stratospheric ozone layer (Taken from web)

ment, plants with stationary lifestyle in open field are maximally subject to UV-induced damage; the redeeming factor in this context is the genetically determined UV protective cell molecular mechanisms particularly operative in plants.

UV-A radiation (315–390 nm) represents a radiation of much lower energy of impact (corresponding to its wavelength). This radiation, though earlier considered as innocuous, is now reported to exhibit cell molecular damaging effect comparable to UV-B – the two radiation types functioning via different mode of action.

1.1.2 Visible Radiation

The visible light fractions comprising of 40 % of the sun's total electromagnetic radiation consist of a spectrum of seven different colors, viz., violet, indigo, blue, green, yellow, orange, and red, with wavelengths ranging from 390 to 700 nm (1 = 1 nm or 1 millionth of a millimeter).

The energy of visible light in sunray is harvested by pigments, viz., chlorophylls, that are capable of harvesting solar radiation by the process of photosynthesis, converting light energy to chemical energy in biochemical compounds. During photo synthesis of these primary and secondary metabolites, plants fix atmospheric CO_2 concomitantly releasing O_2 to be utilized for respiration of all living forms on earth. Plants thus serve as primary producers for life support on earth and have necessarily evolved survival strategies (see Chap. 4) for persistence under ambient UV environment.

1.1.3 Infrared Radiation

At one end of the visible spectrum, beyond red light, sun's radiation presents infrared (IR) radiation that spans from 700 to 1 mm (Liew 2006); this comprises about half (50 %) of the total solar energy fluencing on earth. This range of long wavelength corresponding to a frequency range of approximately 430 THz down to 300 GHz has low energy of impact and thus does not cause harmful effect on life forms. IR radiation is conceptually divided into three types on the basis of wavelength:

Infrared-A: 700–1400 nm
Infrared-B: 1400–3000 nm
Infrared-C: 3000 nm–1 mm

Infrared radiation with low energy and thus incapable of inflicting damaging effects provides heating effect that early in the earth's evolutionary history served to warm up the then supercool earth's atmosphere, thereby facilitating establishment of life-supporting temperature on earth. Ironically, IR radiation that once contributed to the establishment of life supporting temperature in the earth's atmosphere, in recent years, being trapped by enhanced level of greenhouse gases (due to anthropogenic activity) threatens an unprecedented rise of temperature in the troposphere i.e. in earth's atmosphere. This contributes to global warming that adversely affects earth's atmosphere, thereby threatening persistence of life on earth. While causing heating effect in the troposphere, where it adversely affects persistence of life forms, this heating effect on the other hand simultaneously brings about another life destructive situation, i.e., cooling in the stratosphere, thereby facilitating depletion of stratopheric ozone (Ramanathan and Feng 2009) and associated enhanced UV fluence on earth.

1.2 Atmospheric Envelopes Surrounding the Earth

Sunrays, with associated UV radiation, have to pass through five layers of atmosphere surrounding the earth. These are troposphere, stratosphere, mesosphere, thermosphere, and exosphere. These layers vary in thickness, composition, and temperature (Fig. 1.3).

Of these, the two layers nearest to the earth's surface, viz., troposphere and stratosphere, directly influence atmospheric conditions related to life support on earth.

1.2.1 Troposphere

Troposphere, constituting the first layer overlying the earth's surface, extends up to about 9 km (30,000 ft) at the poles and 17 km (56,000 ft) at the equator – thickness of the layer exhibiting some variation due to weather fluctuation in this layer on earth. Since this layer gets heated due to transfer of heat from the earth's surface, the lower part of troposphere is warmer than the upper part, the temperature decreasing with distance away from the earth's surface. Such temperature variation along the height of this layer causes vertical mixing of air (hence the origin of its name in the Greek τρόπος means turn or overturn). The troposphere contains 80 % of the mass of the atmosphere, being composed primarily of nitrogen, (78 %) oxygen (21 %), Carbon dioxide (0.05 %) and only small concentrations of other trace gases.

Fig. 1.3 Layers of the atmosphere with respective characteristics (Taken from web)

Troposphere is the lowermost atmospheric layer over the earth's surface and constitutes earth's weather forming region. Oxygen released by plants in the process of photosynthesis get accumulated in this layer which is consumed for respiration by all life forms. Another potent gaseous accumulation in the troposphere (that has raised international concern since the later part of the last century) constitute greenhouse gases that refer collectively to excessive CO_2, CH_4 (relating to plant growth and agricultural practices), CFCs (relating to industrial emission) as well as some other harmful gases (Sect. 1.7.1) – all originating from adverse anthropogenic activities. The greenhouse gases (see Sect. 1.9) accumulated in the lower region of the troposphere form a "blanket" that restricts heat of fluencing sunshine (including UV and a fraction of IR radiation) from disseminating, thereby simulating greenhouse effect over the earth's surface. As a consequence, increase in green house gases over the past several decades owing to industrial/technological practices has led to temperature rise on earth's surface i.e. lower region of the troposphere i.e. the first envelope on earth's surface.

1.2.2 Stratosphere

Stratosphere is located just above troposphere at a distance of 9–18 km from the earth's surface. The stratosphere is stratified in temperature, with warmer layers higher and cooler layers closer to the earth. The increase of temperature with altitude is a result of the absorption of the sun's ultraviolet radiation by ozone. This is in contrast to the troposphere, near the earth's surface, where temperature decreases with altitude (https://en.wikipedia.org/wiki/Stratosphere). This mediates enhanced cooling at the lower portion of the stratosphere (that overlays troposphere) thereby facilitating ozone depletion at this region. The stratosphere is a region of intense interactions among radiative, dynamical, and chemical processes, in which the horizontal mixing of gaseous components proceeds much more rapidly than does vertical mixing. Air is roughly a thousand times thinner at the top of the stratosphere than it is at sea level. Because of this, jet aircraft and weather balloons reach their maximum operational altitudes within the stratosphere. This layer is of prime importance to the living world because of the presence of a well defined ozone layer within it that serves as a UV shield by selectively absorbing specific UV fraction of solar radiation (UV-B partially and UV-C fully) and restricts penetration of these damaging UV radiation while allowing UV-A (fully unrestricted) into the earth's atmosphere.

1.3 Evolutionary History of Life-Supporting Atmosphere on Earth

During the early Precambrian era, in the absence of any UV restricting barrier (McKenzie et al. 2006), flux of solar UV radiation on the surface of earth was several fold higher than it is today. During such times, early life forms availed of protection from the damaging effects of UV radiation by perpetuating under protection of the oceanic water covering (where UV penetration is largely restricted while visible light passes through and photosynthesis continues). Another predicament for life on earth was that around that time was that young earth's atmosphere lacked molecular oxygen, and thus aerobic life forms were unsuitable for perpetuation on earth's surface; lack of oxygen in the atmosphere presented a condition where only anaerobic life forms could exist.

Around 3.5 billion years ago, during the Archean era when diversification of early anaerobes took place and the first anaerobic photosynthetic bacteria appeared. Sequentially about 3 billion years ago, a group of aquatic organisms, viz., the *Cyanobacteria*, pioneers of the process of oxygenic photosynthesis, began harvesting the energy of solar radiation for conversion to chemical energy in photosynthetically formed organic compounds (published in Scientific American April 2008 issue) – the reaction consisting of combining molecules of H_2O and carbon dioxide (CO_2) producing glucose with evolution of molecular oxygen (O_2) as a by-product. This provided oxygen, necessary for developing an oxygenic atmosphere on earth. With the appearance of other photosynthetic organisms, the photosynthetically evolved oxygen began to accumulate in the atmosphere, initiating an ecological disaster for the early existing anaerobic organisms and simultaneously a boom for the then existing and further evolving aerobic organisms. However despite the enhanced oxygen accumulated in the atmosphere, appearance of aerobic organisms on earth's terrestrial habitat was still "awaited" since the earth's atmosphere was still high in UV radiation. With formation of ozone resulting from reaction between molecular oxygen and atomic oxygen under high energy of UV radiation and its accumulation in the lower region of the stratosphere, a protective ozone layer was formed that restricted UV from fluencing on earth. This layer provided a protective screen for restraining UV radiation from penetrating into the earth's atmosphere. An atmospheric condition was thus developed for appreciably supporting terrestrial life forms – until such times (viz., later half of the last century) when the stratospheric ozone layer began to show signs of disruption.

During evolutionary history by the Ordovician era (by when conditions of atmospheric oxygen and stratospheric ozone layer were life supporting), migration of life forms from underwater to the terrestrial habitat took place. This marked a groundbreaking event in the evolutionary history of earth. By the Silurian period (443–416 million years ago) from when vascular plant fossils are found, O_2 producing life forms appear to have largely occupied the land. By then a balance between protective ozone shield and an appropriately evolved biosphere in an oxygenic atmosphere had been established.

The supercool temperature on earth during the early stages of evolution presented yet another predicament for persistence of life forms on earth. Before the origin of life, average surface temperature of the earth's atmosphere was at about −18 °C (0 °F). This temperature was unsuitable for life on earth. The final step toward making the earth suitable for life support on earth was undertaken by IR radiation trapped under a gaseous "blanket" made up of some gases of nonlife origin (since life was not originated at that time) such as water vapor, carbon dioxide, methane, nitrous oxide. This raised the temperature of earth's atmosphere to a life supporting level. The temperature rise recorded was up to about 15 °C to 22 °C which made the earth livable. In recent times however, these gases together with other gases of anthropogenic origin and even enhancement of CO_2 due to present day deforestation added to the temperature rise that is now related to "greenhouse effect" on earth. This rise of temperature on the earth's surface concomitantly cools the lower part of the stratosphere, thereby creating a condition that brings about depletion in the stratospheric ozone layer (see Sects. 1.8 and 1.9).

1.4 Formation of UV-Restraining Ozone Layer

Ozone is formed at the equator when ultraviolet radiation from the sun strikes the stratosphere dissociating (or "splitting") oxygen molecules (O_2) to atomic oxygen (O). Atomic oxygen thus formed combines with other oxygen molecules to form ozone (O_3). Accumulated ozone forms a layer in the lower part of stratosphere providing a protective screen that differentially restrains UV from penetrating into earth's atmosphere. This protective ozone shield restricts entry of the damaging UV fractions (UV-C fully and UV-B partially) while allowing UV-A to pass unrestricted (Fig. 1.2).

Ozone forming reactions are as follows:

$$O_2 + \text{UV radiation} \rightarrow 2O$$
$$O + O_2 \rightarrow O_3$$

This reaction occurs maximally at the tropics since solar UV is highest at the equator with the sun's zenith angle being the lowest when the sun shines overhead. The ozone formed in this region is drifted by air flow and is distributed around the globe accumulating highest at the poles with gradual reduction at mid latitudes; ozone accumulation is thinnest at the equator. The ozone layer, being thickest at the poles with intermediate thickness at mid latitude, with proportional restraint of UV fluencing on earth is expected to provide proportional UV protection over life forms at these regions. The highest ozone accumulation at the polar regions however began to exhibit disruption during the later part of the last century; this was first recorded as ozone thinning/ozone hole in the Antarctic and subsequently in the Arctic region. In the "post-ozone hole era," since detection of ozone thinning first at the poles and subsequently in mid latitude, associated UV irradiance on earth was recorded high particularly at high and mid latitudes (Madronich and de Gruijl 1994) of both the Southern and Northern Hemispheres (Seckmeyer and McKenzie 1992; McKenzie et al. 1993), during solar solstice associated summer in respective hemispheres. Reports of Norgaard et al. (1998) on UV effect on humans, taking erythema as an index (see Chap. 3, Sect. 3.1) appear corroborated with reports on seasonal variation in UV fluence on earth as a function of solar movement X thickness of stratospheric ozone layer. The authors report increase in erythema (sunburning) development due to UV fluence on earth relative to the values of 1970 was found to be increased by about 7 % at Northern Hemisphere mid latitudes in winter/spring, about 4 % at Northern Hemisphere mid latitudes in summer/fall, about 6 % at Southern Hemisphere mid latitudes on a year-round basis, about 13 % in Antarctica in spring, and about 22 % in the Arctic in spring.

In addition to the stratospheric ozone layer, that provides protection to life forms by restricting UV penetration into earth's atmosphere ground level or "bad" ozone occurring on earth's surface (https://www.epa.gov/ozone-pollution) is created by chemical reactions between oxides of nitrogen (NOx) and volatile organic compounds (VOC) in the presence of sunlight; it is not emitted directly into the air.

$$\text{VOCs} + \text{NOx} + \text{Sunlight} = \text{Ozone}$$

Ground level ozone in addition to having adverse effect on respiratory processes of human may also have harmful effects on sensitive vegetation and ecosystems (https://www.epa.gov/ozone-pollution).

The amount of ozone above a point on the earth's surface is measured in Dobson units (DU) (named after Dr. Gordon Dobson who was a researcher at the University of Oxford). In the 1920s, he built the first instrument to measure total ozone from the ground this instrument is now called the Dobson ozone spectropho tometer. The lowest surface ozone value occurring near the tropics is typically ~260 DU; the value is higher at higher latitudes.

1.5 Factors That Determine Fluence of Solar UV Radiation on Earth

In addition to the ozone layer that restricts fluencing of UV, other factors relating to sun's angle, latitude, altitude, weather condition, and other relevant factors also determine impact level of UV fluencing on earth. UV fluencing on earth is usually measured by ground-based instruments such as a spectrophotometer. However, for global coverage, satellite instruments such as the total ozone mapping spectrometer (TOMS) with accurate calibration (providing corrections for clouds and other atmospheric variables) are essential.

1.5.1 Sun Angle (Zenith Angle) and the Stratospheric Ozone Layer

Differences in sun zenith angle can strongly affect the amount of UV radiation received directly from the sun when not affected by intervening environmental factors such as aerosols/cloud (particularly from biomass burning) (Engelsen 2010). With the sun at its zenith, about 10 % of the total solar radiation and almost 30 % of UV-B and UV-A are diffused. For a solar elevation angle of 20°, about 20 % of the total radiation, 70 % of UV-A, and almost 80 % of UV-B are diffused (Moan et al. 2013). At lower latitudes, the sun is higher in the sky, so because of smaller zenith angle of the sun to the earth's surface, UV rays would travel less distance through the atmosphere and, in turn, expose those latitudes to more UV radiation.

1.5.2 Latitude

Level of spontaneous natural UV radiation on earth depends on several factors (Hader et al. 2003) enumerated above, the most prominent and consistent factors of these being latitude and altitude (Lucas et al. 2006). The sun shifts to the Northern/Southern Hemisphere during summer of the respective hemispheres, and life forms in the lower to mid-latitude in each hemisphere suffer dire consequences of UV radiation caused damage during their respective summers (Diffey 1991). Intensity of ozone layer varies strongly with latitude and is highest at the poles. Variation in UV fluence also depends on sun's zenith angle that is associated with variation in latitude; natural UV fluence that is part of solar radiation is thus subject to the annual solar movement. Sunrays being more inclined at the poles in comparison to that at the equator, the intensity of solar radiation decreases with increasing latitude and UV fluence would be expected to show a similar trend. The fluence of UV at different latitudes in pre ozone depletion era is shown in Fig. 1.4.

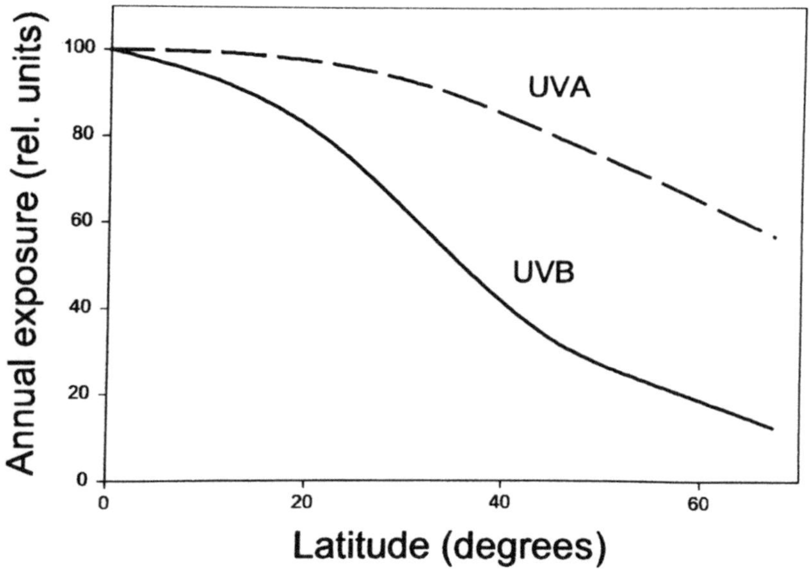

Fig. 1.4 The annual exposure of UV-A at about 360 nm and of UV-B at about 310 nm as functions of the latitude (Taken from an article (Chap. 7) written by Johan Moan – http://www.uio.no/studier/emner/matnat/fys/FYS3610/h04/undervisningsmateriale/Moan7.pdf)

The graph shown in Fig. 1.4 depicts the state of environment prevalent in the pre ozone depletion era, when the ozone layer was naturally thinner in the tropics (allowing higher UV penetration in this region) compared to that in mid and high latitudes. In the tropical region, there is less ozone to absorb the UV radiation as it passes through the earth's enveloping stratospheric region; with increasing latitude, higher ozone accumulation causes lower UV fluence.

In the post ozone depletion era incidence of ozone dents/holes in the polar region and low ozone accumulation at mid latitude resulting from destruction of ozone due to anthropogenic activity has become a common occurrence. Consequently the concept of high ozone concentration at the poles and mid-latitudes does not hold true. Studies in recent times represent a scenario when, as a consequence of depletion of ozone layer during the past several decades, fluence of UV radiation on earth is higher in mid and high latitudes (than earlier recorded). A consequential effect noted is that during the post ozone depletion era UV related cellular damages are greater in the mid and high latitude areas (Madronich and de Gruijl 1994, Rozema et al. 1997). Model calculations incorporating satellite measurements of ozone, aerosol, and cloud, reveal a significant increase in UV-B radiation reaching the earth's surface between 1978 and 2008; this increase occurred at all latitudes except the equatorial zone, with the largest increments taking place at mid- to high latitudes in the Southern Hemisphere (Herman 2010a; Mckenzie et al. 2011) – the Southern Hemisphere being colder than the Northern Hemisphere causes higher ozone depletion in this hemisphere.

Since 1978, there has been a 4–5 % decline in the mid-latitude stratospheric ozone (Herman and Lrko 1994) with associated increase in levels of solar UV-B radiation at the temperate regions (Kerr and McElory 1993; Blumethaler et al. 1994). Ozone layer depletion-related increase in UV fluences involves greater in fluence of UV-B compared to that of UV-A, evidently due to the fact that absorption of UV-B by the ozone layer is appreciably higher compared to that of UV-A; UV-A fluence (being free of absorption by ozone) does not exhibit appreciable correlation with thickness of the ozone layer.

1.5.3 Altitude

UV intensity increases with altitude because there is less atmosphere to absorb the damaging rays at higher altitude. Investigation of the altitude effect revealed a vertical increase of UV fluence approximately 10.7 %/1000 m as a mean value for global and 17.4–19 %/1000 m recorded particularly in the Alps (Blumthaler and Ambach 1990; Blumthaler et al. 1992) with a small general decrease under overcast conditions. In general, an increase in altitude (with increasing air rarefaction) by 1000 m has been recorded to enhance UV irradiance by 6–8 % (Blumthaler et al. 1997). Examples of very high UV exposure have been reported from Cuzco, Peru (13.5°S), and from the South American Andes – in the Southern Hemisphere (in the sparsely populated Atacama Desert in Chile at 4400–5600 m altitude during January, i.e., summer time in the Southern Hemisphere (noon solar zenith angle = 9.5°). High UV fluence has also been reported in Himalayan Mountains – in Northern Hemisphere during July, i.e., summer time in Northern Hemisphere over 100 peaks exceeding 7000 m and also in the Everest peak at 28°N (solar zenith angle = 5.5°) (Herman 2010b). The values presented in these cases were when the summer sun is nearly overhead (including annual cycle of cloud coverage). Such reports indicate the importance of latitude also in altitude related fluence of UV radiation. Variation in UV fluence is further subject to rarefication of air appropriate for the altitudinal level (vide Fig. 1.5 showing UV levels at Assam and Darjeeling, both places being at the same latitude (270 N) but at different altitude (Fig. 1.5). Small solar zenith angles and high elevation may also cause high UV exposure in mountains located even at mid and low latitudes.

1.5.4 Time of Year

The sun's yearly movement determines season; a shift from the equator to 23.5° in each of the Northern and Southern Hemisphere relates to maximum solar radiation (and associated UV fluence) and thus summer in the two hemisphere, i.e., April–June in the Northern Hemisphere and December to February in the Southern Hemisphere, respectively. The solar movement-associated UV fluence exhibits a monthly variation at different altitudes – exhibiting higher UV fluence at high altitude in each hemisphere due to rarefaction of air. UV radiation measured around noon time and given as microwatt/square cm/day in Darjeeling (1300 m a.s.l.) and Assam (130 m a.s.l.), both at 27°N from January to December is shown in Fig. 1.5 (Sect. 4.8 of Chap. 4 and Sect. 6.3.2 of Chap. 6). Reports indicate that intensity of UV radiation over the Qing-Tibetan Plateau region (33°N) is highest from March to September; during this time the amount of UV-B radiation reaching the plateau is higher by 10 % (Wang et al. 2009) than that in the adjacent areas lying at lower altitude (in the same latitude).

1.5.5 Time of Day

When the sun is highest in the sky, i.e., at noon, UV radiation (associated with sun's rays) is highest since the rays have to travel least distance through air that would absorb the radiation. In early morning and late afternoon, since the sun is low in the sky and sun's ray s have to pass through the atmosphere at an angle (thereby covering longer distance), their intensity is greatly reduced. The sun's UV rays are strongest in the 4 h period around local noon when 50–60 % of a summer day UV is received (Diffey 2002). The author reported summer solar spectral irradiance (including UV radiation) in noon on clear days when highest radiation is received; the data includes measurement of UV fluence at all latitudes between tropical and temperate, i.e., 20°–60°, and assumes that solar noon occurs at 1:00 PM.

1.5.6 Water Depth

Dissolved impurities as organic carbon dissolved in water strongly absorb and scatter incident UV radiation reducing intensity of the UV fluence

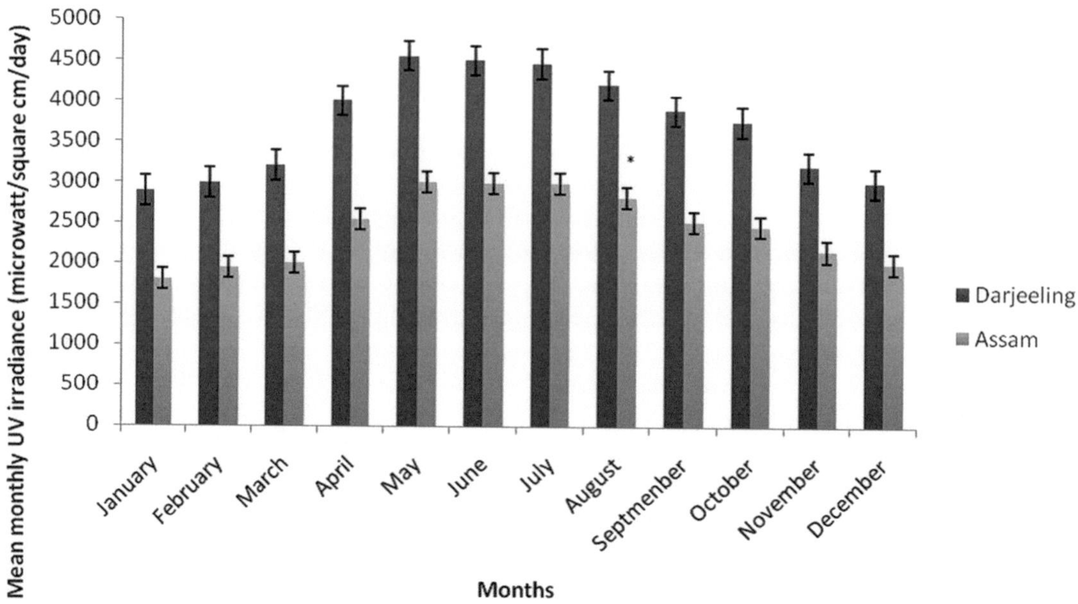

Fig. 1.5 A representation showing UV fluence at Darjeeling at high altitude (1300 m a.s.l) and Assam at low altitude (130 m a.s.l), both at the same latitude, viz., 27°N around noon time. Mean monthly UV irradiance (microwatt/square cm/day) in Darjeeling and Assam (Taken from Shyam Choudhury and Sen-Mandi 2012)

under water; this phenomenon serves to protect aquatic life forms under water. Since different water bodies at different location contain different concentration of dissolved substances and also other particles, effect of UV under water should be evaluated separately for each water body (See Fig. 2.1).

1.5.7 Weather Conditions

UV radiation is hindered by clouds and thus UV fluence varies with thickness of the cloud cover. Frederick and Snell (1990) have shown that clouds can reduce the amount of UV reaching the ground by as much as 90 % on a given day.

1.5.8 Aerosols

Aerosols including dust and smoke, present in the earth's surface, i.e., in the troposphere, absorb UV radiation in addition to scattering the radiation. Although the absorption of UV radiation by aerosols is only partial, in regions of heavy smoke or dust, aerosol particles can absorb more than 50 % of the radiation; aerosols generated from biomass burning are more effective in modifying UV fluence compared to aerosols generated due to automobile exhaust (http://www.ipcc.ch/report/ar5/wg1/). It is of interest to note that while a high latitude ozone loss is expected to increase UV radiation reaching the earth at a region, the effect of enhanced UV fluence could be partially reduced by increase in attenuation by aerosol levels resulting from air pollution (Vogelmann et al. 1992; Tilmes et al. 2012).

1.6 Ultraviolet Index

Evaluation of atmospheric UV radiation is made in terms of an international standard, referred to as ultraviolet index or UV index. The scale ranging from 1 to 10 was developed by Canadian scientists in 1992 and first used in Canada. This UV radiation evaluation was adopted and standardized by the UN's World Health Organiza tion and World Meteorological Organization in

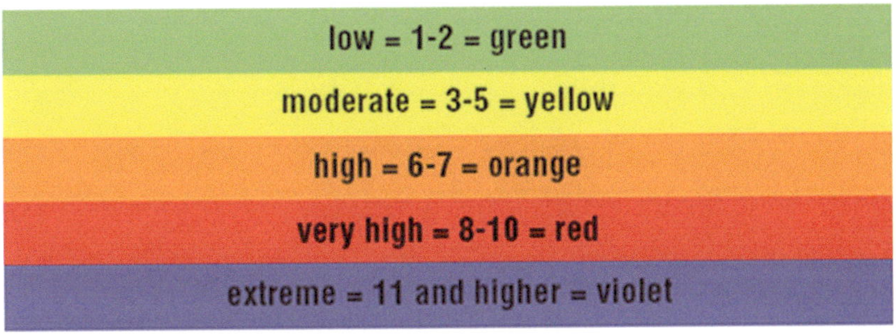

Fig. 1.6 UV index scale (Taken from WHO report 2002)

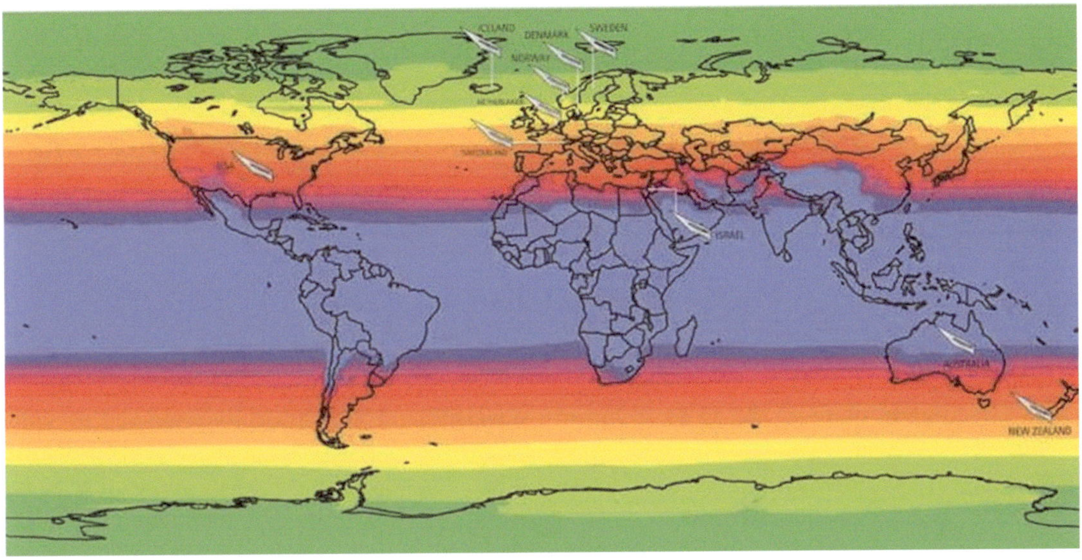

Fig. 1.7 UV fluence on different parts of the world (Taken from web)

1994. According to the WHO recommendations (2002), the UV Index values can be grouped into exposure categories with colour coding (Fig. 1.6). Some notable values of UVI at different places are given below. Throughout the Altiplano region of Peru, Bolivia, Chile, and Argentina, UVI values are much higher, exceeding 20. The highest UVI of 25 was recorded in the grid cell centered on Cuzco, in Southern Peru (13.5°S, 3360 m.s.l.). Total column ozone value was reported to be near 100 Dobson units (DU)) over Antarctica between September and October 1995, the value recorded being much higher compared to the previous ozone amounts in the early 1980s, i.e., near the time when ozone hole was first detected (around 1978 over Antarctica).

UV index is primarily available as daily forecasts for public information and is increasingly available as hourly forecasts as well. The UV power spectrum (strictly expressed in watts per square meter per nanometer of wavelength) is weighted according to a weighting curve known as the erythemal action spectrum. This typically gives a figure of around 250 in midday summer sun and is arbitrarily divided by 25 to generate a convenient index value, which becomes essentially a scale of 0–11 (though ozone depletion is now resulting in higher values). The color-coded UV index scale (Fig. 1.6) gives an indication of the differential UV fluence at different locations on earth (Fig. 1.7) at a given time due to proximity to equator as well as the extended ozone hole.

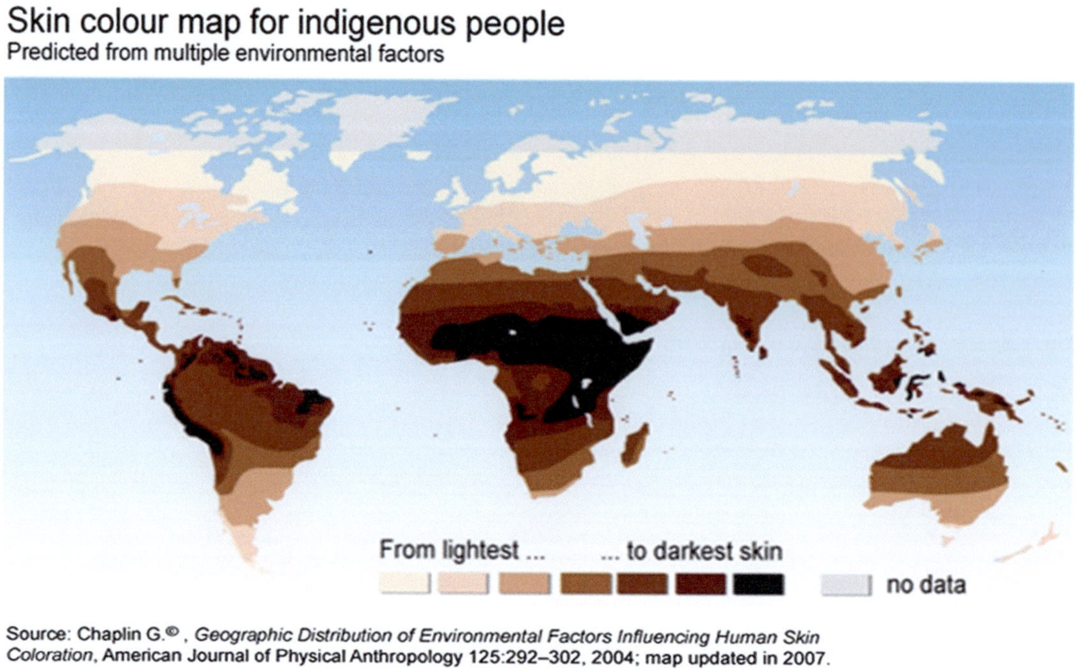

Fig. 1.8 UV index represented as sunburn UV index at different place in the world

The UV index represented as "sunburn UV index" at different place on the globe is shown in Fig. 1.8.

Due to ozone thinning particularly during summer in New Zealand, UVI values regularly exceed 13, even in the south of the country. Mckenzie et al. (2006) showed that peak values of UV index in NZ are about 40 % more than the locations at similar latitudes in North America (Fig. 1.8). This is presumably related to larger ozone depletion in the stratospheric ozone layer in the southern hemisphere.

1.7 Ozone Depletion

Under the restricted entry of UV radiation through the ozone shield into the earth's atmosphere, a state of equilibrium had been established for stable existence of life forms. This state of equilibrium between UV environment and life perpetuation was found to be disrupt when during the 1970s ozone thinning was noted to occur (recorded at the Light Climatic Observatory in Arosa, Switzerland) due to adverse anthropogenic activity-related accumulation of ozone-depleting substances (ODSs) that cause gradual thinning of the protective ozone layer. Subsequently an "ozone hole" was detected in Antarctica (i.e., the coldest region on earth, ideal for ozone breakdown) during the spring of 1978 and subsequently in the Arctic region. Unless recovered, breach in the ozone layer will continue leading to life disruption on earth. In a bid to restore the ozone layer to 1978 level, the United Nations took appropriate measures, embodied in Kyoto Protocol and Montreal Protocol, to reduce release of harmful gases/chemicals by nations of the world (including scope for emission trading – vide Sect. 1.11).

1.7.1 Ozone-Depleting Substances (ODSs)

During the later part of the last century, reports on ozone depletion rang an alarm threatening life to a state of jeopardy. Through subsequent years, temperature and gaseous interactions in the earth's atmosphere due to thoughtless anthropogenic activity-related emission of harmful gases collectively referred to as ozone depleting substances (ODSs) led to a reversion of the process of ozone accumulation in the stratospheric ozone; the ozone depletion thus developed resulted in concomitant enhancement of UV fluence on earth. Breach in the ozone layer, due to presence of ODS in the stratosphere, led to enhanced UV fluence that associated adverse effect on life forms at the troposphere.

Detection of "ozone hole" in the Antarctic region during the later part of the last century initiated widespread concern relating to the increasing UV fluence on earth and associated adverse effect on life forms. This called for urgent studies and publication of reports that established the occurence of a gradual thinning of the stratospheric ozone layer was going on due to careless anthropogenic activity that give out ozone-depleting gases into the atmosphere; the first harmful anthropogenic gas reported in this context was nitrogen oxide, emitted from high altitude aircraft, causing loss of the protective O_3 layer in the stratosphere. Further improved atmospheric model based studies reporting N_2O to be less harmful than initially believed (albeit remaining in the atmosphere for long) discounted the extent of alarm that had been raised. At about the same time, Molina and Rowland (1974) reported chlorofluorocarbons (CFCs) (commonly known by the DuPont brand name of Freon), used mainly as refrigerants and spray propellants, came to be known as more effective as ozone-depleting substance; these compounds undergo photodegradation and release chlorine under solar UV radiation and bring about ozone depletion (Day and Neale 2002).

With growing development of industrialization, a complex interplay of industry-related emission of waste gases, viz., chlorofluorocarbons (CFCs), halons, methyl chloroform, hydrochlorofluorocarbons (HCFC), carbon tetrachloride, methyl bromide, and oxides of nitrogen, all collectively began to accumulate, constituting a group of ozone-depleting substances (ODSs) in the atmosphere. Due to their prolonged persistence in the atmosphere, the CFCs stand out as notoriously effective ozone depleting substance (ODS). Ozone is also destroyed by the free radical catalysts, viz., hydroxyl radical (OH^-), nitric oxide radical (NO^-), chlorine atom (Cl^-), and bromine atom (Br^-) derived from these compounds by energy of sunrays, particularly UV radiation. Ozone is also destroyed when it combines with singlet oxygen atom to form oxygen. Volcanoes also cause tropospheric ozone destruction through volcanic eruption-derived ppb to ppm levels of reactive-free radicals Br, Cl, H, and HO. A series of catalytic cycles involving reactive atoms cause the conversion of ozone molecule (combined with an oxygen atom) to form two molecules of molecular oxygen (O_2), thereby depleting the protective ozone shield in the stratosphere. CFCs are particularly damaging because these cause damage to the atmosphere not only when they are produced, but also because they remain as destructive entities in the atmosphere long after their production. Of these, long-lived molecules like CFC11 and CFC12 having an estimated atmospheric lifetime of 75 and 120 years, respectively. (Ravishankara et al. 1993) contribute to prolonged enhancement of CFC concentration in the stratosphere; these compounds have been reported to be approximately tripled between 1960 and 1985 (Kerr 1988). The CFCs and N_2O are very stable when they are produced in the troposphere; some species of CFCs remain at the ground level participating in global warming, while most of the CFC species are transported to the stratosphere. Some species of chlorofluorocarbons (CFCs) that may find their way to the stratosphere without being destroyed in the troposphere, due to their low reactivity, are represented as the most important ODS under the originally formulated Montreal Protocol.

Once in the stratosphere, the Cl atoms are liberated from the parent compounds, CFCs, by the action of ultraviolet light. Although chlorine is much more abundant in the stratosphere than bromine, bromine atoms are about 60 times more effective than chlorine atoms in chemically destroying ozone in the lower stratosphere. All these chemicals released into the atmosphere gradually lead to conditions conducive to significant ozone loss in the polar regions.

Massive addition of industrial gaseous emission in the troposphere such as N_2O, CFCs (Crutzen 1970) collectively referred to as green house gases added to the natural ground level atmosphere that was already under burden of the then levels of CO_2, CH_4 resulting from massive devegetation and urbanization. During the later part of the last century the expression "Green House Gases" has been extended to include such gases that cause global warming as well as stratospheric ozone depletion. Most of these gases accumulate at the ground level (viz. in the troposphere). Again some species of CFCs drift upwards where they contribute to thinning of the ozone shield in the stratosphere.

1.8 Role of Greenhouse Gas in Global Warming

During the course of evolution the heat energy of incoming sun rays (particularly IR radiation) began to be partially trapped by accumulation of green house gases such as CO_2 and CH_4 that formed a 'blanket' over earth's surface (simulated a closed atmosphere as in a green house) raising the temperature of earth to life supporting level (~ 220 C) while reflecting the remaining radiation (Fig. 1.11) back to space. This helped to create a situation that supported evolution and perpetuation of life on earth until the phenomenon of earth's warming went overboard when adverse anthropogenic activity related emission of harmful industrial gases accumulated over earth's surface. A 'blanket' thus formed by the excessive accumulation of enhanced green house gases that trapped IR radiation resulting a rise of temperature on earth's surface (i.e. in the troposphere – Venkataramanan and Smitha 2011). The extended mass of 'green house gas' globally (Ramanathan and Feng 2009) brought about an unprecedented trend in rise of temperature bringing about global warming and associated climate change that adversely affects present day life forms on earth. Global warming (at the troposphere) brings about a further damaging effect as a reflectory cooling in the adjacent stratosphere – thereby augmenting depletion in the UV protective stratospheric ozone layer. "Green House gases" at the troposphere include a large number of CFC (Chlorofluorocarbon), N_2O, and ofcourse the earlier known CO_2 and CH_4.

Some CFC species are also found at the stratosphere; these are different from the species that remain in the troposphere. The tropospheric CFCs are largely of a nonreactive nature and function as greenhouse gases. When the concentration of the greenhouse gases in the atmosphere increases, the temperature of the earth also increases. This causes more evaporation of water from the earth surface. Water, itself being a greenhouse gas, traps IR radiation, thereby further increasing the temperature. All this leads to global warming – a phenomenon that had a positive effect for life support during early evolution but in recent times threatens sustenance of life due to biologically unsuitable global warming; it is important to appreciate that temperatures will continue to rise over the twenty-first century if greenhouse gas (GHG) emissions continue unabated. To control the anthropogenic emission of greenhouse gases, it is urgently necessary to formulate regulatory mechanisms that would help to maintain biologically conducive temperature on earth. Appropriate regulations have been mandated under Kyoto Protocol; this instrument also allows for emission trading between countries for holding the global emission to an acceptable level.

1.9 Mechanism of Ozone Depletion

The net or overall reaction leading to ozone depletion occurs between atomic oxygen and ozone, forming two oxygen molecules – the reaction occurring at very cold conditions found in lower stratosphere. Heat enhancement on the earth's surface caused by trapping of IR radiation by increasing greenhouse gases results in concomitant cooling at lower stratospheric region, a condition that is conducive for ozone depletion in that region. The cycle may begin with either ClO or Cl. When starting with ClO, the first reaction occurs between ClO and O which form Cl. Then, Cl atom reacts with ozone and reforms ClO, consuming ozone in the process. The cycle then occurs again with another reaction between ClO and O. Chlorine is considered a catalyst for ozone destruction because Cl and ClO are reformed each time the reaction cycle is completed, and ozone is removed. Atomic oxygen (O) is formed when solar ultraviolet radiation (sunlight) reacts with ozone and oxygen molecules. Reactions showing steps involved in ozone depletion due to some ozone-depleting substances such as chlorofluorocarbon are given below.

$CFCl_3$ in the presence of UV light dissociates and liberates nascent Cl atom which then reacts with O_3 to form ClO and O_2. ClO, thus generated, reacts with nascent oxygen to form Cl and O_2. The net result is destruction of ozone molecule leading to ozone hole formation in stratosphere (Stolarski and Cicerone 1984).

$$\begin{gathered} CFCl_3 + \text{UV light} \rightarrow CFCl_2 + Cl \\ Cl + O_{3+} \rightarrow ClO + O_2 \\ \underline{ClO + O \rightarrow Cl + O_2} \\ \text{Net result}: O_3 + O \rightarrow 2O_2 \end{gathered}$$

Being catalytic, free chlorine atom is recycled and attacks another ozone molecule:

$$\begin{gathered} Cl + O_{3+} \rightarrow ClO + O_2 \\ ClO + O \rightarrow Cl + O_2 \end{gathered}$$

which is repeated in cyclic reactions.

Reactions of bromine oxide derived from methyl bromide (a commonly used fertilizer) are also an important ozone-depleting compound (Foster et al. 2001).

$$\begin{gathered} ClO + BrO \rightarrow Cl + Br + O_2 \\ Cl + O_3 \rightarrow ClO + O_2 \\ Br + O_3 \rightarrow BrO + O_2 \\ \text{Net result}: 2O_3 \rightarrow 3O_2 \end{gathered}$$

Significant destruction of ozone occurs in polar region because of ClO abundance as a resultant of repeated cycles of ozone depletion occurring in the cold polar stratosphere. The reaction between these two halogen monoxides, viz., ClO and BrO, efficiently destroys ozone. The net reaction in both cases is two ozone molecules forming three oxygen molecules.

Oxides of nitrogen (NO and NO_2) react catalytically with ozone accelerating the rate of depletion of ozone (Crutzen 1970). Nitrogen oxides catalytically destroy ozone as shown in the following reactions:

$$\begin{gathered} NO + O_3 \rightarrow NO_2 + O_2 \\ O + NO_2 \rightarrow NO + O_2 \\ \underline{O_3 + \text{UV light} \rightarrow O_2 + O} \\ \text{Net result}: 2O_3 + O \rightarrow 3O_2 \end{gathered}$$

The reaction of Br with O_3 followed by the self-reaction of the BrO produces atomic Br, and in due course the photolysis of Br_2 molecule regenerates Br atom which again attacks ozone molecule leading to destruction of ozone (Foster et al. 2001).

$$\begin{gathered} 2Br + 2O_3 = 2BrO + 2O_2 \\ BrO + BrO = 2Br + O_2 \\ Br_2 + h\upsilon = 2Br \end{gathered}$$

In winter (starting in April–May, with the setting of sun), the stratosphere above the Antarctic continent gets colder than it does anywhere else on earth with temperature frequently dropping below −45 °C. Since ozone-depleting reactions brought about by ODS is augmented at low temperature, "ozone hole" due to ozone thinning occurred initially and maximally at the Antarctic pole. Also being one of the windiest places on earth, in May and June, howling stratospheric winds begin to blow clockwise around the Antarctic continent which gradually forms an enormous ring of moving air – called the stratospheric polar vortex. The very low temperature inside the Antarctic polar vortex causes the formation of extremely small icy particles that in turn make up polar stratospheric clouds (PSC). Chlorine molecules remain attached to these icy particles which in this bound form do not attack ozone molecules. With the onset of sunshine in late August, UV-B strikes the reservoir molecules, thereby releasing chlorine gases inside the vortex. The molecules dissociate and release billions of chlorine atoms that begin attacking ozone molecules. Due to this phenomenon, massive ozone destruction occurs resulting ozone hole in the stratosphere above Antarctica. A change for filling up the ozone hole usually begins in early October when the continent and air above begin to warm up. By early November, the strong stratospheric winds circling Antarctica die down and the polar vortex breaks up. As it does, ozone rich air from outside the vortex flows in and much of the ozone that was destroyed is replaced. Usually at the end of November, the amount of ozone in the stratosphere over Antarctica almost returns to normal level. In the next winter, the cycle will begin again. Each year the ozone layer is subject to increasing thinning. From 1978 to 1980, ozone depletion in the stratosphere became a topic of increasing concern. Catalytic cycle of ozone destruction involving chlorine and ozone became evident around 1978. In 1985, loss of large amount of ozone over Halley Bay, Antarctica, was reported (Farman et al. 1985) associated with "large losses of total ozone in Antarctica due to seasonal ClOx/NOx interac tion." Ozone destruction is reported to be most intense over the Southern Hemisphere (Crutzen 1992) evidently because of the much colder atmosphere (that is conducive to ozone depletion) compared to that in the Arctic.

In the presence of UV radiation, ODSs form many intermediates such as hydrogen chloride, chlorine nitrate, bromine nitrate, etc. which remain as reservoir gases in polar stratospheric clouds (PSC). These reservoir gases in turn form chlorine monoxide, bromine monoxide, and also nascent chlorine and bromine which are reactive in nature. The ozone depleting mechanism is schematically represented in Fig. 1.9. Hydrochloric acid reacts with chlorine nitrate to form nitric acid which accumulates in the PSC, and simultaneously chlorine gas is liberated (Reaction 1 of Fig. 1.9) which is broken down into its nascent form by the energy of UV radiation (Reaction 2 of Fig. 1.9). Chlorine atoms then combine with O_3 atoms to produce ClO and O_2 gas (Reaction 3 of Fig. 1.9). ClO in turn photodissociates and forms oxygen atoms and chlorine atoms (Reaction 4 of Fig. 1.9). Two oxygen atoms will combine to create O_2, while the two chlorine ions will continue to react with other O_3 molecules.

Till recently overall 50 % losses of ozone layer have been recorded over Antarctica, representing 5 % of global ozone. This high level of loss at the Antarctica is due to the very low temperature at the lower level of stratosphere where ozone depletion takes place each year. Though refilled every year with onset of summer, some ozone loss is carried over; this loss of each year accumulated over the

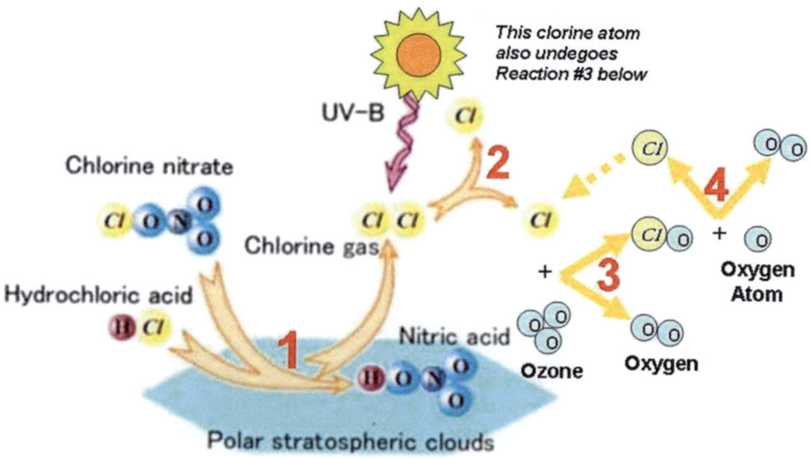

Fig. 1.9 Schematic diagram showing different steps in ozone depletion in the stratosphere (Taken from web)

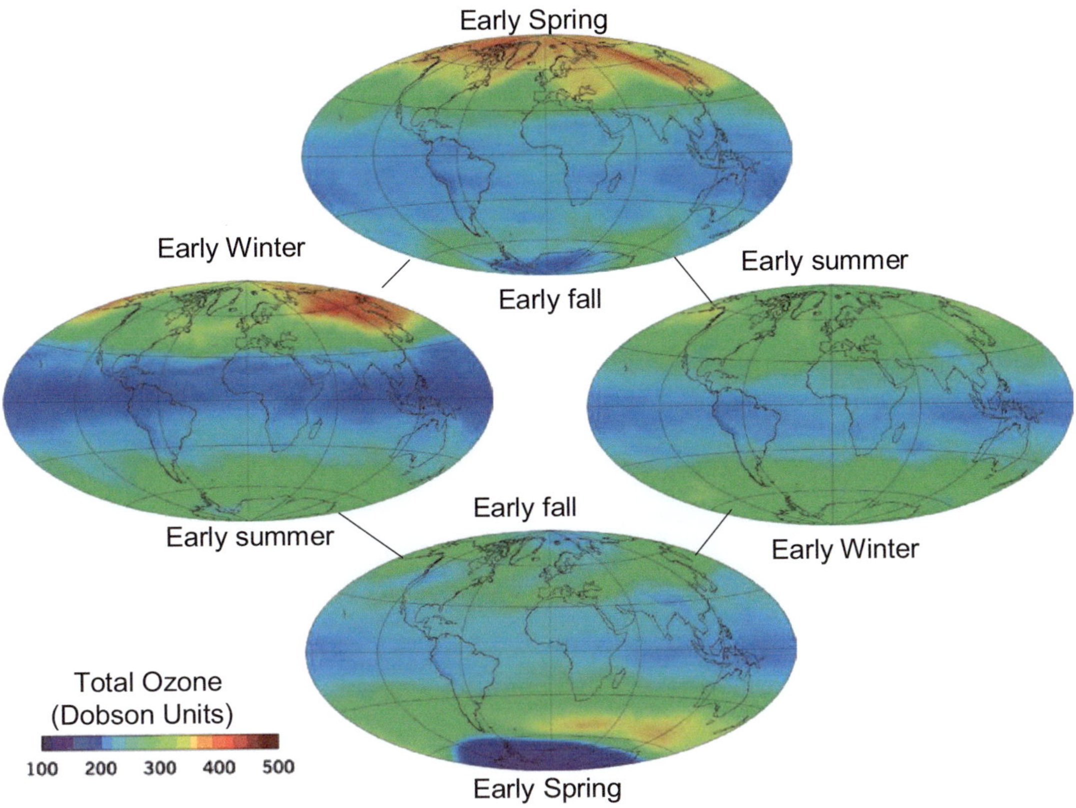

Fig. 1.10 Schematic diagram showing how ozone layer formation and depletion varies with different seasons throughout the year over the poles (Taken from web)

years resulting in effective ozone loss throughout the globe. Although initially observed over the Antarctic, formation of "dents"/ozone holes in the stratospheric ozone layer over the Arctic has also been reported. It may be noted that in the Arctic region, due to atmospheric conditions not being as suitable for the ozone destruction, the depletion of ozone over the Arctic has been less dramatic each year (Brune et al. 1991; Austin et al. 1992).

Since the detection of ozone layer depletion (due to anthropogenic activities) during the later part of the last century, harmful effects of UV-B are noticed from mid latitude to high latitude areas (Madronich and de Gruijl 1994). Stratospheric ozone has decreased over the period 1978–1991 (during the ozone depletion era) by 3–5 % in mid-latitudes and 6–8 % at higher latitudes, with no losses near the low ozone zone, viz., the equator (Gleason et al. 1993).

The damage (in terms of ozone depletion) that had already reached up to 2008 is represented in Figs. 1.10 and 1.11.

1.10 Kyoto Protocol

The Kyoto Protocol is an international treaty which extends the 1992 United Nations Framework Convention on Climate Change (UNFCCC) that commits state parties to reduce greenhouse gas emissions, based on the premise that (a) global warming exists and (b) man-made CO_2 emissions have caused it (http://unfccc.int/kyoto_protocol/items/2830.php). The Kyoto Protocol was adopted in Kyoto, Japan, on 11 December 1997 and entered into force on 16 February 2005. First agreed in 1997, it took 8 years for participating countries to ratify the

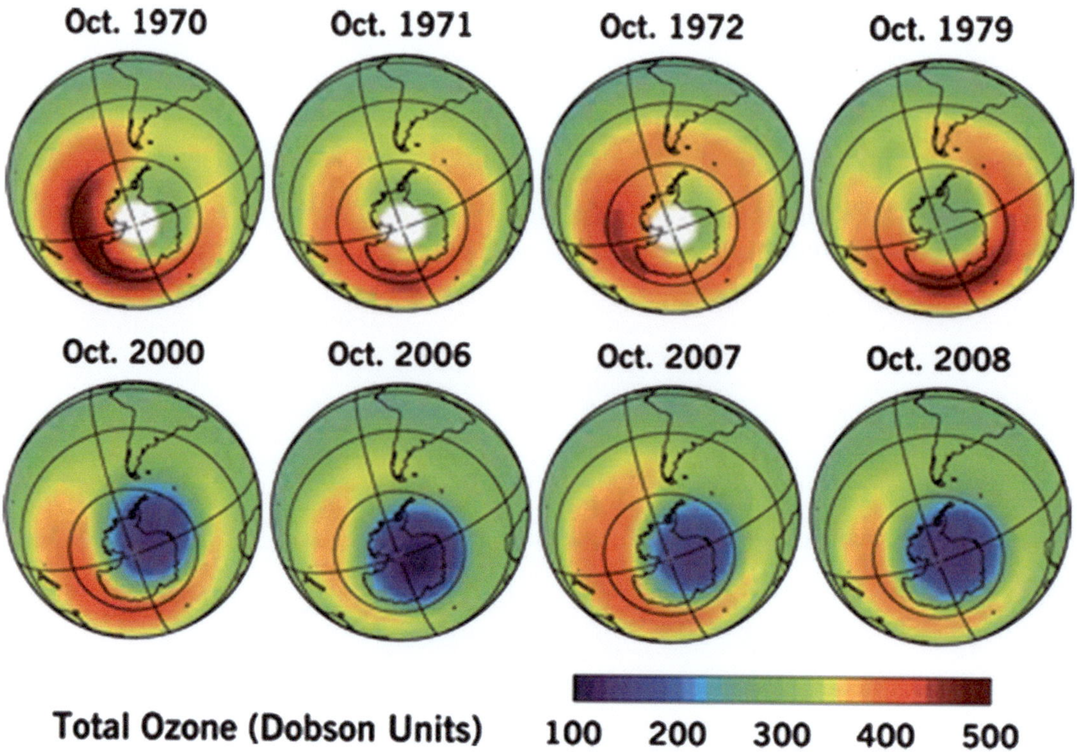

Fig. 1.11 The picture depicts a season-wise depletion of ozone in 2009 (Taken from web)

Kyoto Protocol. Being linked to the United Nations Framework Convention on Climate Change, this arrangement commits its parties by setting internationally binding emission reduction targets. Under the Protocol, countries must meet their targets primarily through national measures. However, the Protocol also offers them an additional means to meet their targets by way of three market-based mechanisms.

The Kyoto mechanisms are:

- International emissions trading
- Clean development mechanism (CDM)
- Joint implementation (JI)

Appreciating that greenhouse gases that disrupt atmospheric balance basically relates to global atmospheric quality, Kyoto Protocol allows for several "flexible mechanisms" such as "emissions trading." The clean development mechanisms (CDM) allows joint implementation by allowing Annex I countries, viz., the industrialized countries (that were members of the Organisation for Economic Co-operation and Development – OECD) to meet their greenhouse gas (GHG) emission limitations by purchasing GHG emission reduction credits from any other non-industrialized countries that would not be producing industrially related hazardous gases through financial exchanges for projects that would help to collectively reduce global emissions of harmful gases. The mechanisms help to stimulate green investment and help parties meet their emission targets in a cost-effective way. Under the Protocol, countries' actual emissions have to be monitored, and precise records have to be kept of the trades carried out. Recognizing that developed countries are principally responsible for the current high levels of GHG emissions in the atmosphere as a result of more than 150 years of industrial activity, the Protocol places a heavier burden on developed nations under the principle of "common but differentiated responsibilities."

1.11 The Montreal Protocol

Chlorofluorocarbons and other ozone-depleting substances are now globally recognized as the main cause of the observed depletion of the ozone layer. Molina and Rowland (winners of 1995 Nobel Prize in Atmospheric Chemistry) in 1974 reported the potential of CFCs to deplete stratospheric ozone. This provided an early warning, and their scientific report led to inclusion in citizen action plan and national regulation for reduction in ODS emission into the atmosphere. A decade later referring to the discovery of the ozone hole over Antarctica and this being attributed to ozone-depleting substances further heightened the concern. Scientists with the British Antarctic Survey announced the discovery of a huge hole in the ozone layer over Antarctica and also reported a substantial thinning of the ozone layer through successive years as ozone hole has been opening every spring and filling up in summer every year since 1970 with associated minor loss of ozone each year. Recognizing the significant threat of several industrially emitted gases in damaging the UV restraining stratospheric ozone layer, and the fact that concentration of CFCs in the stratosphere was rapidly increasing since 1960, in 1981 UNEP initiated action on a proposal to develop a Vienna Convention on the ozone layer. The draft convention was first presented to the international community that year. A lack of understanding about the extent of the environmental risks from ozone depletion made early acceptance of the convention difficult. Countries could not agree on specific control measures making the Vienna Convention merely a framework treaty for developing mechanisms to reduce and phase out the global production of ozone-depleting substances and also facilitate cooperation on research. As on April 2014, 114 countries have signed this treaty. Although calling for international monitoring and scientific assessment, this treaty does not include legally binding reduction goals. The binding regulations are laid out in an accompanying Montreal Protocol. The Montreal Protocol provided a mechanism to reduce and phase out the global production and consumption of ozone depleting substances (referred to as ODSs). This was opened for signature on 16 September 1987 and entered into force on 1 January 1989 and was followed by a meeting in Helsinki, May 1989. By September 2009, it had been signed and ratified by 196 countries, achieving universal participation.

Since the universal acceptance of the Montreal Protocol, in 1987 there has been growing evidence for the recovery of the stratospheric ozone layer. As per Montreal Protocol's provisions, the ozone layer should return to its pre-1980 levels by 2050–2075. This will also require cooperation of the Kyoto Protocol that calls for global reduction in anthropogenic production of greenhouse gases. It is important to appreciate of course that although the damaging CFCs are being phased out, they will remain a significant problem for some time yet, as they have accumulated in the atmosphere to very high levels. These CFCs will continue to penetrate slowly into the stratosphere well into the later part of this century; new emissions thus need to be meticulously monitored and restricted. Caution is also called for with respect to "HFCs," "the super greenhouse gases" that are coming into widespread use as replacements of Montreal Protocol's recognized CFCs such as CFC-11, CFC-12, CFC-113, CFC-114, and CFC-115.

While most chemicals involved in the reactions for ozone destruction have been banned under the Montreal Protocol, two of the chemicals viz. methyl bromide and nitrous oxide (byproduct of fuel burning viz. aircraft exhaust) had been left out from the list of banned chemicals under the Montreal Protocol. The reasons being: Methyl bromide is commonly used as a cheap pesticide by farmers in developing countries that could not be suitably replaced; methyl bromide is also released in small amount by some aquatic organisms and brassicaceae group of plants. N_2O that constitutes aircraft exhaust is not very reactive. Thus despite being the first detected chemical responsible for ozone depletion (Portmann et al. 2012) and in spite of the similarities between N_2O and other ODSs in causing ozone depletion N_2O has not been considered to be an ODS in the same sense as chlorine- and bromine-containing source gases; this is because of the comparatively low O_3 damaging impact of N_2O (Day and Neale 2002). However recent information that N_2O may remain present in the atmosphere for prolonged period has raised concern and is being considered by Montreal Protocol for inclusion among banned chemicals (Ravishankara et al. 2009). Limiting future N_2O emissions (from aircraft emission and also by agricultural related nitrification-denitrification process) would enhance the recovery of the ozone layer from its depleted state not only by functioning to restrict ODS in the stratosphere but also by restricting global warming in the troposphere. Hence limiting N_2O in the atmosphere is of primary concern. This would reduce adverse anthropogenic activity related change in climate by reducing N_2O related heating effect (of green house gases) in the troposphere and simultaneously restriction in cooling effect associated ozone loss in the stratosphere. Realization that enhancement of green house gases in the troposphere in turn causes chilling in the stratosphere, a condition that promotes ozone depletion reactions, calls for urgent restriction in production of excessive green house gases; this has been mandated by Montreal Protocol.

The Montreal Protocol also provides scope for addressing issues of dealing with new ODS discovered in scientific research. The protocol helps to establish ways for countries to periodically review the latest technical information on the substances in question by processes such as the ozone depletion potential-weighted anthropogenic emissions.

Though bromine is a potent ozone-depleting component, the use of methyl bromide, a major source of Br gas in atmosphere, was not banned under Montreal Protocol for many years because of its use as fertilizers in developing and under developed countries. It has been decided that methyl bromide use in Article 5(1) countries will be phased out in 2015. It may be noted however that many countries, both industrialized and developing, had committed to phase out methyl bromide well in advance of the Montreal Protocol schedule. Further, nitrous oxide (N_2O – released to the atmosphere due to aircraft fuel burning) which remains unregulated by the Montreal Protocol, because of its comparatively low O_3 damaging impact (Day and Neale 2002), is the single most important ozone depleting substance today and is predicted to remain such for the rest of the twenty-first century. Thus, thinning of the stratospheric ozone and related increase in UV radiation fluencing on earth is predicted to continue (despite measures undertaken to control) in the future Caldwell et al. 2003; McKenzie et al. 2003.

Bibliography

Austin J, Butchart N, Shine KP (1992) Possibility of an Arctic ozone hole in a doubled-CO_2 climate. Nature 360:221–225

Blumethaler M, Ambach W, Silbernagl R, Staehelin J (1994) Erythemal UV-B irradiance under ozone deficiencies in winter/spring 1993. Photochem Photobiol 59:657–659

Blumthaler M, Ambach W (1990) Indication of increasing solar ultraviolet-B radiation flux in alpine regions. Science 248(4952):206–208

Blumthaler M, Ambach W, Rehwald W (1992) Solar UV-A and UV-B radiation fluxes at two alpine stations at different altitudes. Theor Appl Climatol 46:39–44

Blumthaler M, Ambach W, Ellinger R (1997) Increase in solar UV radiation with altitude. J Photochem Photobiol 39:130–134

Bornman JF, Barnes PW, Robinson SA, Ballaré CL, Flint SD, Caldwell MM (2015) Solar ultraviolet radiation and ozone depletion-driven climate change: effects on terrestrial ecosystems. Photochem Photobiol Sci 14:88–107

Brune WH, Anderson JG, Toohey DW, Fahey DW et al (1991) The potential for ozone depletion in the Arctic polar stratosphere. Science 252:1260–1266

Caldwell MM, Ballare CL, Bornman JF, Flint SD, Bj€orn LO, Teramura AH, Kulandaivelu G, Tevini M (2003) Terrestrial ecosystems increased solar ultraviolet radiation and interactions with other climatic change factors. Photochem Photobiol Sci 2:29–38

Chaplin G (2004) Geographic distribution of environmental factors influencing human skin coloration. Am J Phys Anthropol 125:292–304

Crutzen PJ (1970) The influence of nitrogen oxides on the atmospheric ozone content. Q J R Meteorol Soc 96:320–325

Crutzen PJ (1992) Ultraviolet on the increase. Nature 356:104–105
Day TA, Neale PJ (2002) Effects of UV-B radiation on terrestrial and aquatic primary producers. Ann Rev Ecol Syst 33:371–396
Diffey BL (1991) Solar ultraviolet radiation effects on biological systems. Rev Phys Med Biol 36(3):299–328
Diffey BL (2002) Sources and measurement of ultraviolet radiation. Methods 28:4–13
Engelsen O (2010) The relationship between ultraviolet radiation exposure and vitamin D status. Nutrient 2:482–495
Farman JC, Gardiner BG, Shanklin JD (1985) Large losses of total ozone in Antarctica reveal seasonal ClOx/NOx interaction. Nature 315:207–210
Foster KL, Plastridge RA, Bottenheim JW, Shepson PB, Finlayson-Pitts BJ, Spicer CW (2001) The role of Br2 and BrCl in surface ozone destruction at polar sunrise. Science 291:471–474
Frederick JE, Snell HE (1990) Tropospheric influence on solar ultraviolet radiation: the role of clouds. J Clim 3:373–381
Gleason JF, Bhartia PK, Herman JR, McPeters R (1993) Record low global ozone in 1992. Science 260: 523–526
Hader DP, Kumar HD, Smith RC (2003) Aquatic ecosystems: effects of solar ultraviolet radiation and interactions with other climatic change factors. Photochem Photobiol Sci 2:39–50
Herman JR (2010a) Use of an improved radiation amplification factor to estimate the effect of total ozone changes on action spectrum weighted irradiances and an instrument response function. J Geophys Res 115, D23119
Herman JR (2010b) Global increase in UV irradiance during the past 30 years (1978–2008) estimated from satellite data. J Geophys Res 115
Herman JR, Lrko D (1994) Low ozone amounts during 1992–93 from nimbus 7 and meteor 3 total ozone mapping spectrometers. J Geophys Res 98:12783–12793
http://unfccc.int/kyoto_protocol/items/2830.php
http://www.ipcc.ch/report/ar5/wg1/
http://www.uio.no/studier/emner/matnat/fys/FYS3610/h04/undervisningsmateriale/Moan7.pdf
https://www.epa.gov/ozone-pollution
https://en.wikipedia.org/wiki/Stratosphere
https://www.skepticalscience.com/water-vapor-greenhouse-gas.htm
Kerr RA (1988) Stratospheric ozone is decreasing. Science 239:1489–1491
Kerr JB, McElory (1993) Evidence for large upward trends of ultraviolet –B radiation linked to ozone depletion. Science 262:1032–1034
Liew SC (2006) Electromagnetic waves. Centre for Remote Imaging, Sensing and Processing. Retrieved, pp 10–27
Lucas R, McMichael T, Smith W, Armstrong B (2006). Solar ultraviolet radiation. Global burden of disease from solar ultraviolet radiation World Health Organization, Environmental Burden of Disease Series, 13
Madronich S, de Gruijl FR (1994) Stratospheric ozone depletion between 1978 and 1992: implications for biologically active ultraviolet-B radiation and non-melanoma skin cancer incidence. Photochem Photobiol 59(5):541–546
McKenzie RL, Kotkamp M, Seckmeyer G, Erb R, Gies R, Toomey S (1993) First southern hemisphere intercomparison of measured solar UV spectra. Geophys Res Lett 20(20):2223–2226
McKenzie RL, Bjo¨rn LO, Bais A, Ilyas M (2003) Changes in biologically active ultraviolet radiation reaching the Earth's surface. Photochem Photobiol Sci 2:5–15
McKenzie RL, Bodeker GE, Scott G, Slusser J (2006) Geographical differences in erythemally-weighted UV measured at mid-latitude USDA sites. Photochem Photobiol Sci 5(3):343–352
McKenzie RL, Aucamp PJ, Bais AF, Bj¨orn LO, Ilyas M, Madronich S (2011) Ozone depletion and climate change: impacts on UV radiation. Photochem Photobiol Sci 10:182–198
Moan J, Grigagavicius M, Dahlback A, Baturaite Z, Juzeniene A (2013) UV-radiation and health optimal time for sun exposure. In: Reichrath J (ed) Sunlight, vitamin D and skin cancer. Landes Bioscience and Springer science + business media, pp 1–6
Molina MJ, Rowland FS (1974) Stratospheric sink for chlorofluoromethanes: chlorine atom catalyzed destruction of ozone. Nature 249:810–812
Norgaard MA, Anderson CB, Pettersson G, Caldwell MM (1998) Changes in biologically active ultraviolet radiation reaching the Earth's surface. J Photochem Photobiol B Biol 46(1):21–31
Portmann RW, Daniel JS, Ravishankara AR (2012) Stratospheric ozone depletion due to nitrous oxide: influences of other gases. Phil Trans R Soc B 367:1256–1264
Ramanathan V, Feng Y (2009) Air pollution, greenhouse gases and climate change: global and regional perspectives. Atmos Environ 43:37–50
Ravishankara AR, Solomon S, Turnipseed AA, Waren RF (1993) Science 259:194–199
Ravishankara AR, Daniel JS, Portmann RW (2009) Nitrous oxide (N2O): the dominant ozone-depleting substance emitted in the 21st century. Science 326:123–125
Rozema J, Staaij J, Bjorn LO, Caldwell M (1997) UV-B as an environmental factor in plant life: stress and regulation. Trends Ecol Evol 12:22–28
Seckmeyer GRL, McKenzie RL (1992) Increased ultraviolet radiation in New Zealand (45øS) relative to Germany (48øN). Nature 359:135–137
Shyam Choudhury S, Sen-Mandi S (2012) Natural ultra violet irradiance related variation in antioxidant and aroma compounds in tea (Camelia sinensis L. Kuntze) plants grown into different altitudes. Int J Environ Biol 2(1):1–6
Stolarski RS, Cicerone RJ (1984) Stratospheric chlorine: a possible sink for ozone. Can J Chem 52:1610–1615
Tilmes S et al (2012) Impact of very short-lived halogens on stratospheric ozone abundance and UV radiation in

a geo-engineered atmosphere. Atmos Chem Phys 12:10945–10955
Venkataramanan M, Smitha (2011) Causes and effects of global warming. Indian J Sci Technol 4:226–229
Vogelmann AM, Ackerman TP, Turco RP (1992) Enhancements in biologically effective ultraviolet radiation following volcanic eruptions. Nature 359: 47–49
Wang CT, Cao GM, Wang QL, Jing ZC, Ding LM, Long RJ (2009) Changes in plant biomass and species composition of alpine Kobresia meadows along altitudinal gradient on the Qinghai-Tibetan Plateau. Sci China Life Sci 51:86–94
WHO (2002) Global solar UV index: a practical guide. A joint recommendation of the World Health Organization, World Meteorological Organization, United Nations Environmental Programme, and the International Commission on Non-Ionizing Radiation Protection. World Health Organization, Geneva. ISBN 92 4 159007 6

Effect of UV Radiation on Life Forms 2

2.1 Introduction

Early in evolution, when life originated on earth, its atmosphere had almost no free oxygen (http://www.scientificamerican.com/article/origin-of-oxygen-in-atmosphere/), and flux of solar UV radiation in the earth's atmosphere was several-fold higher than it is today (McKenzie et al. 2007). During that time terrestrial habitat was not suitable for life support. About 4 billion years ago, life originated, as low oxygen requiring/anaerobic forms (in deep ocean near a hydrothermal vent (Maher and Stephenson 1988)). Consequently, early life forms perpetuated under protection of oceanic water (Nisbet and Sleep 2001) where growth and evolution of life forms continued. Availability of oxygen in the earth's atmosphere was marked by appearance of the *Cyanobacteria* group of organisms (during the Archean era) that initiated the process of oxygenic photosynthesis; this resulted in accumulation of molecular oxygen in the earth's atmosphere. Subsequent arrival of other oxygen-evolving organisms served to enrich the oxygen content in troposphere that in addition to providing respiratory oxygen for life forms also contributed toward formation of an ozone layer in the stratosphere, thereby restricting entry of UV radiation into the earth's atmosphere. At this point of evolution, the earth's atmosphere with restricted UV fluence as well as enhanced oxygen concentration in the troposphere together with development of life supporting temperature brought about by trapping of small levels of IR radiation under the then (low) concentration of greenhouse gases gradually established an environmental condition that was conducive for life support in terrestrial habitat. Subsequently by the Ordovician era, migration of life forms from aquatic to the terrestrial habitat became a large scale event. For maintaining a balance among life forms then evolved along multiple evolutionary lines, plants, enabled by their evolutionary development, providing food and shelter for other trophic levels and oxygen for maintaining atmospheric balance on earth, qualify as "primary producers" on earth. Perpetuation of plants that is central to perpetuation of life forms on earth is maintained by environment-related adjustment of cell molecular events made effective by molecular plasticity specifically evolved in plants. This established an equilibrium among life forms that was maintained till indications of thinning in the ozone layer followed by detection of an ozone hole (during the later part of the past century) raised an alarm. The first "ozone hole" was detected in Antarctica where ozone destruction is most intense (Crutzen 1992), evidently because of the much lower temperature in this region compared to that in the Northern Hemisphere. Subsequently, ozone loss was also recorded in the Arctic region (Proffitt et al. 1990) as well as in other latitudes worldwide (Brasseur et al. 1990). The potential impacts of

S. Sen Mandi, *Natural UV Radiation in Enhancing Survival Value and Quality of Plants*,
DOI 10.1007/978-81-322-2767-0_2

an increase in solar UV-B radiation reaching the earth's surface due to stratospheric ozone depletion have been investigated by several research groups since detection of ozone hole. Herman and Larko (1994) reported a 4–5 % decline in the mid-latitude stratospheric ozone layer. This widespread breach in the ozone layer resulted in increasing penetration of UV into the earth's atmosphere. Report of Kerr and McElory (1993) indicates that locational disruption of the ozone layer has led to proportionally enhanced UV fluence at different location. Since the later part of the last century when ozone thinning had begun, reports on damaging effects on life forms are becoming available in the literature, more than before. Reports on UV effects on interactive relation between components in communities of plants and lower organisms, thereby affecting their population survival on earth, are also available (Hader et al. 2007). A commonly observed damaging effect of UV on humans particularly in the post-ozone hole era includes increase in skin melanoma frequency. Such adverse effects are particularly observed among white immigrants that do not have built in tolerance developed through UV preexposure based development of UV protective compounds, viz., melanin in skin (cf. development of phenolic compounds in plants), as a means of acclimation strategy under high UV environment in Australia (Southern Hemisphere where ozone thinning is intense). This underscores the predicament faced by life forms under enhanced UV fluence in the post ozone hole era (Erdei and Torres 2013). Tevini (1993) have reviewed the effect of enhanced UV-B radiation on terrestrial plants. UV induced macromolecular damage in embryonic axis of orthodox seed (where due to lack of moisture, synthesis, and function of repair enzymes and protective compounds, viz., flavonoids, is precluded) renders the seed nonviable – this results in loss of rich biodiversity on earth (Andersen et al. 2004). Loss of biodiversity through UV-induced cell damage in (dehydrated) seed system is thus a faster event compared to that in whole plant (hydrated system). Plants in the hydrated state are particularly endowed with UV acclimation processes that include macromolecular repair as well as molecular rearrangements for the synthesis of UV protective compounds. Plants growing in open field usually display low damaging effects.

Literature accumulated till date provides information on a wide range of UV-induced effects on plants as well as other life forms. Potential impacts of increased solar UV-B on global plant productivity in open field have been reported by Teramura and Sullivan (1991). The authors in 1994 have reported on UV effects of UV-B radiation on photosynthesis and growth of terrestrial plants (Teramura and Sullivan 1994). Intraspecific differences in growth and yield of soybean exposed to ultraviolet B radiation under greenhouse and field conditions have been reported by Teramura and Murali (1986). Sensitive plants often exhibit reduced growth (plant height, dry weight, leaf area, etc.), photosynthetic activity, and flowering (Tevini and Teramura 1989). Adverse effects of UV radiation on different plants have been reported by many authors (studied in maize, Teramura (1983), and in rice, Carletti et al. (2003) and Hidema et al. (2007)). Intraspecific variation in response to UV radiation has been established from studies on physiological, morphological, and leaf optical properties studied on two barley varieties, viz., Barke and Bonus (Klem et al. 2012); the authors have also demonstrated that in addition to varietal difference, UV effect also exhibited variation with plant age/maturity in determining plants' sensitivity to UV.

Study of Paul and Gwynn-Jones (2003) provides a comprehensive understanding of the ecological role of life forms vis-à-vis damaging effect of UV. In closed chamber/laboratory experiments, the authors have reported that increases in UV radiation can damage many organisms that on the other hand may exhibit amelioration effect under combination of radiation of different wavelengths (Deckmyn and Impens 1997). While plants growing under artificial UV lamps exhibit major damaging effect, supplement of photosynthetically active radiation available in open air exhibits moderation of the damaging effects (Kakani et al. 2003). Evidence to establish that open air experiments provide realistic information on UV effect on plants is evident from the following experiments: high levels of white light as well as UV-A/blue light available in open field have been reported to mediate photorepair mechanisms, thereby ameliorating the UV-B damage (studied in Soybean – Warner and Caldwell 1983; Mirecki and Teramura 1984).

Paul and Gwynn-Jones (2003) have emphasized the need for studying the influence of solar UV not only on (isolated) plants alone but also in association with other coexisting organisms, viz., insects, pests, and pathogens for better understanding and future application for developing useful interactive outcomes.

While effect of UV on individual life forms has often been studied, research into the effects of ultraviolet (UV) radiation on communities in terrestrial ecosystems remains a relatively new discipline. Effect of UV-B have sometime been reported with regard to status of survival between crop plants and weed as also with respect to components of phytoplankton exhibiting impact of UV on the extensive marine ecosystem.

Generally representing a damaging radiation, solar UV (UV-B and UV-A) have often been reported to be effectively used by microbes, plants, and animals as source of information about their environment (Hader et al. 2007).

Studies on natural adaptations under variation of UV fluence in specific life forms particularly in plants relate to challenges for survival under conditions of ozone depletion (compared to the reference year 1978) – this calls for understanding on a wider perspective, vis-à-vis reports provided in ACIA (2005), taking into consideration adaptations and abilities for coexistence among communities under enhanced UV radiation. This radiation might adversely affect certain species of insects and pests that coexist, through a net counterbalancing effect where the direct negative effects of increased UV radiation may appear adjusted (Raviv and Antignus 2004). A system where UV-induced damage may be countered within the same system is that of plants that due to their unique attribute of cellular plasticity exhibit greater adaptability for survival under variation of UV radiation (Weinig et al. 2004).

2.2 Effect of UV on Terrestrial Ecosystem

Terrestrial ecosystems comprising a major component of the biosphere face the threat of major consequences under ravages of enhanced UV radiation. While organisms in aquatic ecosystems have the depth of water as partial protective covering, the terrestrial ecosystem components, viz., plants, animals, microbes, and humans, have to cope with more drastic ravages of UV radiation. Many of the most important, yet less predictable consequences of UV radiation on life forms in terrestrial ecosystem are manifestation of cell molecular rearrangements induced by UV radiation; variation in chemical composition (higher flavonoid content) of plants, in addition to conferring UV acclimation (and thus survival value) in plants, may also bring about variation in interaction with associated life forms; this determines survival capacity of other entities in the ecosystem (Caldwell et al. 1998). UV might change the ability of life forms for coexisting with other species (Aarssen 1989). Impairment of growth and productivity of plants (correlated with varietal difference) under UV radiation have been discussed by Greenberg et al. (1997) in the context of UV upregulated flavonoid synthesis in leaves; this may adversely affect plant interaction with insects that feed on leaves (Hartley and Gange 2009). Thus, insects or even microorganisms may stand to lose not only for physical association but may also be unable to adjust to changed metabolism of plants under enhanced UV-related adaptations. UV effects, with regard to altered plant chemistry and pigmentation, which can influence plant–animal interactions, have received wide attention. Terrestrial plant habitats that represent the major community of plants include agricultural lands, forests, grasslands, marsh lands, savannahs, deserts, and tundra at various latitudes and altitudes (Teramura and Sullivan 1994). Functioning of these ecosystems involves many attributes that could potentially be affected by increased solar UV-B radiation, including plant biomass production, seed production, plant consumption by herbivores including insects, and disease incidence of plants and animals. These factors, in isolation or collectively, bring about population fluctuations including changes in species composition as well as balance between crops and weeds (Barnes et al. 1988) involving species variation and plant biomass development.

At the ecosystem level, the effects are often not immediately clear since interaction between the component entities brings about variation in relationship within and between the communities (Whitlock 2014). Among the terrestrial organisms, plants which maximally endure UV radiation exhibit development of epidermal wax, deposition, and pigmentation that serve as UV screens thereby providing UV acclimation. Another means of UV acclimation involves antioxidative protection conferred by UV upregulated compounds viz. flavonoids (see Chap. 4). Being developed through evolution to counter damaging effects of UV, the extent of these protective mechanisms is found to correlate with the UV fluence exhibiting adaptation/amelioration at the respective UV location (Filella and Peñuelas 1999). Such adaptive modifications are seen to vary as a function of varietal/genetic difference (Bieza and Lois 2001). Despite such evolutionary developments, all life forms in different ecosystems undergo correlated UV fluence caused adverse effects.

Experiments on food crops under high UV environment have shown lower yields for several key crops such as rice, soy beans, and sorghum (Xiong and Day 2001; Krizek et al. 1997, 1998); such effect could be due to macromolecular damage in the embryonic axis of seeds during post harvest "dry" storage of orthodox seed under ambient UV environment; the lower yield could relate to variable extent of seed viability loss depending on maturity status of seeds. Due to uniform status of seed viability in a seed lot due to adverse effect of UV on seed during post harvest "dry" storage in open field (farmers' seed storage), crop stand establishment, and thus crop productivity would be adversely affected. Provided that percentage of crop stand establishment is not adversely affected (due to reduction in seed vigor), crop productivity will not be reduced (Ganguli et al. 1992). The potential impact of increased UV radiation on plants and animals is of particular concern because they form the basis of food supply at various trophic levels.

Specific studies in this regard reveal that competitive balance between wheat and major weeds could be altered under elevated UV-B conditions. If species competition relationship changes occur on a broad scale, it will be one of the most important ecosystem level consequences of stratospheric ozone reduction, especially in non-agricultural ecosystems. From the ecological point, species competition, involving even subtle decreases in plant height correlated with solar UV-B radiation, might be of eventual significance if different species were affected to different extent by fluencing UV radiation.

2.3 Effect of UV on Aquatic Ecosystems

In aquatic ecosystems, phytoplanktons are at the start of the aquatic food chain providing primary food source for the zooplankton, fish, and other aquatic organisms that account for 30 % of the world's intake of animal protein. Accounting for 30 % of the world's intake of animal protein (Tevini 1993), productivity of this group is restricted to the upper layers of the water, and thus a small increase in atmospheric UV-B would significantly reduce the size the plankton populations (Hader et al. 2007), the extent of UV reaching into depth of water where phytoplankton grow being important for survival of the group. It is thus important to assess adverse effect of UV on photosynthetic aquatic organisms particularly phytoplanktons and to develop strategies for counteracting the adverse effects of enhanced ambient UV.

UV radiation directly damages fish, insects, shrimps, crabs, amphibians, and other animals during their early development (Browman et al. 2003; Heyer 2003; Buck and Callaghan 1999). Pollution of the water by toxic substances may heighten the adverse effects of UV radiation, working its way up the food chain; less plankton means less food for the animals that feed on them

– this has particular implication on fish stock that already stands depleted by overfishing.

The aquatic life forms, though protected to some extent by the surface water layer, are still subject to adverse effects of UV to the extent that penetrate up to the relevant water layer.

Factors that cause variation in UV penetration through water bodies include turbidity, concentration of colorless or colored dissolved organic matter (DOM/CDOM), phytoplankton concentration, and suspended particles that affect water's optical properties (Tedetti and Sempèrè 2006). Light intensity decreases exponentially with depth of the water column, while biologically damaging levels of UVR, subject to variation in dissolved matter (Fig. 2.1), have been reported to penetrate up to 30 m (Shick et al. 1995). The penetration of UVR wavelengths into water is highly dependent on the depth and color of the water body. Color of the water bodies is defined by the amount of colored dissolved organic matter (CDOM) and is estimated from the concentration of dissolved organic carbon (DOC); this value can be used to predict UVR transparency, except in the lowest CDOM environments such as in some alpine lakes (Laurion et al. 2000; Sommaruga 2001; Sommaruga and Augustin 2006) where suspended particles/turbidity also appear to play a role (albeit often not significant) in UVR attenuation. The large arctic rivers are relatively more protected from UV radiation because of their high CDOM concentration (Gibson et al. 2000). Such reports suggest that under conditions of low CDOM, i.e., when there is less organic matter in the upper layers of the water, UV radiation can penetrate deeper into the water and adversely affect plants and animals living there. In the

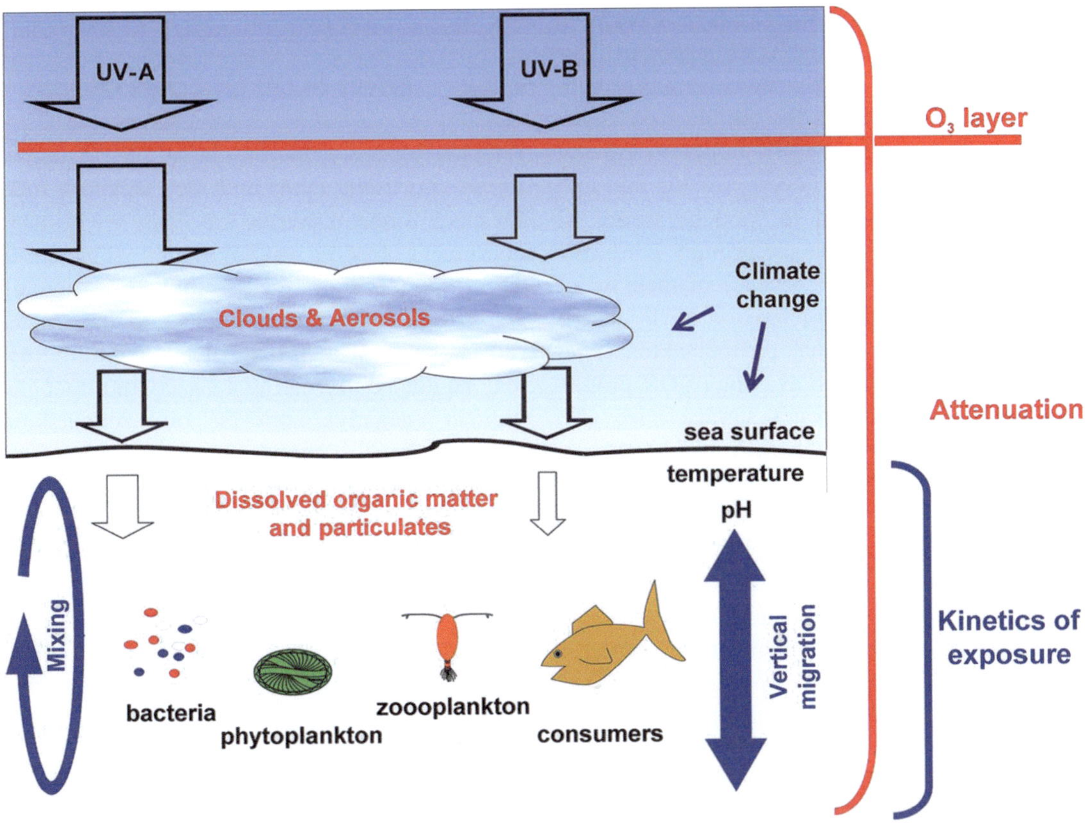

Fig. 2.1 Main factors affecting the quantity and quality of UV radiation received by aquatic organisms (Taken from Goncalves et al. 2010)

hyper-oligotrophic waters of the South Pacific, the depth at which UV-B radiation (at 305 nm) is reduced to 10 % of the surface value (expressed as $Z_{10\ \%}$) was recorded as maximum at 28 m (Tedetti et al. 2007), whereas in Northern European coastal waters, $Z_{10\ \%}$ (at 310 nm) was maximum at 0.08–10.4 m (Aas and Høerslev 2001).

Aquatic ecosystem consists of both freshwater and marine ecosystems. Both these ecosystems consist of planktons (including the phytoplanktons, zooplanktons, and bacterioplanktons) along with nektons (fish). The UV radiation affects each of these trophic levels to different extent depending on their genetically determined susceptibility to UV related damages.

2.3.1 Effect of UV on Phytoplanktons

UV-B levels at the upper ocean layer have been reported to inhibit phytoplankton activities. This can have an impact on the extensive marine ecosystem, thereby affecting the marine food chain. Phytoplanktons, due to their ability to capture sunlight and convert the energy of the sun into chemical energy storing in food by means of photosynthesis, constitute the primary producers in the oceans. Evidently because of their major role in producing a major fraction of the global food chain, cyanobacteria, phytoplanktons, and macroalgae in addition to availing of UV protective measures of the water surface have also evolved internal UV screens such as scytonemin, mycosporin-like amino acids (cf. plants see Chap. 4), and other UV-absorbing substances as a means of (additional) protection from UV radiation (Rozema et al. 2002). These simple aromatic organic molecules representing the early UV screens in aqueous environments in the course of evolution subsequently developed into specialized UV absorbers found in terrestrial plants, eukaryotic photosynthetic organisms, as well as in cyanobacteria (UNEP-2006).

Equatorial regions contain densities of phytoplankton approximately 10–100 times smaller than the circumpolar regions (UNEP 1994). Additional concentrations of phytoplankton occur in upwelling areas along the continental shelves. Investigations in Antarctica indicate that high UV-B radiation levels limit phytoplankton productivity (UNEP 1994; Tevini 1993). While they harvest longer wavelength radiation for converting solar energy into chemical energy in food and other biomolecules, phytoplanktons are also exposed to solar UV radiation, albeit to a lesser extent compared to terrestrial plant. Such reports call for studies on the role of phytoplanktons as primary producers on earth even in the ozone depletion era. Solar UV-B radiation (280–320 nm) affects motility and orientation in motile organisms in the aqueous ecosystems and also affects photosynthesis (assayed by monitoring oxygen production) in cyanobacteria, phytoplankton, and macroalgae (Hader et al. 2007).

2.3.2 Effect of UV on Zooplanktons

Zooplanktons are marine larval or immature forms of various kinds and they serve as food for many forms of marine life. They are primary consumers, feeding mainly on phytoplankton. The importance of this assemblage in aquatic ecosystems makes zooplanktons important material for study of the effect of UV radiation that may be done by assessing a variety of direct and indirect negative effects on zooplankton populations. Most commonly reported UV effect on zooplanktons include increased death rates (Zagarese et al. 2003); other deleterious effects of UV radiation includes DNA damage (Malloy et al. 1997), reduced growth rates (de Lange et al. 1999), decreased fecundity (Williamson et al. 1994; Zellmer 1996, 1998; Huebner et al. 2006), and gut damage in zooplankton when fed with UVR preexposed algae (Zellmer et al. 2004). These damages however depend on the physiological condition, the habitat, and the inherent capacity of the organism to protect itself from deleterious effects of UV-B.

2.3.3 Effect of UV on Bacterioplanktons

Bacterioplanktons contribute a significant biomass component in aquatic ecosystems and play a key role in biogeochemical processes, recycling, and remineralization of materials and energy within the food chain. Inhibition of bacterial production (Bailey et al. 1983; Aas et al. 1996; Pakulski et al. 1998) and bacterial biomass (Sierachi and Sieburth 1986; Helbling et al. 1995; Gustavson et al. 2000) has often been studied with respect to UV effect. They do not produce any UV-related screening pigments but overcome solar radiation stress by rapid cell division and effective molecular repair mechanisms. Up to the extent that the repair mechanism in the bacterioplanktons can counteract the damage induced by UV radiation, there occurs no considerable effect on their population; however, when amount of CPD (cyclobutane pyrimidine dimers – a DNA damage parameter) accumulation becomes significantly high under high solar radiation, DNA repair efficiency is reduced and the population decreases.

2.3.4 Effect of UV on Fish

A large number of studies have reported damaging effects of UV-B on fish eggs and larvae. These damages include DNA damage (Browman et al. 2003; Malloy et al. 1997) and increased mortality of larvae (Hunter et al. 1978; Kouwenberg et al. 1999; Steeger et al. 1999). Malformation (Dong et al. 2007); lesion of the skin, eyes, and brains (Blazer et al. 1997; McFadzen et al. 2000); retarded growth (Jokinen et al. 2008); and immune depression (Markkula et al. 2005, 2006, 2009) of fish have also been reported. In correlative studies Blazer et al. (1997) have reported that although increased UV radiation is known to have damaging effects such as severe sunburn on some species of fish such as *Oncorhynchus clarki*, fish populations by and large also benefit under UV fluence that through proportional penetrations into water bodies affect slowing down of infection rates and/or inhibit the spread of some parasites. On the other hand, UV-induced immunosuppression in host fish in turn exacerbates the effects of parasitism, disease, and contaminant load on individual fish. This will possibly lower fish population productivity by decreasing survival. Other modes of UV caused reduction in fish population have been demonstrated in experiments with gobies where, in the UV-treated larvae, both escape rate and escape distance (36 %, 3867.5 mm, respectively) were significantly lower than those of control larvae (75 %, 6964.7 mm, respectively). There was also a significant difference in survival (UV, 35 %, control, 63 %) (Fukunishi et al. 2012).

2.4 Effect of UV on Microbes

Soil microorganisms have significant role in carbon and nutrient cycling. So their response to elevated UV-B radiation in regard of the community structure, biomass, and functioning is essential for fully understanding the effects of global climate change on ecosystems. UV-B radiation can only penetrate up to about 100 μ under the soil surface. The soil microbial community thus has often been considered to remain unaffected by stratospheric ozone depletion. Although some subsequent studies (Bornman et al. 2015) indicate better soil penetration of UV, comprehensive understanding on effect of UV on global climate change and its effect on ecosystems requires further studies.

Due to lack of significant surface penetration by UV, only epigeous microorganisms (that have adapted to life on the phylloplane, with nutrition-based grouping into biotrophic, pathogenic, or saprotrophic types) are maximally affected (Johnson 2003). Cell molecular damages occur in these organisms by impact of UV-B radiation that causes direct damage to DNA or indirectly via active oxygen species formed due to absorption of the UV by other macromolecules (cf. plants – see Chap. 3). Results of ecological studies of the phyllosphere clearly indicate that UVR has a negative impact on individual microbial species as well as on complex microbial communities encompassing a number of trophic levels (Newsham et al. 1997; Paul 2000). UV-induced damaging effects in fungi are particularly observed in the reproductive aspects (Moody et al. 1999). In case of a forest tree (such as *Quercus robur*), it has been reported that supplemental UV radiation for 6 months causes reduction in fungal growth on the adaxial surfaces (Newsham et al. 1997). Ayres et al. (1996) have demonstrated that two species of yeast, viz., *Aureobasidium pullulans* and *Saccharomyces roseus*, are more affected by UV than the other species because they have less amount of dark pigmentation in their skin and are thus less capable of restricting UV entry into cells; these organisms are thus more vulnerable to UV radiation-caused damage. Other studies have reported similar responses to increased UV-B radiation by phylloplane ballistosporous yeasts, found in *Q. robur*, and a number of crop and horticultural species (Ayres et al. 1996; Rasanayagam et al. 1995).

In fungi population, the reproductive aspects have been reported to be particularly sensitive to UV-induced damage (Moody et al. 1999). Moody et al. (1999) measured spore germination in a range of fungi isolated from decomposing *Rubus chamaemorus* L. and *Calluna vulgaris* L. litter collected from Moor House National Nature Reserve, England, and from the phylloplane of *Q. robur* saplings. The spores were exposed to elevated UV-B radiation (corresponding to a 15 % ozone depletion) for up to a total of 6 h within two 12 h photoperiods. Interestingly, germination was only significantly reduced in fungi isolated from litter. The species most affected were *Penicillium hordei*, *Aspergillus fumigatus*, and *P. spinulosum* in which the germination percentage was found to be reduced up to 38–44 %. Additionally, spore production in many of the litter species tested was greatly reduced, and, in *A. fumigatus*, *P. spinulosum*, and *Trichoderma viride*, spore production ceased altogether. A possible mechanism for the adaptation by fungi is a lower surface area-to-volume ratio of their spores, which may have greater importance than the degree of pigmentation in *Hyphomycetes* and conidiomata production in *Coelomycetes* (Moody et al. 1999). Studies of Gunasekera et al. (1997) have demonstrated that phylloplane yeasts cultured from plants in high altitudes (receiving more natural UV-B) were less sensitive to UV-B exposure than yeasts isolated from plants in low altitudes (receiving less natural UV-B). Similar responses have also been demonstrated for other organisms including higher plant species (Sullivan and Teramura 1988). Such observation suggests that prior exposure to UV preconditions cellular mechanism for UV tolerance.

2.5 Effect of UV on Cyanobacteria

Cyanobacteria, the major biomass producers both in aquatic and terrestrial ecosystems, represent more than 50 % of the biomass in many aquatic ecosystems (Sinha and Hader 2004). In addition to being the pioneer in oxygen production on earth, importance of *Cyanobacteria* also relates to their nitrogen fixing capacity that causes them to serve as important fertilizers both in aquatic and in terrestrial plant habitats such as tropical rice fields. Several studies have demonstrated adaptation to UV stress expressed as increased resistance to UV-caused damaging effect in *Cyanobacteria* (Prabha and Kulandaivelu 2002; Rajagopal et al. 2005; Helbling et al. 2006). Norris et al. (2002) demonstrated adaptive mutagenesis associated with increasing resistance to UV-B in *Cyanobacteria*.

2.6 Effect of UV on Plants

The importance of interactions among plant species in determining the structure and dynamics of plant communities is widely recognized (Tilman 1988; Loreau et al. 2001). Wide inter and intra-specific differences in growth, production of dry matter, and physiological and biochemical changes in response to UV-B radiation have often been reported (Kramer et al. 1991;

Mpoloka 2008; Fedina et al. 2010; Klem et al. 2012).

In the terrestrial ecosystem, UV-B fluence in open field is potentially detrimental to all living forms but is particularly harmful to plants. This is because terrestrial plants have a stationary lifestyle and an obligatory requirement of sunlight (Strid et al. 1994) for harnessing solar energy to produce utilizable biochemical compounds (by photosynthesis). Thus, while fulfilling the beneficial role as primary producer in the ecological cycle, plants suffer UV induced damages, such damaging effects being largely reduced in open field through interaction of UV with other radiation particularly the PAR. For realistic understanding of UV effect on plants it is thus important to study plants in open air. The effectiveness of UV-B varies both among species and among cultivars of a given species. UV-sensitive plants often exhibit reduced growth (plant height, dry weight, leaf area, etc.), photosynthetic activity, and flowering.

It is important to appreciate that damaging effects of UV appear particularly exaggerated when recorded under greenhouse/closed chamber conditions where plants are exposed to UV radiation without regard to interaction of the UV wavebands with the photosynthetically active radiation (PAR) in the experimental setup (Kolb et al. 2001). This is evident from the fact that while (soybean) plants were found to be moderately susceptible to UV radiation when grown in open field, the same plants exhibit higher sensitivity to UV when grown in greenhouse under low levels of non-UV radiation including PAR (Teramura and Sullivan 1991). This demonstrates that greenhouse experiments do not provide information on natural effect of UV on plants; interaction with other natural factors and life forms in the open field in an ecosystem is also important in bringing about the final manifestation of UV effects on plants, the primary producers. This brings about a focus on studies of the earth's natural ecosystem. For developing a realistic understanding into the effects of UV on plants, it is important to consider studies in open air or at least make appropriate corrections on data obtained from studies in closed chambers.

In plants, UV induced effects are discernible widely at morphological, physiological, biochemical, and molecular levels. Exposure to UV-B has often been reported to cause decreased leaf area, plant height, and reduced dry weight; such changes are often correlated with somewhat increased axillary branching (Greenberg et al. 1997; Furness et al. 1999). Dai et al. (1995) reported that in rice plants UV dose accumulated after a few weeks of UV-B exposure is sufficient to significantly reduce leaf area and plant dry weight. Such studies demonstrate that UV effect is a function of duration of exposure and age of the plant. If the UV-B dosage exceeds the limits of tolerance, plant leaf anatomy is changed and biomass is decreased (Coleman and Day 2004; Kakani et al. 2003; Zhao et al. 2003). UV has been reported to adversely affect the most important physiological processes, such as photosynthesis by damaging D1 and D2 proteins (of PS-II) and reducing Rubisco (Bjorn et al. 1997). UV-B radiation may also decrease the penetration of PAR causing reduction in photosynthetic and accessory pigment formation as well as impairment of stomatal function and altered canopy morphology and thus retard photosynthetic carbon assimilation (Teramura and Sullivan 1994).

Bjorn et al. (1997) have reported positive responses of UV in many morphological and physiological processes such as marked increase in flowering and a stimulation of photosynthetic processes in some species of plants; the authors also reported that UV-B treatment caused enhancement in drought tolerance of *Pinus pinea* and *Pinus halepensis* by increasing leaf cuticle thickness, a trait that by restricting UV entry into cells also helps to acclimatize plants to ambient UV radiation on location. The basic means (genetic/epigenetic) of restriction of UV entry, thereby conferring UV tolerance in plants, is the upregulation of gene expression for synthesis of UV-absorbing compounds, viz., flavonoids (see Chap. 4), a compound that is upregulated in synthesis as a function of UV intensity-based dose variation resulting from weather fluctuations. From studies at different altitude locations (at the same latitude), Choudhury and

Sen Mandi (2012) have demonstrated that accumulation of UV screening compound (viz., flavonoid) is correlated with altitude-related variation in UV intensity as a function of genetic variation among plants (see Chap. 6). Variation among species and also between varieties and cultivars of a species for tolerance to natural UV radiation has also been reported in soybean – (Biggs et al. 1981; Teramura and Murali 1986). Other studies have also reported marked variability in sensitivity to UV-B radiation among plant species/cultivars – exhibiting genetic variation (Tevini et al. 1981; Teramura 1983). Plants native to low elevations have been found to be sensitive when exposed to high UV-B, but plants from higher elevations where UV-B fluence is high, exhibit tolerance to experimental condition of high UV-B radiations (Teramura and Sullivan 1991). Such studies indicate the development of UV tolerance, exhibiting UV-adapted traits developed during preexposure to high UV fluence; such plants appear to have evolved specific adaptations that protect them from the deleterious effects of UV-B (Antonovics 1975). This may be exemplified by the findings of Caldwell et al. (1982) who demonstrated that in the arctic region, with naturally high accumulation of the protective ozone layer, arctic ecotypes, variants of the same species, viz., *Oxyria digyna*, were consistently more sensitive to high UV radiation due to their preexposures under low UV environment than their counterparts collected from mountains in lower latitudes (with high UV fluence). This demonstrates that the plant's tolerance to UV is determined by the extent of UV previously experienced by the plant. Thus, plants growing under high UV (e.g., at high altitudes) would exhibit sensitivity to UV if it had earlier grown under lower UV fluence compared to plants preexposed to high UV radiation. Thus, UV protective mechanism, e.g., flavonoid accumulation during preexposures to high UV, would serve as protective compounds in plants even in changed habitat. The presence of secondary compounds such as flavonoids which may act as UV screens in the epidermal layer restricting entry of UV to sensitive tissue layers has been shown to vary greatly among various plant species collected along a latitudinal gradient (Robberecht and Caldwell 1978). Other studies indicate that a simple correlation between UV sensitivity and epidermal flavonoid concentrations does not seem to exist in wheat and wild oat (Barnes et al. 1987). Studies of Caldwell et al. (1982) and Barnes et al. (1987) indicate cofunctioning of other UV-related cell protective mechanisms such as resistance of the photosynthetic apparatus and DNA damage repair.

To acclimatize under UV radiation, plants have also developed enzymatic antioxidants (e.g., superoxide dismutase) (Surplus et al. 1998) as different from the nonenzymatic antioxidants (viz., flavonoid; see Sect. 4.4.2 of Chap. 4) that prevent ROS (including UV-induced ROS) mediated macromolecular damage (Heijde and Ulm 2012). The cell molecular mechanisms involved in developing DNA repair process constitute an important UV acclimation strategy in plants (see Chap. 4). Thus, high-intensity UV-induced damaging effect in plants, in turn countered by low-dose UV radiation, supports persistence of plants and thus life on earth. This would explain survival of plant population through the years under the predicament of locational/seasonal/diurnal variation in UV fluence even in the post-"ozone hole" era.

Studies till date on the effect of UV on plants have provided the following information (detailed in Chaps. 4, 5, 6, and 7):

- Plants exhibit genotype/species/cultivar specific variation as UV-induced effect demonstrating a gene X environment effect.
- UV radiation though generally known to be damaging (particularly when studied in closed chamber) is also known to confer positive effects (when studied under natural open-field conditions) on plants.
- In addition to morphological adaptations, cell molecular events that confer UV adaptation,

induced under low-dose UV fluence, and that include DNA repair and also upregulation of biosynthesis of phenolic compounds for providing internal UV screens and also non-enzymatic antioxidants evidently to counter UV effects in plants, are known.

2.6.1 Effect of UV on Crops

Till date most of the studies available in the literature relates to studies on crop species in temperate environments. It is important to appreciate however that UV fluence in the tropics are much higher than in the temperate regions and a lot of agriculture is carried out in tropical countries. For successful plant growth and development in cultivation sites under reduced ozone and concomitant enhancement in UV fluence, it is important to (a) select appropriate UV-adapted/UV acclimated germplasm through molecular screening (Talai and SenMandi 2010) and (b) develop UV tolerant plants through marker assisted breeding based incorporation of UV-tolerant traits into otherwise desirable plants by the process of breeding/plant transformation. Understanding gene involvement and related secondary metabolites in plant protection should help toward developing improved plants not only with respect to plant survival but also with respect to plant quality. Such studies would further be useful in appropriate selection of cultivation sites under the threat of a changing UV environment on earth. For comprehensive understanding on survival under variation of UV fluence at different geographic location and different times of the year, it would be particularly useful to examine the types of adaptations observed in plants growing naturally in regions of the world with variation in solar UV-B flux. In this context, case studies on leaf of tea plants, carried out in the subtropical region of the Northern Hemisphere at two different altitudes but on same latitude (27°N), are presented in Case Study II of both Chaps. 4 and 6. Both these studies, conducted during summer in natural habitat (Bhattacharya and Sen Mandi 2012), demonstrate development of higher flavonoid content that confers proportional aroma and antioxidant value in tea leaf at high altitude. Such studies provide useful information for developing strategies toward exploiting natural phenomenon on plant quality and survival under variation of UV fluence. Interesting information on plant tolerance to UV have been generated from studies of Teramura and Sullivan (1991) on seeds collected from 132 native and introduced plant species growing over a 3000 m elevational gradient in Hawaii (23°N latitude). The studies demonstrate that extent of UV radiation (proportional to altitude of habitat) during early stages of growth determine performance of later growth.

Experiments on food crops under high UV environment have shown lower yields for several key crops such as rice, soy beans and sorghum (Xiong and Day 2001; Krizek et al. 1997, 1998); such effect could be due to macromolecular damage in the embryonic axis of seeds during post harvest 'dry' storage of orthodox seed under ambient UV environment; the lower yield could relate to variable extent of seed viability loss depending on maturity status of seeds. Due to uniniform status of seed viability in a seed lot due to adverse effect of UV on seed during post harvest 'dry' storage in open field (farmers' seed storage), crop stand establishment and thus crop productivity would be adversely affected. Provided that percentage of crop stand establishment is not adversely affected (due to reduction in seed vigour), crop productivity will not be reduced (Ganguli et al. 1992). Studies of Ganguli et al. (1992) have demonstrated that orthodox (rice) seeds adversely affected during post-harvest seed storage under ambient UV in the tropical region bring about adverse effect on crop productivity due to reduced plant establishment under adverse field condition. However, provided that the storage period is not too long and thereby accumulation of damage is not high, seeds may repair such damages incurred during early imbibition associated preemergence stage of the embryonic axis (cf. Biever et al. 2014). Provided that all seedlings may establish in the field, albeit delayed, crop productivity will not be adversely affected. Field validation studies conducted over several growing seasons are crucial in any UV-B impact assessment of agricultural productivity. Few studies have documented the effects of UV-B on total plant yield under field conditions. One notable exception is a 6-year study with soybean demonstrating harvestable yield reductions under a simulated 25 % ozone depletion (Tevini and Teramura 1989). These effects are found to be further modified by prevailing microclimatic conditions. Plants tend to be less sensitive to UV-B radiation under drought or mineral deficiency, while sensitivity increases under low levels of visible light. Field experiments conducted on two soybean cultivars demonstrated intraspecific adverse effect of UV on growth of the sensitive soybean variety, Essen. Solar UV-B exposure is known to reduce biomass production of some plant species by 10–35 % (Krizek et al. 1997, 1998; Mazza et al. 1999; Xiong and Day 2001). Most studies conducted under field conditions show responses varying in extent between species and cultivars to increasing UV-B radiation.

It is important to appreciate that seeds that are not of high viability status soon lose viability in farmers' storage particularly in tropical region where under effect of UV, if not countered by protective compounds (phenolics/flavonoids) synthesized (during secondary metabolism in field-grown plants) and stored in embryonic axis of orthodox quiescent state (at <8 %mc) in post-harvest seed storage, seeds succumb to the damaging effect of UV radiation, thereby bringing about loss of seed viability and thus crop productivity – the extent of UV effect being a function of genetic difference among the plants (see Sect. 6.3.1 of Chap. 6). This results in loss of valuable plant genetic resource. Comparative study on effect of solar UV radiation on seedlings were conducted also on seedlings of wheat (using a German cultivar and Chilean cultivars both summer and winter cultivars) exposed to high solar radiation at a high mountain site in the North Chile Atacama desert (2300 m above sea level). The data revealed effect of both UV-A and UV-B radiation on plant length, leaf area, and dry weight as well as root area and dry weight; the differences (except in leaf length) were more pronounced in the German cultivar than in the Chilean cultivars that were already adapted due to preexposures to high ambient irradiances prevalent in their site of origin. Induction of synthesis of screening pigments in all cultivars appeared induced by both UV-A and UV-B; the induction was more obvious in the German cultivar that required more protection (from UV) than the Chilean cultivars that were already adapted to high level of UV. These changes in plant chemistry also have other important consequences (in addition to shelf life/seed viability protection) evident in other trophic processes, including herbivory, disease, decomposition, and nutrient cycling.

Phenolic compounds, synthesized in field grown plants under UV radiation in addition to providing cellular protection, also affect plant quality. Such UV effects persist long after the end of exposure and is realized in (a) quality of crop yield as has been shown in a few species including tomato, potato (Biggs and Kossuth 1978), and sugar beet (Ambler et al. 1978) and (b) agronomic quality of seed, viz., rice (see Sect. 6.3.1 of Chap. 6).

Flavonoids also have other ecological roles, for example, in defense against herbivores and pathogens. UV-B radiation would also have indirect effects on rice production through effects on other components of rice ecosystem such as weeds, diseases, and nitrogen-fixing cyanobacteria.

2.6.2 Effect of UV on Forests

In contrast to crops, few studies have been undertaken on the effect of UV on woody perennials (trees) that occupy extensive land area and account for up to 80 % of global net primary productivity (Whittaker 1975; Solomon and West 1986). Again, whereas study reports are available that demonstrate deleterious effects of UV-B radiation on tree growth and physiology (Kossuth and Biggs 1981; Sullivan and Teramura 1988, 1989), many of these studies do not provide information beyond a single growing season. Further, lack of long-term effects of UV-B radiation on trees in open field makes it difficult to develop comprehensive understanding the effect of UV on forest trees. Due to their economic importance and widespread global distribution, conifers have been selected for study in over half of the studies on trees. Some 15 species of conifers have been tested to date for susceptibility to UV-B radiation. Of these, seven were deleteriously affected, five were resistant, and three were favored by UV-B radiation (Sullivan and Teramura 1988). Of these, loblolly pine, a leading commercial species with pulp producing capacity (Harlow and Harbar 1969; Walker 1980), exhibited most susceptibility as manifested in reduction of biomass and height. Such studies are particularly important since there would be enormous economic consequences if and when increasing levels of UV-B substantially reduce loblolly pine productivity. Experimental demonstration of the protective role of low dose UV-related flavonoid accumulation in these trees studied through long-term field experiments under fluctuations of natural UV radiation would be useful for understanding effect of UV on growth and survival of forest trees. There exists considerable variation in plant species sensitivity to UV-B radiation (Tevini et al. 1981; Teramura 1983). Some species show sensitivity to present levels of UV-B radiation (Bogenrieder and Klein 1978), while others are apparently unaffected by UV enhancements (experiments conducted at high altitude, i.e., 3000 m – Becwar et al. 1982) – evidently a reflection of the plants' original location, i.e., extent to which the plants have been preexposed. Plants have several mechanisms to ameliorate or repair UV-induced adverse effects and may acclimate to a certain extent to increased UV-B radiation, subject to genetically controlled acclimation potential in the plants under study (Frohnmeyer and Staiger 2003).

2.7 Effect of UV on Animals

Animals avail of protection from overexposure to harmful (UV) rays by fur covering their body; the radiation may nevertheless damage uncovered portion of their body, viz., the nose, paws, and skin around the muzzle. Without fur, human stands susceptible to UV-induced cancer, particularly of the skin. UV radiation may damage eyesight in animals.

Due to recently enhanced UV fluence on earth (as a consequence of stratospheric ozone depletion), frequency of malignancy, eyesight defects, and immunosuppression has increased in animals including humans. The skin of most animals however remains protected from overexposure to harmful rays by their exoskeletons, e.g., fur coat, plumage, hair, feather, scales, etc. However, there are reports that UV-B radiation damage their nose, paws, and skin around the muzzle that are not covered by exoskeletal features (http://www.grida.no/publications/vg/ozone/page/1389.aspx).

2.7.1 Effect of UV on Amphibians

Exposure to UV can affect amphibians at all life stages; this can kill amphibians and induce sublethal effects in embryos, larvae, and adults. As the amphibians lay their eggs in open, shallow water, their embryo and larvae are particularly susceptible to UV-B damage. Although there exist considerable controversies regarding the extent of damage by UV-B radiation (Heyer 2003; Blaustein and Kats 2003; Licht 2003; Leinaas 2002), amphibians such as frogs and salamanders are more susceptible to UV radiation than other vertebrates. Blaustein et al. (1995) have reported that hatching of the long toed salamander embryos was varied in absence or presence of UV-B exposure. It was found that percentage of hatching is 95 % in case of shielded embryos whereas only 14.5 % embryos (without such protective shield) survived after UV-B exposure. It has also been reported that 91.9 % embryos which are exposed to UV-B suffered from morphological deformities in comparison to shielded embryos in which the amount of damage was found to be only 0.5 % (Blaustein et al. 1995). Survival of amphibian species has been found to be correlated with in vivo levels of DNA photolyase. Species with high DNA photolyase activity show lower mortality rates of embryos. Apart from these, UV-B also affects defense mechanism of eggs and embryos, thereby allowing pathogens to infect the eggs. Kiesecker et al. (2001) reported a pathogen (*Saprolegnia ferax*) outbreak event in amphibian (*Bufo boreas*) populations in the western USA; the author linked this occurrence to climate-induced changes in UV-B exposure. UV-B radiation has also been reported to accelerate the decline of frog population (events causing concern for biodiversity loss) by reducing food supply (Whittaker et al. 2013). Although the mechanisms leading to major declines in many amphibian populations are complex, it seems that solar UV radiation is a significant factor, particularly the subtle changes in UV that are caused by emissions of manmade ozone-depleting substances and El Niño oscillations.

2.7.2 Effect of UV on Birds

Birds can perceive the near ultraviolet part (320–400 nm) of the incident solar radiation. Some birds of prey use UV reflectance in their feeding strategy to detect the proper location of prey. UV perception and its application in foraging have also been reported in many herbivorous bird species. Birds have also been shown to use their UV perception ability for recognition of their own eggs (Rajchard 2009). Edwards Jr. (2003) have reported that chickens which are vitamin D3 deficient when exposed to UV light had increased body weight, bone ash, and plasma Ca and decreased incidence of rickets and tibial dyschondroplasia (TD). The knowledge of all specific aspects of bird physiology and role of UV radiation would be immensely useful for the breeders (Rajchard 2009).

2.7.3 Effect of UV on Harmful Insects

Direct damaging effect of UV on insects includes increased mortality of eggs or larvae, reduction in growth, and decrease in the fecundity of adults (Buck and Callaghan 1999). The immune system of insects might also be affected adversely by elevated UV-B as is also reported in case of mammalian immune systems (Paul and Gwynn-Jones 2003). Enhanced UV-B radiation affects growth and survival (Veteli et al. 2003) and behavior (Mazza et al. 1999) of many harmful insects. For example, insects are found to feed underside of leaves where UV-B penetration is low (Paul et al. 1997) or avoid areas of the plant where secondary metabolite produced by plants for protection against harmful UV radiation accumulate (McCloud et al. 1992). Behavioral responses to solar UV-B radiation have also been reported for thrips, the leaf-eating insects which cause severe damage to a variety of commercially important crops (Braga et al. 2001). Thrips preferred leaves of plants that were not exposed to solar UV-B in field choice experiments. This is speculated thus that they have special mechanism for sensing and avoiding exposure to solar UV-B.

UV radiation has been reported to cause significant decrease in survival of *Dysdercus koenigii*, one of the most destructive cotton pests. Increased time of exposure delayed molting into adults. Availability of such information should help to use UV radiation for pest control. It is pertinent to mention however that although UV-B effects have been reported to be useful in the plant–pathogen interaction as a means of crop protection, only a few data are available to infer general rule in plant–pathogen system (Paul 2000).

2.7.4 Effect of UV on Beneficial Insects

In pollinating insects, UV affects growth and development as well as insects' selection of breeding grounds. The caterpillars of subarctic moths have cuticles that help in absorption of UV (Buck and Callaghan 1999), thereby causing cellular damage. UV-B radiation may affect both the autumnal moth and its host and the mountain birch, because of a high adaptation to UV-B radiation of these high-latitude species. Increased UV light affects the growth and development in caterpillar of painted lady butterflies (*Vanessa cardui*) (http://www.bio.miami.edu/ecosummer/eco2015/caterpillar_macaque.pdf). Some flowers, like *Brassica rapa*, have a UV guide for aiding nectar search for the butterfly where the petals reflect near UV light whereas the center of the flower absorbs UV light, creating a visible dark center in the flower when seen in UV condition. This UV guide plays a significant role in *P. rapae* foraging. Studies at ecosystem level reveal that enhanced ultraviolet B (UV-B) radiation may have multiple effects on both plants and animals and affect plant–herbivore interactions directly and indirectly by inducing changes in host plant quality.

2.8 Effect of UV on Humans

In addition to indirect effects of UV on humans through effects on related entities, viz., plants, animals, and microbes, that constitute the biosphere, UV-B can cause direct beneficial as well as harmful effect on humans.

2.8.1 Beneficial Effects of Direct UV Exposure on Humans

UV-B triggers production of vitamin D that is required for bone and muscle development and also for proper functioning of immune system. It also protects human being from the risk of onset of some kinds of cancer, viz., colon cancer. Ninety percent of all requisite vitamin D is produced in the skin that is exposed to sun (Holick 2004). This effect occurs under the influence of UV radiation. It is interesting to note that UV radiation helps to cure skin conditions such as psoriasis (condition when the skin sheds its cells too quickly and develops itchy, scaly patches). Limited exposure to UV allows growth of new skin cells and relieves the symptoms (Holick 2004).

2.8.2 Harmful Effects Due to Direct UV Exposure on Humans

UV radiation at high intensity affects the skin, eye, and immune system (Fig. 2.2). Of these, sunburn (erythema) is the most common acute effect of excessive UV radiation exposure. Prolonged exposure to UV radiation induces damage to the keratinocytes, fibrous tissue, and blood vessels causing premature skin aging photodermatoses and actinic keratoses. Eye is another target of enhanced UV radiation, and cataract incidents have found to be increased by 0.5 % per 1 % depletion of ozone layer (UNEP 1994; Tevini 1993).

Laboratory and epidemiological studies demonstrate that UV-B induces non-melanoma skin cancer and plays a pivotal role in development of malignant melanoma. A lot of research has been performed to understand the relationship between various forms of skin cancer and

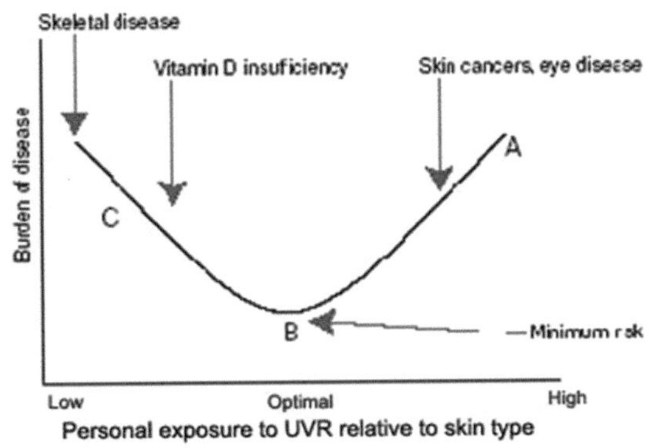

Fig. 2.2 Relationship of exposure to UVR and burden of disease (Taken from WHO report)

increased UV-B radiation (Anna et al. 2007). The United States Environmental Protection Agency (USEPA) has used the results of these studies to conclude that recently enhanced UV radiation results in increased incidences of cutaneous malignant melanoma (potentially mortal skin cancers). High UV exposure (due to the Antarctic ozone hole) to the inhabitants in the southern regions of Australia (Fig. 1.9) appears to be the cause of high frequency of skin cancer among the fair-skinned immigrants in Australia; the local inhabitants of Australia exhibit higher tolerance to UV attributable to genetic difference between these two populations. In addition to this report by USEPA, other analyses have also acknowledged the adverse relationship between reduced stratospheric ozone and increased cancer incidences (Shea 1988; Van Der Leun 1986).

Bibliography

Aarssen LW (1989) Competitive ability and species coexistence: a 'plant's-eye' view. Oikos 56:386401. Copenhagen 1989

Aas E, Høerslev NK (2001) Attenuation of ultraviolet irradiance in North European coastal waters. Oceanologia 43:139–168

Aas P, Lyons MM, Pledger R, Mitchell DL, Jeffrey WH (1996) Inhibition of bacterial activities by solar radiation in nearshore waters and the Gulf of Mexico. Aquat Microb Ecol 11:229–238

ACIA (2005) Arctic climate impact assessment. Cambridge University Press, Cambridge

Ambler JE, Rowland RA, Maher NK (1978). Response of selected vegetable and agronomic crops to increased UV-B radiation under field conditions. In: UV-B biological and climatic effects research (BACER), Final report. EPA-IAG-D6-0168. EPA, Washington, DC, SIRA File No. 142.22

Andersen MK, Hauggaard-Nielsen H, Weiner J, Steen JE (2004) Competitive dynamics in two- and three-component intercrops. J Appl Ecol 44:545–551

Anna B, Blazej Z, Jacqueline G, Andrew CJ, Jeffrey R, Andrzej S (2007) Mechanism of UV-related carcinogenesis and its contribution to nevi/melanoma. Expert Rev Dermatol 2(4):451–469

Antonovics J (1975). Predicting evolutionary response of natural populations to increased UV radiation. In: Nachtwey DS, Caldwell MM, Biggs RH (eds) Climatic ipact assessment program (CIAP), monograph 5. Chap. 8, pp. 3–6. U.S. Dept. Transport, Report No. DOT-TST-75-55, National Techn. Infor. Serv., Springfield, Virginia

Ayres PG, Gunasekera TS, Rasanayagam MS, Paul ND (1996) In: Frankland JC, Magan N, Gadd GM (eds) Fungi and environmental change. Cambridge University Press, Cambridge, pp 32–50

Bailey CA, Neihof RA, Tabor PS (1983) Inhibitory effect of solar radiation on amino acid uptake in Chesapeake Bay bacteria. Appl Environ Microb 46:44–49

Barnes PW, Flint SD, Caldwell MM (1987) Photosynthesis damage and protective pigments in plants from a latitudinal arctic/alpine gradient exposed to supplemental UV-B radiation in the field. Arct Alp Res 9:21–27

Barnes PW, Jordan PW, Gold G, Flint SD, Caldwell MM (1988) Competition, morphology and canopy structure in wheat (*Triticum aestivum* L.) and wild oat (*Avena fatua* L.) exposed to enhanced ultraviolet-B radiation. Funct Ecol 2:319–330

Becwar MR, Moore FD III, Burke MJ (1982) Effects of deletion and enhancement of ultraviolet-B (280–315 nm) radiation on plants grown at 3000 m elevation. J Am Soc Hortic Sci 107:771–779

Bhattacharya S, Sen-Mandi S (2012) Natural ultra violet irradiance related variation in antioxidant and aroma compounds in tea (*Camelia sinensis* L. Kuntze) plants grown in two different altitudes. Int J Environ Biol 2 (1):1–6

Biever JJ, Brinkman D, Gardener G (2014) UV-B inhibition of hypocotyl growth in etiolated *Arabidopsis thaliana* seedlings is a consequence of cell cycle arrest initiated by photodimer accumulation. J Exp Bot 65 (11):2949–2961

Bieza K, Lois R (2001) An Arabidopsis mutant tolerant to lethal ultraviolet-B levels shows constitutively elevated accumulation of flavonoids and other phenolics. Plant Physiol 126:1105–1115

Biggs RH, Kossuth SV (1978) Effects of ultraviolet-B radiation enhancement under field conditions on potatoes, tomatoes, corn, rice, southern peas, peanuts, squash, mustard and radish. UV-B Biological and Climatic Effects Research (BACER), Final Report, EPA, Washington, DC

Biggs RH, Kossuth SV, Teramura AH (1981) Response of 19 cultivars of soybeans to ultraviolet-B irradiance. Physiol Plant 53:19–26

Bjorn LO, Callaghan TV, Johnsen I, Lee JA, Manetas Y, Paul ND, Sonesson M, Wellburn AR, Coop D, Heide-Jørgensen HS, Gehrke C, Gwynn-Jones D, Johanson U, Kyparissis A, Levizou E, Nikolopoulos D, Petropoulou Y, Stephanou M (1997) The effects of UV-B radiation on European heathland species. Plant Ecol 128:253–264

Blaustein AR, Kats LB (2003) Amphibians in a very bad light. Bioscience 3:1028–1029

Blaustein AR, Edmond B, Kiesecker JM, Beatty JJ, Hokit DG (1995) Ambient ultraviolet radiation causes mortality in salamander eggs. Ecol Appl 5:740–743

Blazer VS, Fabacher DL, Little EE, Ewing MS, Kocan KM (1997) Effects of ultraviolet-B radiation on fish: histologic comparison of a UV-B-sensitive and a UV-B-tolerant species. J Aquat Anim Health 9:132–143

Bogenrieder A, Klein R (1978) Die abhangigkeit der UV-empfindlichkeit von der lichtqualitat bei der aufzucht (*Lactuca sativa* L.). Angew Bot 52:283–293

Bornman JF, Barnes PW, Robinson SA, Ballaré CL, Flint SD, Caldwell MM (2015) Solar ultraviolet radiation and ozone depletion-driven climate change: effects on terrestrial ecosystems. Photochem Photobiol Sci 14:88–107

Braga GUL, Flint SD, Miller CD, Anderson AJ, Roberts DW (2001) Both solar UV-A and UV-B radiation impair conidial culturability and delay germination in the entomopathogenic fungus *Metarhizium anisopliae*. Photochem Photobiol 74:734–739

Brasseur GP, Granier C, Waiters S (1990) Future changes in stratospheric ozone and the role of heterogeneous chemistry. Nature 348:626–628

Browman HI, Vetter RD, Rodriguez CA, Cullen JJ, Davis RF et al (2003) Ultraviolet (280–400 nm)-induced DNA damage in the eggs and larvae of *Calanus finmarchicus* G. (Copepoda) and Atlantic cod (*Gadus morhua*). Photochem Photobiol 77:397–404

Buck N, Callaghan TV (1999) The direct and indirect effects of enhanced UV-B on the moth caterpillar *Epirrita autumnata*. Ecol Bull 47:68–76

Caldwell MM, Robberecht R, Nowak RS, Billings WD (1982) Differential photosynthetic inhibition by ultraviolet radiation in species from the arctic-alpine life zone. Arct Alp Res 14:195–202

Caldwell MM, Bjorn LO, Bomman JF, Flint SD, Kulandaivelu G, Teramura M, Tevini M (1998) Effect of increased solar ultraviolet radiation on terrestrial ecosystem. J Photochem Photobiol 46(5):40–52

Carletti P, Masi A, Wonisch A, Grill D, Tausz M, Ferretti M (2003) Changes in antioxidant and pigment pool dimensions in UV-B radiation maize seedlings. Environ Exp Bot 50:149–157

Coleman RS, Day TA (2004) Response of cotton and sorghum to several levels of subambient solar UV-B radiation: a test of the saturation hypothesis. Physiol Plant 122:362–372

Crutzen PJ (1992) Ultraviolet on the increase. Nature 356:104–105

Dai Q, Peng S, Chavez AQ, Vergara BS (1995) Effects of UV-B radiation on stomatal density and opening in rice (*Oryza sativa* L.). Ann Bot 76:65–70

Deckmyn G, Impens I (1997) Combined effects of enhanced UV-B radiation and nitrogen deficiency on the growth, composition and photosynthesis if rye. Plant Ecol 128:235–240

de Lange HJ, Verschoor AM, Gylstra R, Cuppen JGM, van Donk E (1999) Effects of artificial ultraviolet-B radiation on experimental aquatic microcosms. Freshw Biol 42:545–560

Dong Q, Svoboda K, Tiersch TR, Monroe WT (2007) Photobiological effects of UV-A and UV-B light in zebrafish embryos: evidence for a competent photorepair system. J Photochem Photobiol B Biol 88:137–146

Edwards Jr (2003) Effects of UV-B radiation of very young chickens on growth and bone development. Br J Nutr 90:151–160

Erdei E, Torres SM (2013) A new understanding in the epidemiology of melanoma. Expert Rev Anticancer Ther 10(11):1811–1823

Fedina I, Hidema J, Velitchkova M, Georgieva K, Nedeva D (2010) UV-B induced stress responses in three rice cultivars. Biol Plant 54(3):571–574

Filella I, Peñuelas J (1999) Altitudinal differences in UV absorbance. UV reflectance and related morphological traits of *Quercus ilex* and *Rhododendron ferrugineum* in the Mediterranean region. Plant Ecol 145:157–165

Frohnmeyer H, Staiger D (2003) Ultraviolet-B radiation-mediated responses in plants. Balancing damage and protection. Plant Physiol 133:1420–1428

Fukunishi Y, Browman HI, Durif CMF, Bjelland RM, Skiftesvik AB (2012) Effect of sub-lethal exposure to Ultraviolet radiation on the escape performance of Atlantic Cod Larvae (*Gadus morhua*). PLoS One 7 (4):1–6

Furness N, Upadhyaya MK, Ormrod DP (1999) Seedling growth and leaf surface morphological responses of three rangeland weeds to ultraviolet-B radiation. Weed Sci 47:427–434

Ganguli S, Das G, Sen-Mandi (1992) Plant emergence and productivity of different vigour of wheat (Triticum aestivum) seeds. Ind J Agr Sci 62 (3):224–227

Gibson JAE, Vincent WF, Nieke B, Pienitz R (2000) Control of biological exposure to UV radiation in the Arctic Ocean: comparison of the roles of ozone and riverine dissolved organic matter. Arctic 53:372–382

Gonçalves RJ, Souza MS, Aigo J, Modenutti B, Balseiro E, Villafañe VE, Cussac V, Helbling EW (2010) Responses of plankton and fish from temperate zones to UVR and temperature in a context of global change. Ecol Aust 20:129–153

Greenberg BM, Wilson MI, Huang X-D, Duxbury CL, Gerhaddt KE, Gensemer RW (1997) The effects of ultraviolet-B radiation on higher plants. In: Wang W, Goursuch J, Hughes JS (eds) Plants for environmental studies. CRC Press, Boca Raton, pp 1–35

Gunasekera TS, Paul ND, Ayres PG (1997) The effects of ultraviolet-B (UV-B: 290–320 nm) radiation on blister blight diseases of tea (*Camellia sinensis*). Plant Pathol 46:179–185

Gustavson K, Garde K, Wängberg SA, Selmer JS (2000) Influence of UV-B radiation on bacterial activity in coastal waters. J Plankton Res 22(8):1501–1511

Hader DP, Kumar HD, Smith RC, Worrest RC (2007) Effects of solar UV radiation on aquatic ecosystems and interactions with climate change. Photochem Photobiol Sci 6:267–285

Harlow WM, Harbar ES (1969) Textbook of dendrology. McGraw-Hill, Springer Publication, New York, pp 87–90. ISBN 07-026569-0

Hartley SE, Gange AC (2009) Impacts of plant symbiotic fungi on insect herbivores: mutualism in a multitrophic context. Annu Rev Entomol 54:323–342

Heijde M, Ulm R (2012) UV-B photoreceptor-mediated signaling in plants. Trends Plant Sci 17:230–237

Helbling EW, Marguet ER, Villafane VE, Holm-Hansen O (1995) Bacterioplankton viability in antarctic waters as affected by solar ultraviolet radiation. Mar Ecol Prog Ser 126:293–298

Helbling EW, Gao K, Ai H, Ma Z, Villafane VE (2006) Differential responses of Nostoc *sphaeroides* and *Arthrospira platensis* to solar ultraviolet radiation exposure. J Appl Phycol 18:57–66

Herman JR, Larko D (1994) Low ozone amounts during 1992–93 from Nimbus 7 and Meteor 3 total ozone mapping spectrometers. J Geophys Res 98:12783–12793

Heyer WR (2003) Ultraviolet-B and amphibian. Bioscience 53:540–541

Hidema J, Taguchi T, Ono T, Teranishi M, Yamamoto K, Kumagai T (2007) UV-B resistance and CPD photolyase activity in rice. Plant J 50:70–79

Holick MF (2004) Vitamin D: importance in the prevention of cancers, type 1 diabetes, heart disease, and osteoporosis. Am J Clin Nutr 79:362–371

http://www.bio.miami.edu/ecosummer/eco2015/caterpillar_macaque.pdf

http://www.grida.no/publications/vg/ozone/page/1389.aspx

http://www.scientificamerican.com/article/origin-of-oxygen-in-atmosphere/

Huebner JD, Young DLW, Loadman NL, Lentz VJ, Wiegand MD (2006) Age-dependent survival, reproduction and photorepair activity in Daphnia magna (Straus, 1820) after exposure to artificial ultraviolet radiation. Photochem Photobiol 82:1656–1661

Hunter JR, Taylor JH, Moser HG (1978) Effect of ultraviolet radiation on eggs and larvae of the northern anchovy, Engraulis mordax, and the pacific mackerel, Scomber japonicus, during the embryonic stage. Photochem Photobiol 29:325–338

Johnson D (2003) Response of terrestrial microorganisms to ultraviolet-B radiation in ecosystems. Res Microbiol 154:315–320

Jokinen IE, Markkula ES, Salo HM, Kuhn P, Nikoskelainen S et al (2008) Exposure to increased ambient ultraviolet B radiation has negative effects on growth, condition and immune function of juvenile Atlantic salmon (*Salmo salar*). Photochem Photobiol 84:1265–1271

Kakani VG, Reddy KR, Zhao D, Mohammed AR (2003) Effects of ultraviolet-B radiation on cotton (*Gossypium hirsutum* L.) morphology and anatomy. Ann Bot 91:817–826

Kerr JB, McElory (1993) Evidence for large upward trends of ultraviolet – B radiation linked to ozone depletion. Science 262:1032–1034

Kiesecker JM, Blaustein AR, Belden LK (2001) Complex causes of amphibian decline. Nature 410:681–684

Klem K, Alexander AC, Holuba P, Kovca D, Spundaa V, Robsonb M, Urbana O (2012) Interactive effect of PAR and UV radiation on the physiology, morphology and leaf optical properties of two barley varieties. Environ Exp Bot 75:52–64

Kolb CA, Käser MA, Kopecký J, Zotz G, Riederer M, Pfündel EE (2001) Effects of natural intensities of visible and ultraviolet radiation on epidermal ultraviolet screening and photosynthesis in grape leaves. Plant Physiol 127(3):863–875

Kossuth SV, Biggs RH (1981) Ultraviolet-B radiation effects on early seedling growth of Pinaceae species. Can J For Res 11:243–248

Kouwenberg JHM, Browman HI, Cullen JJ, Davis RF, St-Pierre JF et al (1999) Biological weighting of ultraviolet (280–400 nm) induced mortality in marine zooplankton and fish. I. Atlantic cod (*Gadus morhua*) eggs. Mar Biol 134:269–284

Kramer GF, Norman HA, Krizek DT, Mirecki RM (1991) Influence of UV-B radiation on polyamines, lipid peroxidation and membrane lipids in cucumber. Phytochemistry 30:2101–2108

Krizek DT, Mirecki RM, Britz SJ (1997) Inhibitory effects of ambient levels of solar UV-A and UV-B radiation on growth of cucumber. Physiol Plant 100:886–893

Krizek DT, Britz SJ, Mirecki RM (1998) Inhibitory effects of ambient levels of solar UV-A and UV-B radiation on growth of cv. new red fire lettuce. Physiol Plant 103:1–7

Laurion I, Ventura M, Catalan J, Psenner R, Sommaruga R (2000) Attenuation of ultraviolet radiation in mountain lakes: factors controlling the among- and within-lake variability. Limnol Oceanogr 45 (6):1274–1288

Leinaas HP (2002) UV tolerance, pigmentation and life forms in high Arctic Collembola. In: Hessen DO (ed) UV radiation and arctic ecosystems. Springer, Berlin, pp 123–134

Licht LE (2003) Shedding light on ultraviolet radiation and amphibian embryos. Bioscience 53:551–561

Loreau M, Naeem S, Inchausti P, Bengtsson J, Grime JP, Hector A, Hooper DU, Huston MA, Raffaelli D, Schmid B, Tilman D, Wardle DA (2001) Ecology – biodiversity and ecosystem functioning: current knowledge and future challenges. Science 294: 804–808

Maher KA, Stephenson DJ (1988) Impact frustration of the origin of life. Nature 331(6157):612–614

Malloy KD, Holman MA, Mitchell D, Detrich HW (1997) Solar UV-B-induced DNA damage and photoenzymatic DNA repair in Antarctic zooplankton. Proc Natl Acad Sci U S A 94:1258–1263

Markkula SE, Salo HM, Immonen AK, Jokinen EM (2005) Effects of short- and long-term ultraviolet B radiation on the immune system of the common carp (*Cyprinus carpio*). Photochem Photobiol 81:595–602

Markkula SE, Salo HM, Rikalainen AK, Jokinen EI (2006) Different sensitivity of carp (*Cyprinus carpio*) and rainbow trout (*Oncorhynchus mykiss*) to the immunomodulatory effects of UV-B radiation. Fish Shellfish Immunol 21:70–79

Markkula E, Salo HM, Rikalainen K, Jokinen IE (2009) Long-term UV-B radiation affects the immune functions of Carp (*Cyprinus carpio*) and Rainbow Trout (*Oncorhynchus mykiss*). Photochem Photobiol 85:347–352

Mazza CA, Battista D, Zima AM, Szwarcberg-Bracchitta-M, Giordano CV, Acevedo A, Scopel AL, Ballaré CL (1999) The effects of solar ultraviolet-B radiation on the growth and yield of barley are accompanied by increased DNA damage and antioxidant responses. Plant Cell Environ 22:61–70

McCloud ES, Berenbaum MR, Tuveson RW (1992) Furanocoumarin content and phototoxicity of rough lemon (Citrus jambhiri) foliage exposed to enhanced ultraviolet-B (UV-B) radiation. J Chem Ecol 18:1125–1137

McFadzen I, Baynes S, Hallam J, Beesley A, Lowe D (2000) Histopathology of the skin of UV-B irradiated sole (*Solea solea*) and turbot (*Scophthalmus maximus*) larvae. Mar Environ Res 50:273–277

McKenzie RL, Aucamp PJ, Bais AF, Bjorn LO, Ilyas M (2007) Changes in biologically active ultraviolet radiation reaching the Earth's surface. Photochem Photobiol Sci 2:5–15

Mirecki RM, Teramura AH (1984) Effects of ultraviolet-B irradiance on soybean. The dependence of plant sensitivity on the photosynthetic photon flux density during and after leaf expansion. Plant Physiol 74:475–480

Moody SA, Newsham KK, Ayres PG, Paul ND (1999) Variation in the responses of litter and phylloplane fungi to UV-B radiation (290–315 nm). Mycol Res 103(11):1469–1477

Mpoloka SW (2008) Effect of prolonged UV-B exposure in plants. Afr J Biotechnol 7(25):4874–4883

Newsham KK, Low MNR, McLeod AR, Greenslade PD, Emmett BA (1997) Ultraviolet-B radiation influences the abundance and distribution of phylloplane fungi on pedunculate oak (*Quercus robur*). New Phytol 136:287–297

Nisbet EG, Sleep NH (2001) The habitat and nature of early life. Nature 409:1083–1091

Norris TB, McDermott TR, Castenholz RW (2002) The long term effects of UV exclusion on the microbial composition and photosynthetic competence of bacteria in hot-spring microbial mats. FEMS Microbiol Ecol 39:193–209

Pakulski JD, Aas P, Jeffrey W et al (1998) Influence of light on bacterioplankton production and respiration in a subtropical coral reef. Aquat Microb Ecol 14:137–148

Paul ND (2000) Stratospheric ozone depletion, UV-B radiation and crop disease. Environ Pollut 108: 343–355

Paul ND, Gwynn-Jones D (2003) Ecological roles of solar UV radiation: towards an integrated approach. Trends Ecol Evol 18:48–55

Paul ND, Rasanayagam S, Moody SA, Hatcher PE, Ayres PG (1997) The role of interactions between trophic levels in determining the effects of UV-B on terrestrial ecosystems. Plant Ecol 128:296–308

Prabha GL, Kulandaivelu G (2002) Induced UV-B resistance against photosynthesis damage by adaptive mutagenesis in *Synechococcus* PCC 7942. Plant Sci 162:663–669

Proffitt MH, Margitan JJ, Kelly KK, Lowenstein M, Podolske JR (1990) Ozone loss inferred from high-altitude aircraft measurements. Nature 347:31–33

Rajagopal S, Sicora C, Varkonyi Z, Mustardy L, Mohanty P (2005) Protective effect of supplemental low intensity white light on ultraviolet-B exposure-induced impairment in cyanobacterium *Spirulina platensis*: formation of air vacuoles as a possible protective measure. Photosynth Res 85:181–189

Rajchard J (2009) Ultraviolet (UV) light perception by birds – a review. Vet Med 54(8):351–359

Rasanayagam MS, Paul ND, Royle DJ, Ayres PG (1995) Variation in responses of *Septoria triciti* and *S. nodorum* to UV-B radiation in vitro. Mycol Res 99:1371–1377

Raviv M, Antignus Y (2004) UV radiation effects on pathogens and insect pests of greenhouse-grown crops. Photochem Photobiol 79(3):219–226

Robberecht R, Caldwell MM (1978) Leaf epidermal transmittance of ultraviolet radiation and its implication for plant sensitivity to ultraviolet-radiation induced injury. Oecologia 32:277–287

Rozema J, Bjorn OL, Bornman JF et al (2002) The role of UV-B radiation in aquatic and terrestrial ecosystems – an experimental and functional analysis of the evolution of UV-B absorbing compounds. J Photochem Photobiol B Biol 66:2–12

Shea CP (1988) Protecting life on earth: steps to save the ozone layer, Worldwatch Paper 87. Worldwatch Institute, Washington, DC

Shick JM, Lesser MP, Dunlap WC, Stochaj WR, Chalker BE, Won JW (1995) Depth-dependent response to solar ultraviolet-radiation and oxidative stress in the zooxanthellate coral *Acropora microphthalma*. Mar Biol 122:41–51

Shyam Choudhury S, Sen Mandi S (2012) Natural ultra violet radiation on field grown rice (Oryza sativa L.) plants confer protection against oxidative stress in seed during storage under subtropical ambience. Environ Pollut 1:2

Sierachi ME, Sieburth JM (1986) Sunlight induced growth delay of planktonic marine bacteria in filtered seawater. Mar Ecol Prog Ser 33:19–27

Sinha RP, H¨ader D-P (2004) Ultraviolet screening compounds in algae: role in evolution. In: Britto SJ (ed) Molecular systematics. The Rapinat Herbarium & Centre for Molecular Systematics, Tiruchirappalli, pp 91–117

Solomon AM, West DC (1986) Simulating forest responses to expected climate change in eastern North America: applications to decision-making in the forest industry. In: Shands WE (ed) Climate change and future forest management in the United States. The Conservation Foundation, Washington, DC

Sommaruga R (2001) The role of UV radiation in the ecology of alpine lakes. J Photochem Photobiol B Biol 62:35–42
Sommaruga R, Augustin G (2006) Seasonality in UV transparency of an alpine lake is associated to changes in phytoplankton biomass. Aquat Sci 68:129–141
Steeger HU, Wiemer M, Freitag JF, Paul RJ (1999) Vitality of plaice embryos (Pleuronectes platessa) at moderate UV-B exposure. J Sea Res 42:27–34
Strid A, Chow WS, Anderson JM (1994) UV-B damage and protection at the molecular level in plants. Photosynth Res 39:475–489
Sullivan JH, Teramura AH (1988) Effects of ultraviolet radiation on seedling growth in the Pinaceae. Am J Bot 75:225–230
Sullivan JK, Teramura AH (1989) The effects of ultraviolet-B radiation on Iobelly pine: 1. Growth, photosynthesis and pigment production in greenhouse-grown seedlings. Physiol Plant 77: 202–207
Surplus SL, Jordan BR, Murphy AM, Carr JP, Thomas B, Mackerness SA-H (1998) UV-B induced response in Arabidopsis thaliana: role of salicylic acid and ROS in the regulation of transcripts and acidic PR proteins. Plant Cell Environ 21:685–694
Talai S, Sen-Mandi S (2010) Seed vigour-related DNA marker in rice shows homology with Acetyl CoA Carboxylase gene. Acta Physiol Plant 32:153–167
Tedetti M, Sempèrè R (2006) Penetration of ultraviolet radiation in the marine environment. A review. Photochem Photobiol 82:389–397
Tedetti M, Sempe're' R, Vasilkov A, Charrie're B, Ne'rini D, Miller WL, Kawamura K, Raimbault P (2007) High penetration of ultraviolet radiation in the southeast Pacific waters. Geophys Res Lett 34: L12610. doi:10.1029/2007GL029823
Teramura AH (1983) Effects of ultraviolet-B radiation on the growth and yield of crop plants. Physiol Plant 58:415–427
Teramura AH, Murali NS (1986) Intraspecific differences in growth and yield of soybean exposed to ultraviolet-B radiation under greenhouse and field conditions. Environ Exp Bot 26:89–95
Teramura AH, Sullivan JH (1991) Potential impacts of increased solar UV-B on global plant productivity. In: Riklis E (ed) Photobiology. Plenum Press, New York, pp 625–634
Teramura AH, Sullivan JH (1994) Effects of UV-B radiation on photosynthesis and growth of terrestrial plants. Photosynth Res 39(3):463–473
Tevini M (1993) Effects on enhanced UV-B radiation on terrestrial Plants. In: Tevini M (ed) UV-B radiation and ozone depletion. Effects on humans, animals, plants, microorganisms and materials. Lewis Publishers, Boca Raton, pp 125–153
Tevini M, Teramura AH (1989) UV-B effect on terrestrial plants. Photochem Photobiol 50:479–487
Tevini M, Iwanzik W, Thoma U (1981) Some effects of enhanced UV-B radiation on the growth and composition of plants. Planta 153:388–394
Tilman D (1988) Plant strategies and the dynamics and structure of plant communities. Princeton University Press, Princeton
United Nations Environment Programme (UNEP) (1994) Environmental effects of ozone depletion: 1994 assessment
United Nations Environment Programme (UNEP) (2006) Climate change and its effect on ecosystem: assessment
Van Der Leun JC (1986). Health effects of ultraviolet radiation. Draft report to the United Nations Environmental Protection Coordinating Committee on the Ozone Layer. Bilthoven. 19–21 November
Veteli TO, Tegelberg R, Pusenius J, Sipura M, Julkunen-Tiitto R, Aphalo PJ, Tahvanainen J (2003) Interactions between willows and insect herbivores under enhanced ultraviolet-B radiation. Oecologia 137:312–320
Walker LC (1980) The southern pine region. In: Barrett JW (ed) Regional silviculture of the United States. Wiley, New York. ISBN 0471-05645-6
Warner CW, Caldwell MM (1983) Influence of photon flux density in the 400–700 nm waveband of inhibition of photosynthesis by UV-B (280–320 nm) radiation in soybean leaves: separation of indirect and immediate effects. Photochem Photobiol 38:341–346
Weinig C, Gravuer KA, Kane NC, Schmitt J (2004) Testing adaptive plasticity to UV: costs and benefits of stem elongation and light-induced phenolics. Evolution 58:2645–2656
Whitlock R (2014) Relationships between adaptive and neutral genetic diversity and ecological structure and functioning: a meta-analysis. J Ecol 102(4): 857–872
Whittaker RH (1975) Communities and ecosystems. MacMillan Co., New York
Whittaker K, Koo MS, Wake DB, Vredenburg VT (2013) Global declines of amphibians. Encyclopedia of biodiversity, vol 3. Elsevier
Williamson CE, Zagarese HE, Schulze PC, Hargreaves BR, Seva J (1994) The impact of short-term exposure to UV-B radiation on zooplankton communities in north temperate lakes. J Plankton Res 16:205–218
Xiong FS, Day TA (2001) Effect of solar ultraviolet-B radiation during springtime ozone depletion on photosynthesis and biomass production of Antarctic vascular plants. Plant Physiol 125:738–751
Zagarese HE, Tartarotti B, Añón Suárez A (2003) The significance of ultraviolet radiation for aquatic organisms. In: Ambasht RS, Ambasht NK (eds) Modern trends in applied aquatic ecology. Kluwer Academic/Plenum Press, New York, pp 173–200
Zellmer ID (1996) The impact on UV-B tolerance and recovery from UV-B damage in Daphnia pulex. Hydrobiologia 319:87–92

Zellmer ID (1998) The effects of natural UV-A and UV-B on subarctic Daphnia pulicaria in its natural habitat. Hydrobiologia 379:55–62

Zellmer ID, Arts MT, Abele D, Humbeck K (2004) Evidence of sublethal damage in Daphnia (Cladocera) during exposure to solar UV radiation in subarctic ponds. Arct Antarct Alp Res 36:370–377

Zhao D, Reddy KR, Kakani VG, Read JJ, Sullivan JH (2003) Growth and physiological responses of cotton (*Gossypium hirsutum* L.) to elevated carbon dioxide and ultraviolet-B radiation under controlled environmental conditions. Plant Cell Environ 26(5):771–782

3 UV Radiation-Induced Damage at Molecular Level

3.1 Introduction

Sun's emission of UV radiation, conceptually divided into three wavebands (viz., UV-A, UV-B, UV-C) on the basis of wavelengths and associated energy of impact, is represented in the earth's atmosphere (after passing through the stratospheric ozone layer) only by UV-B and UV-A, together constituting only 0.5 % of total solar UV (UV-C is totally restricted from entering the earth's atmosphere –see Fig. 1.2). Even this small fraction of solar UV (A+B) radiation fluencing on earth threatens to jeopardize life processes; this presents the highest challenge to survival of life forms on earth.

Of the 0.5% of solar radiation entering the troposphere, UV-B constitutes a small proportion (5–10% – Saewan and Jimtaisong 2013) while UV-A constitutes 90%. Even though present at a low level, UV-B represents the most damaging UV component in earth's atmosphere; UV-A, despite its high proportion in the troposphere, was long believed to be innocuous apparently because of its low energy of impact (correlated with its low wavelength). Recent studies (Girard et al. 2011) have however established that cell molecular damages caused by UV-A are comparable to the damaging effects of UV-B – albeit effective through varied mode of action.

Damage of DNA is the most hazardous effect of UV relating to persistence of life since this causes disruption in the cell cycle process (Kunz et al. 2005) and if not repaired in time would lead to programmed cell death/apoptosis. UV-B, by direct impact, causes damage to macromolecules such as DNA, proteins, and phospholipids (Beggs et al. 1986), UV-A related direct impact induced damage has been reported only for DNA (Mouret et al. 2006); other macromolecules viz. proteins and phospholipids incur UV-A induced damage indirectly, generally via a specific ROS species, viz., singlet oxygen, that is particularly produced by UV-A evidently because of its low energy (Baier et al. 2007). UV-A related production of other ROS species production and their role in macromolecular damage induction have also been reported (Moldau 1999). UV-B, in addition to causing direct damage, also produces ROS of the type commonly known in cells. ROS mediated UV-B damage is thus a function of nonspecific ROS, although recently available information report that UV-B also produces singlet oxygen as reactive oxygen species (Regensburger et al. 2012).

The ubiquitous presence of UV radiation spontaneously fluencing in the earth's atmosphere causes damaging impact on all living forms, the effect being particularly hazardous for plants because of their stationary habit in open field. This presents threat of dire consequence for survival of life on earth. This is of particular concern with respect to plants that being primary producers on earth, are pivotal for maintenance of life on earth. This calls for developing

S. Sen Mandi, *Natural UV Radiation in Enhancing Survival Value and Quality of Plants*,
DOI 10.1007/978-81-322-2767-0_3

comprehensive understanding on UV induced damages in plants. Data from experiments often conducted in greenhouse exhibit exaggerated damaging effects in plants (Gao et al. 2004) compared to data obtained from experiments in open field (Kakani et al. 2003). This establishes that in presence of other environmental radiations (viz. Photosynthetically Active Radiation- PAR) in open atmosphere, the damaging effects of UV appear largely moderated (Kolb et al. 2001). Teramura and Sullivan (1991) reported that while (soybean) plants were found to be moderately susceptible to UV radiation when grown in open field, the same plants exhibit higher sensitivity to UV when grown in green house under low levels of non-UV radiation including PAR (Teramura and Sullivan 1991). Such studies demonstrate the importance of ambient radiation status for developing realistic understanding on natural effect of UV on plants. It is important also to note that solar and artificial UV-B radiation sources have different spectral energy distributions, and, thus in attempts at deriving realistic data from closed chamber experiments, it is important that the effect of artificial radiation be weighted for determining its biological effectiveness (Hunter et al. 1981).

UV radiation is reported to induce damaging effect on plants growing in open air (Bjorn 1996). It is important to note that UV induced adverse effect evident in plants is not similar among different species and cultivars (Teramura and Murali 1986; Mohammed and Tarpley 2009). This indicates that variation in UV induced effects in plants is manifested as a resultant of genetically controlled UV acclimation processes. Studying the effects of elevated UV-B radiation on native and cultivated plants of Southern Africa Musil et al. (2002) established the role of genetic variation on manifested UV effect. This determines the resultant survival status of plants on earth. Wide inter- and intraspecific differences in growth, production of dry matter and physiological and biochemical changes in response to UV-B radiation have often been reported (Kramer et al. 1991; Mpoloka 2008; Fedina et al. 2010). Strid et al. (1990) and Battag and Brennan (2000) showed that following UV-B treatment, the concentrations of Chlorophyll a and b in cucumber (*Cucumis sativus* L.) were not affected by UV-B treatment. Studies on effect of enhanced ultraviolet radiation on metabolite composition and growth of plants such as barley, maize, kidney beans and radishes (Tevini et al. 1981) indicate that genetically regulated molecular rearrangements involved in diversion of primary metabolism to secondary metabolism for synthesis of flavonoids (i.e. UV screening compounds) is basic to development of UV acclimation that ensures survival of plants, the primary producers on earth.

When supraoptimal UV-B doses overweigh genetically controlled UV acclimation capacity, plants exhibit damaging effects (Bartholic et al. 1975; Biggs and Basiouny 1975; Caldwell et al. 1978; Mpoloka 2008) manifested in different processes such as in: inhibited plant growth (Sullivan and Teramura 1989; Mohammed and Tarpley 2009), reduced biomass production (Tosserams et al. 2001; Tsukaguchi and Iida 2008), reduced crop yield (Teramura and Sullivan 1991; Corlett et al. 1997; Mohammed and Tarpley 2009) and reduced capacity of photosynthetic primary reaction (Ilwanzik et al. 1983; Hidema et al. 1991; Teramura and Sullivan 1994). Reports that ambient UV radiation in the post ozone hole era is taking its toll often in the way of varietally regulated loss of biodiversity are also reported in the literature (Bornman et al. 2015).

With realization of stratospheric ozone thinning and detection of an "ozone hole" in the Antarctica during spring of 1978. In the post-ozone hole era, thus for monitoring UV fluence in the earth's atmosphere and its associated biological effect, organized systems of monitoring UV radiation flux at different location on earth were established in 1981 at Jungfraujoch (3576 m a.s.l) and at Innsbruck (577 m a.s.l) in Switzerland; these centers record daily totals of UV-A and UV-B radiation fluxes and global radiation. A long-standing series of records at the Light-Climatic Observatory in Arosa, Switzerland, revealed that the content of total ozone of the atmosphere (largely accumulating in the stratospheric ozone layer) has been decreasing by 0.23 % per year since the beginning of the 1970s (Staehilin and Dütsch 1989; Staehilin et al. 1990), and this was correlated with an increase in solar UV-B radiation flux by 0.5–1 % per year since 1981, as recorded in the course of seasonal recording at the High Alpine Research Station, Jungfraujoch (Blumthaler and Ambach 1990) and Insbruck. For determining UV effect on humans, biologically effective doses of erythema reaction (reflecting variation in UV dose) have been recorded as an index of biological effect.

For a comprehensive assessment of biological effects under environment-related variable UV fluence, development of biological sensors for monitoring the impact of solar UV radiation under various environmental conditions would be a useful endeavor (Yagura et al. 2011). Exemplary field experiments of Schuch et al. (2012) using a DNA dosimeter together with photometric assessment of solar UV-B/UV-A radiation at different latitudes of South America showed that UV-related DNA lesion pattern largely reflects the relative incidence of UV-A and UV-B at a specific latitude. Such spectrophotometric data matches with data derived from biochemical studies on specific DNA repair enzymes and antibodies, using biochemical and immunological methods for evaluating solar radiation associated UV-induced DNA damage. Such studies established that the genotoxic potential of UV varies according to latitude related variation of UV intensity (Schuch et al. 2012).

3.2 Adverse Effect of UV Radiation on Living Cells

UV radiation at high intensity, even in brief exposure, have been reported to exhibit macromolecular damage by a) direct impact on DNA, proteins, membrane lipids, and associated inhibitory effect on plant growth and development, protein synthesis, and photosynthetic reactions (Bjorn 1996; Casati and Walbot 2004; Jansen et al. 1998) and b) indirect effect resulting from oxidative damage mediated by ROS. Such macromoleculer damage cause disruption of important cell molecular events that, under extreme conditions, lead to irreversible cell damages such as tissue necrosis that may ultimately kill the plants (Girotti 2001; Rebeiz et al. 1988).

3.2.1 Indirect Effect

Indirect effect of UV-B radiation is mediated by a range of reactive oxygen species (ROS), viz., H_2O_2 (hydrogen peroxide), OH^- (hydroxyl ion), O_2^- (superoxide) (Agrawal and Rathore 2007; Du et al. 2011; Singh et al. 2011), and $_1O^2$ (singlet oxygen) (Fig. 3.1) (Moldau 1999). The form of reactive oxygen particularly produced by UV-A is singlet oxygen (Baier et al. 2007). UV-B-mediated ROS cause oxidative damage

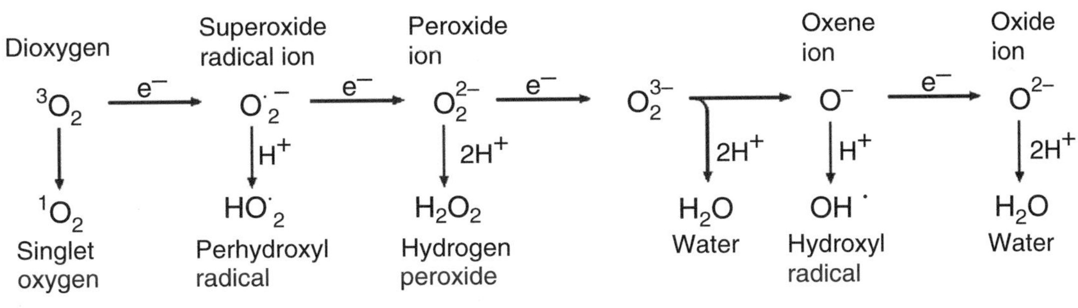

Fig. 3.1 Generation of different ROS by energy transfer or sequential univalent reduction of ground-state triplet oxygen (Taken from Apel and Hurt 2004)

of macromolecules, viz., nucleic acids, membrane lipids, and proteins including enzymes (Roleda et al. 2006a, b). While UV-B gives rise to a wide range of ROS species, UV-A being lower in energy, compared to UV-B (Sutherland and Griffin 1981), generates particularly singlet oxygen ($_1O^2$) (Baier et al. 2007); these damage DNA via photosensitizing reactions (Alscher et al.1997). Singlet oxygen is the major ROS involved in UV-A-mediated cell damage (Tyrrell and Pidoux 1989, Basu-Modak and Tyrell 1993). In recent years however, UV-B has also been shown to produce singlet oxygen by its effect on fatty acid and vitamins (Regensburger et al. 2012).

3.2.1.1 Generation of different types of ROS by energy transfer or electron transfer from ground state oxygen (Taken from Apel and Hurt 2004)

ROS constitute a group of free radicals, reactive molecules, and ions that are derived from ground state oxygen (O_2) either during cell molecular reactions or by direct impact of UV. Conversion of oxygen to its reactive state results from energy transfer or by electron transfer reactions; while sequential reduction of oxygen to superoxide, hydrogen peroxide, and hydroxyl radical results from elecron transfer, singlet oxygen is formed by energy transfer mechanism. Steps in electron/energy transfer for generation of different species of ROS is shown in Fig. 3.1.

The reaction between photoexcited sensitizer and other molecules involves a variety of mechanisms for energy transfer that involve electron/ hydrogen shift between reactants and products. Singlet oxygen ($_1O^2$), an excited state molecule and mediator of cellular oxidation evident in skin photodamage (Davies 2004; Karin 1998; Grether-Beck et al. 2000), is formed by direct energy transfer between the excited sensitizer and ground-state triplet oxygen (Fig. 3.2 reaction II A). Formation of superoxide radical anion (O_2^-) occurs via electron transfer with production of a sensitizer radical cation (Fig. 3.2 reaction II B) or after intermediate reduction of the sensitizer by a substrate with subsequent single electron reduction of oxygen (Fig. 3.2 reaction I). Superoxide radical anions formed by type I or II mechanism can then give rise to H_2O_2 by spontaneous dismutation.

Photosensitization is terminated by spontaneous or quencher-induced return of the photoexcited sensitizer chromophore to the electronic ground state; this is potentially followed by another cycle of photoexcitation. Irreversible termination may occur upon chemical sensitizer destruction (e.g., dye bleaching by $_1O^2$ attack) (Wondrak et al. 2006).

Various environmental (biotic and abiotic) stresses cause formation of ROS by energizing ground level oxygen (Dat et al. 2000; van Breusegem et al. 2001; Blokhina et al. 2003) including UV radiation (Apel and Hirt 2004; Mittler et al. 2004). Such reaction is mediated by small cellular molecules. Superoxide radical

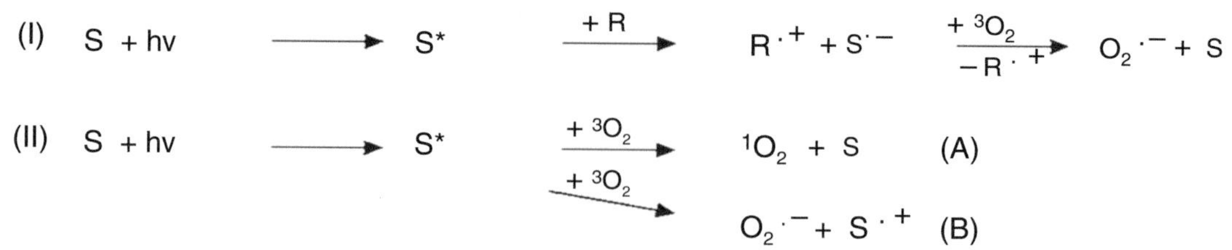

Fig. 3.2 Formation of ROS and organic free radicals by photosensitization reactions type I and type II. *S* sensitizer chromophores in the electronic ground state, *R* substrate molecules, *S** sensitizer chromophores in the excited state (Taken from Wondrak et al. 2006)

generated by UV-B either directly or through the production of H_2O_2 induces several genes. Pharmacological experiments demonstrate that NADPH oxidase, remaining attached to plasma membrane is responsible for ROS production thereby inducing UV-B-related defense genes (Mittler 2002; Apel and Hirt 2004; A-H-Mackerness et al. 2001).

About 1 % of O_2 consumed by plants is converted to ROS (Asada and Takahashi 1987) through UV radiation-induced activation of small molecules in the cytosol such as riboflavin, tryptophan, bilirubin, chlorophyll, and porphyrin that function as photosensitizers (Apel and Hurt 2004). The reactive oxygen species (ROS) thus generated constitute singlet oxygen ($_1O^2$), perhydroxyl radical (HO_2^-), hydrogen peroxide (H_2O_2), and hydroxyl radical (OH^-). Of these, singlet oxygen ($_1O^2$) was initially known to be related to UV-A (Wondrak et al. 2006). In recent years, association of singlet oxygen has also been reported as produced by UV-B (Regensburger et al. 2012). A large fraction of ROS are produced in various subcellular organelles such as chloroplasts, mitochondria, peroxisomes, endoplasmic reticulum, apoplast, and also in plasma membrane and cell wall studied in UV-induced cell death in plants (Nawkar et al. 2013). From studies on mammalian tissues, Lee and Wei (2000), Jackson and Loeb (2001), and Loft and Poulsen (1996) have reported that UV-induced ROS is produced by redox reactions involving electron transfer groups such as quinolines, their phenolic precursors, metal complexes, aromatic nitro compounds, and conjugated imines.

Disturbance in the CO_2 assimilation process, due to UV-B-induced damage of photosynthetic machinery, leads to excessive production of ROS in the chloroplast; such oxidative stress inhibits the repair of photodamage to the photosynthetic machinery (Nishiyama et al. 2001). Rao et al. (1996) reported that UV-B exposure related generation of activated oxygen species occurs also in mitochondria by increasing NADPH oxidase activity. ROS produced in mitochondria and chloroplasts is aggravated particularly during various biotic and abiotic stresses (Asada and Takahashi 1987; Del Rio et al. 1992) and aging related oxidative stress (Elstner 1987; Lee et al. 2004). Leakage of electrons (generated as a by-product in the electron transport chains of chloroplasts and of mitochondria), on tagging onto cellular O_2, gives rise to additional ROS. ROS produced in subcellular organelles are released into cytosol due to UV induced disruption of organellar membrane as an outcome of photoabsorption-mediated peroxidation of unsaturated fatty acids by UV (Murphy and Vu 1996). These, referred to as "unaccounted ROS," add to the population of ROS in the cytoplasm. Thus formed ROS population (combination of cellular along with organeller) functions as non specific ROS that causes a) direct cell molecular damage and also (b) damage countering by signal transduction-mediated upregulation (Casati and Walbot 2004) in synthesis of repair enzymes, viz., DNA photolyase.

UV induced (in the presence of intracellular factors, e.g., Ca^{2+}) increase in intracellular ROS levels may occur immediately or after sometime. Masaki et al. (2009) have reported increase in intracellular ROS after 7 h. The immediate response was found to be triggered by Ca^{2+}; this is demonstrated by experiments where BAPTA (a specific Ca^{2+}-chelating agent) completely abolished the elevation of UV-induced immediate ROS population; calcium has also been found to regulate elevation of ROS after 7 h of UV-B exposure. Experiments on HaCaT human keratinocytes have demonstrated that UV-B radiation is responsible for intracellular production and release of ROS in a dose-dependent manner (Fig. 3.3) (Beak et al. 2004).

UV-B-induced ROS, adding to the metabolically derived ROS population, induces detrimental effect on enzyme activities and gene expression; this eventually leads to cellular injury and programmed cell death (PCD) (Mackerness et al. 2001). ROS causes oxidative base damage of DNA resulting in formation of 8-hydroxyguanine (8OH-G) and thymine glycol; such induced modification of DNA bases may even result in strand breaks.

Excessive ROS (UV induced) oxidizes free nucleotides such as 8-hydroxydeoxygunanosine triphosphate (8OH-dGTP) and thereby adversely

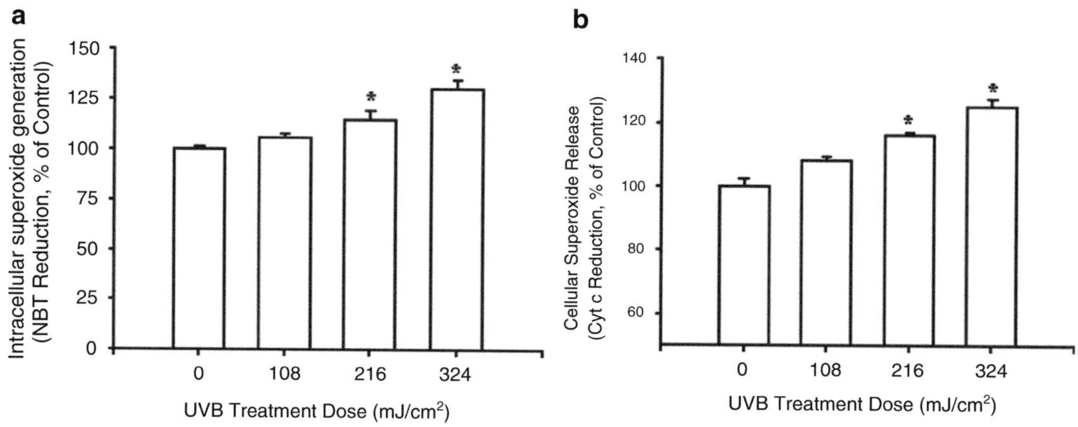

Fig. 3.3 UV-B radiation-induced ROS generation in a dose-dependent manner in HaCaT cells. ROS level was measured by NBT reduction (**a**) and cytochrome c reduction assay (**b**) (Taken from Beak et al. 2004)

affect DNA synthesis. UV has also been reported to induce oxidative stress-mediated mutations in the cellular genome through an indirect mechanism involving catalase in an unexpected role. Catalases that are generally known to be antioxidative enzymes, (and expected to lose integrity under UV radiation linear strand break) , apparently play a direct role in UV-B-induced generation of reactive oxygen species (Heck et al. 2003); the ROS and other reactive chemical intermediates produced by catalase after UV-B absorption are reported to be quenched by other antioxidants present in cells.

3.2.2 Direct Effect

UV radiation at high intensity brings about macromolecular damages directly by absorption into (a) DNA, thereby adversely affecting transcription and associated cell cycle processes (Kunz et al. 2005); (b) proteins, thus adversely affect enzyme activity as well as various physiological processes including photosynthesis (Bjorn 1996; Casati and Walbot 2004; Jansen et al. 1998); and (c) phospholipids, thereby causing membrane lipid destabilization (Jansen et al. 1998).

Both UV-A and UV-B radiation, by direct hit, may cause double-strand breaks (DSB) and single strand breaks (SSB) in DNA by disrupting phosphodiester bond between adjacent nucleotides (Greinert et al. 2012). In addition to strand breaks (SSB and DSB), UV radiation also induces photodimer formation by covalent linkage formation of double bonds between adjacent pyrimidine bases (C=C). These are considered as major DNA lesions induced by UV radiation (Whitmore et al. 2001). The major types of photodimers formed under UV radiation include cyclobutane pyrimidine dimers (CPD – Fig. 3.5) and (6-4) pyrimidine-pyrimidone (6-4PP – Fig. 3.5). Cyclobutane pyrimidine dimer (CPD) thus formed may cause DNA DSB (Fig. 3.4) as well as SSB (Sileman and Nicholson 2000; Girard et al. 2011).

Proteins strongly absorbing UV radiation at ~280 nm or higher wavelengths constitute direct targets of UV radiation and act as chrompphores. By directly targeting aromatic amino acids such as tyrosine, phenylalanine, and tryptophan, UV-B causes photomodification of proteins through photooxidation reactions that cause tertiary structure destabilization and consequent inactivation of proteins (Khoroshilova et al. 1990). UV-B also affects cell membranes by photo absorption-mediated peroxidation of unsaturated fatty acids, thereby affecting membrane lipid composition (Jansen et al. 1998). Studying UV-exposed maize plants Carletti et al. (2003) have demonstrated that tocopherol content is significantly reduced in high UV fluence location, indicating that cell membrane

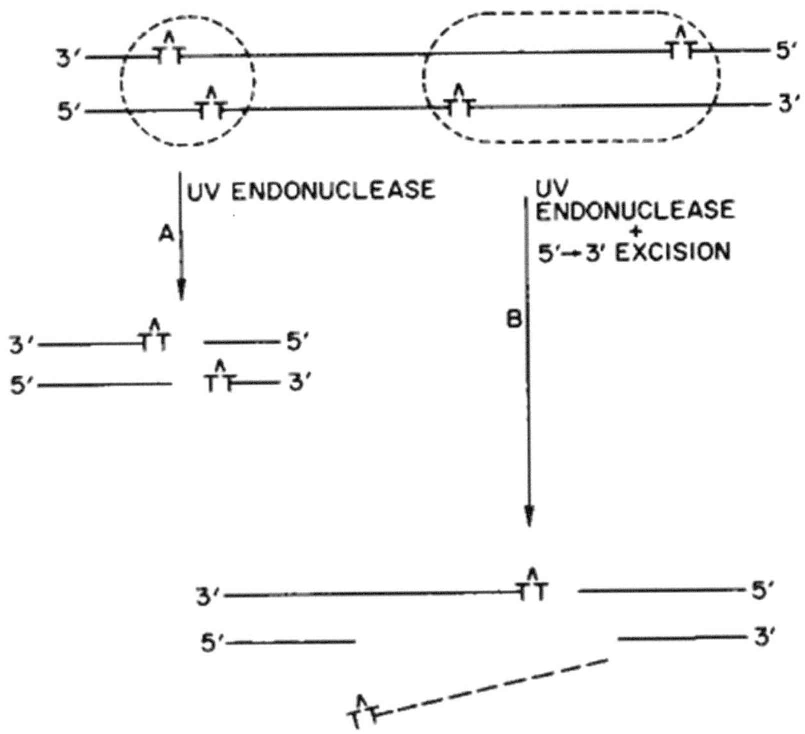

Fig. 3.4 Schematic diagram showing formation of DNA DSB by UV radiation (Taken from Bonura and Smith 1975)

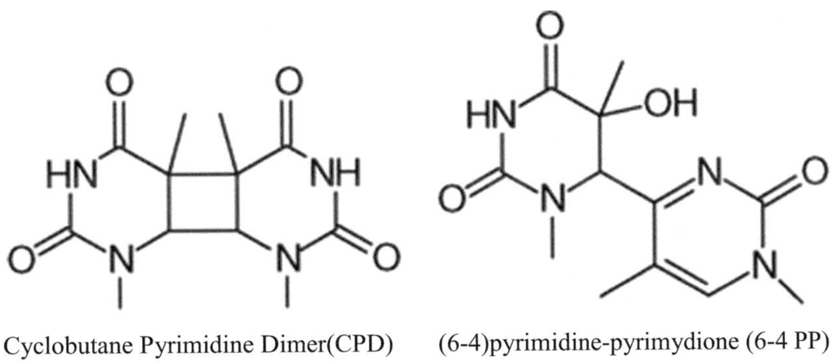

Fig. 3.5 Chemical structure of the UV-induced photodimers (Taken from Douki and Cadet 2003)

proteins/phospholipids also act as primary targets for UV-B radiation.

3.3 DNA Damage by UV Radiation

DNA, with its constituent bases serving as chromophores, is one of the key targets for UV-induced damage in both prokaryotic and eukaryotic cells. Both UV-B (280–315 nm) and UV-A (315–390 nm) either directly (see Sect. 3.2.2) or indirectly via reactive oxygen species (ROS) (see Sect. 3.2.1.1) bring about massive damage to cell molecular entities particularly DNA (Mazza et al. 2000; Brenner and Hearing 2008), thereby adversely affecting transcription and cell cycle progression (Kunz et al. 2005). Although both UV-A and UV-B cause DNA damage, distinctly different effects have often been noted for UV-A (315nm-390nm) and UV-B (280 nm to 315 nm). UV radiation has been classified as a Class I carcinogen (El Ghissassi et al. 2009) by the International Agency for the Research on Cancer. UV radiation-induced damaging effects on DNA have been studied in a large number of life forms: in bacteria (Peak and Peak 1982; Peak et al. 1984), cyanobacteria (Sinha et al. 2001), phytoplankton (Buma et al. 2001), macroalgae (Pakker et al. 2000), plants (Quatie et al. 1992), animals, and humans (Stein et al. 1989; Sinha and Hader 2002; Lima-Bessa et al. 2008). Two proteins, viz., ATM (ataxia-telangiectasia mutated) and ATR (ataxia-telangiectasia and Rad3-related), detected in *Arabidopsis* have been reported to act as sensors for DNA damage. It has been reported that ATR-deficient plants show hypersensitivity to UV-B radiation and exhibit altered G2-phase cell cycle checkpoints (Culligan et al. 2004; Garcia et al. 2003).

Formation of CPD, the most commonly occurring photodimer of DNA, has been shown to vary with time of day, altitude, and season being higher at noon or on sunny days indicating a direct correlation of high-intensity UV with photodimer formation and related DNA damage. Wang et al. (2014) from their studies on two species, viz., *Polygonum sachalinense* and *Plantago asiatica*, at two elevations (1700 m a.s.l. and 300 m a.s.l.) reported formation of CPD to be severe at 1700 m a.s.l. than at 300 m a.s.l., thereby establishing that DNA damage in plants is greater at higher elevations where UV fluence is higher. Although both UV-A and UV-B cause DNA damage distinctly, different effects have often been noted for UV-A (315–390 nm) and UV-B (280–315 nm).

3.3.1 UV-B-Induced DNA Damage

UV-B is known to cause four types of DNA damage, viz., (a) DNA double-strand break, (b) DNA single-strand break, (c) photodimer formation, and (d) oxidative damage by ROS production.

3.3.1.1 DNA Double-Strand Break by UV-B

UV-B radiation causes formation of CPD that results in a replication-blocking lesion; this causes stalling of DNA polymerase enzyme at the site of the lesion resulting in formation of a Y-shaped DNA structure that is found to be recognized by a specific endonuclease. The endonuclease creates a nick at the site of dimer formation, and, as a result of nick formation in both the strands (where CPD is formed), double-strand break (DSB) occurs in the DNA (Fig. 3.4) (Ohnishi et al. 2009). Free radicals (UV generated or otherwise) may also cause DSB (Box et al. 2001) by preventing the topoisomerase II (Top2)-mediated DNA disruption (Ohnishi et al. 2009; Banath and Olive 2003). Recently, Harper et al. (2010) have shown that UV radiation induced SSBs and non-DSB DNA damages contribute to the formation of replication-induced DSB.

3.3.1.2 DNA Single-Strand Breaks by UV-B

Single-strand breaks in DNA is also caused by UV-B-generated free radicals, viz., $\bullet OH^-$, via a pathway involving hydrogen abstraction at C3, C4, and C5 of the 2-deoxyribose moiety (Von Sonntag 2006; Chatgilialoglu et al. 2011).

3.3.1.3 Photodimer Formation by UV-B

In addition to single strand break (SSB) and double-strand break (DSB), direct impact of UV radiation also induces formation of various kinds of photodimers such as CPD and 6-4PP, thereby affecting structure and function of DNA (Fig. 3.5) (Landry et al. 1997). Formation of photodimers at bipyrimidine sites occurs in a strong sequence dependent manner leading to formation of CPDs. In addition, UV radiation produces, although at significantly low levels, purine dimers, a photoproduct at TpA sequences, and also pyrimidine monoadducts such as photohydrates (Sage 1993). Formation of CPD, the most commonly occurring photodimer of DNA, has been shown to vary with time of day, altitude and season being higher at noon or on sunny days indicating a direct correlation of high intensity UV with photodimer formation and related DNA damage. Wang et al. (2014) from their studies on two species viz. *Polygonum sachalinense* and *Plantago asiatica* at two elevations (1700 m a.s.l. and 300 m a.s.l.) reported formation of CPD to be severe at 1700 m a.s.l. than at 300 m a.s.l.; thereby establishing that DNA damage in plants is greater at higher elevations where UV fluence is higher.

Photodimers are formed by UV-B radiation in the following decreasing order: c,s Thy-Thy >6-4PP at TC sequence > Thy-Cyt. In contrast, CT and, to a lower extent, CC sites exhibit a low photoreactivity as inferred from the low yield of formation of both CPDs and 6-4PPs at these sequences (Douki and Cadet 2003). The main photochemical reactions within cellular DNA involve mostly the direct absorption of UV-B photons by the bases, mostly pyrimidine nucleotides such as thymine, cytosine, and minor 5-methylcytosine (Ravanat et al. 2001); purine bases may also be the target of UV-B radiation although to a smaller extent, as a result of the direct absorption of photons within the wavelength range 280–315 nm. In addition to CPD formation which constitutes approximately 75 % of the total DNA damage by UV radiation (Hutchinson et al. 1987), other types of DNA damage may also occur through other photoproducts such as 6-4PPs (Hutchinson et al. 1987) and 8-oxoG formation (Molinier et al. 2008). That UV-B is also able to induce the formation of 8-oxo-7,8-dihydroguanine (8-oxoG) in the DNA has been shown in experiments with mouse epidermis (Liu et al. 2000) and several cell systems that include mouse keratinocytes (Dubest et al. 2002; Ahmad et al. 1997) and Chinese hamster ovary cells (Nakajima et al. 1998). As different from direct impact of UV in the formation of photodimers, the mechanism involving •OH radical in the oxidation of the guanine base has also been reported (Ahmad et al. 1997).

Damaging effects of UV have been extensively studied in *Arabidopsis* in which availability of mutants that are incapable of DNA photorepair made it possible to study the deleterious effect of UV more intensely. The UV radiation-sensitive *uvr1* mutant of *Arabidopsis* exhibits failure to repair photodamage in DNA and cannot photorepair (6-4)-pyrimidine-pyrimidone photodimers (Britt et al. 1993). Landry et al. (1997) reported another *Arabidopsis* UV radiation-sensitive mutant, *uvr2-1*, that is deficient in photorepair of CPDs (as different from the 6-4PP repair deficiency in uvr-1); cloning and sequencing of the photolyase genes from UV radiation-sensitive (uvr2-1) and wild-type strains indicated alterations of DNA sequence that lead to defective photolyase protein (Ahmad et al. 1997).

3.3.2 UV-A-Induced DNA Damage

UV-A-caused DNA damage include formation of CPDs, 8-oxoG, oxidatively damaged pyrimidines (6-4PP), SSB, and DSB (Girard et al. 2011). In 2012, Greinert et al. additionally reported the presence of UV-A-induced DSB as a detectable DNA lesion. From studies on mammalian cells and human skin, Douki et al. (2003) and Mouret et al. (2006) have also reported CPDs as the major lesions induced by UV-A. Proportion of different types of UV-A-induced DNA damage are represented in Fig. 3.6 – depicting extent of different types of UV-A induced DNA damage recorded till 2011. Singlet oxygen generated by the type II photosensitization mechanism (Fig. 3.3) is a major contributor to formation of 8-oxoGua, with a small contribution coming from hydroxyl radicals.

3.3.2.1 DNA Double-Strand Break by UV-A

It was believed in the past that UV-A, being of lower energy compared to UV-B, is innocuous for inducing double-strand breaks (DSB) in DNA. Earlier, Rizzo et al. (2011) had demonstrated that no DSB is produced when cells are irradiated with UV-A. Cadet and Douki (2011) also supported this concept since their experiments revealed that UV-A induces more than twice as much 8-oxodG compared to SSBs or alkali-labile sites. Despite these arguments, in recent years, Greinert et al. (2012), in view of studies on various end points, viz., (a) micronucleus formation, (b) clonogenic survival (Phillipson et al. 2002; Shorrocks et al. 2008), (c) gH2AX formation (He et al. 2006; Lu et al. 2006; Rapp and Greulich 2004), (d) basic

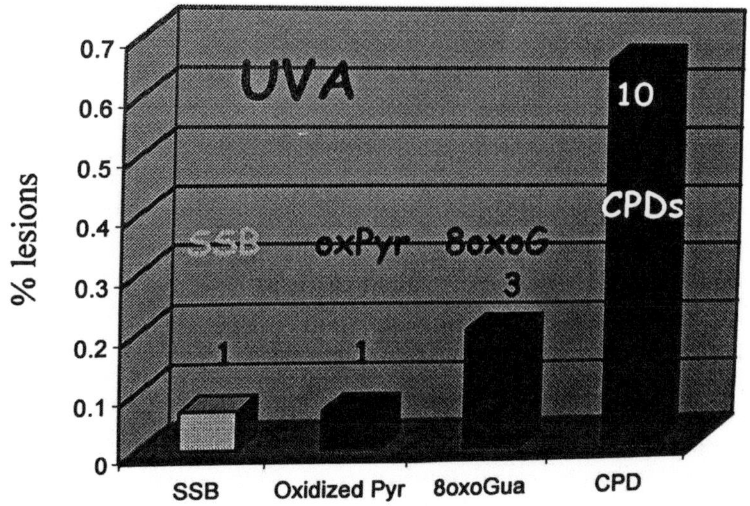

Fig. 3.6 Comparative representation of UV-A-induced CPDs, 8oxoguanine (8oxoG), oxidatively damaged pyrimidines (oxPy), single-strand breaks (SSB) formation in studies on Chinese hamster cells (Taken from Girard et al. 2011). DSB is not shown in the figure

studies on cell lines deficient in DSB repair pathways (Fell et al. 2002), and (e) their own analytical studies using the comet assay to assess DNA integrity and use of cH2AX as markers for DSB formation on the dose-dependent DSB induction by UV-A (studied in G1-synchronized human keratinocytes (HaCaT) and primary human skin fibroblasts), have concluded that UV-A does indeed induce DSB in a replication-independent manner through clustered single-strand breaks albeit via locally generated reactive oxygen species.

3.3.2.2 DNA Single-Strand Break by UV-A

Single strand break caused by UV-A exposure requires oxygen like other UV-A-mediated biological events such as DNA damage which is mediated by UV-A induced formation of singlet oxygen (type II photosensitization), electron abstraction (type I photosensitization) as well as by hydroxyl radicals (Girard et al. 2011). UV-A induced single strand break in bacterial and phage DNA has been reported by Tyrrell et al. (1974). Induction of DNA single-strand break (SSB) and DNA-protein crosslinks has been demonstrated in mammalian cells upon UV-A radiation (Peak and Peak 1990). UV-A radiation-induced formation of DNA single-strand breaks in human blood lymphocytes exposed to 365 nm have also been reported by Osipov et al. (2014). The authors also suggest a significant role of oxygen radicals, in particular OH^- and $_1O^2$, in the mechanism of UV-A-induced DNA damage in human lymphocytes. In such studies, DNA single-strand breaks have been demonstrated by using sensitive methods such as comet assays, commercially available specific antibodies, and DNA repair enzymes. Several reports have established that in mammalian cells, UV-A radiation also induces formation of oxidized pyrimidines, and oxidized purines (Zhang et al. 1997; Kvam and Tyrrell 1997; Douki et al. 1999; Besaratinia et al. 2008). Among these, the most abundant oxidative DNA damage is found to be 8-oxodG (an oxidized derivative of deoxyguanosine) (Kielbassa et al. 1997; Pouget et al. 2000; Douki et al. 2003) in addition to UV-A-induced CPD formation (Douki et al. 2003; Courdavault et al. 2004; Mouret et al. 2006).

3.3.2.3 Photodimer Formation by UV-A

Like UV-B, UV-A radiation in addition to single-strand and double-strand break also induces photodimer formation. The low energy of UV-A excites DNA, thereby forming bipyrimidine photoproducts such as CPDs and 6- 4PPs and its derivatives such as Dewar valence isomers though the amount is less in comparison to UV-B. Most of the photodimer formation by UV-A is mediated by photosensitizers with an exception of 6-4PPs which are only produced by direct absorption. The CPDs are formed more at TT sites due to the fact that thymine has the lowest triplet-state energy among the four bases found in DNA. In contrast, 8-oxodG is essentially formed by photosensitization reaction mediated by singlet oxygen. Plasmid DNA after UV-A radiation gives rise to CPDs by a mechanism that seem not to involve photosensitization reactions; this is confirmed from the fact that no photosensitizer was present in the radiation mixture. The distribution of the different types of DNA damage induced by UV-A is similar in plasmid or isolated DNA and in cells (Kuluncsics et al. 1999; Jiang et al. 2009; Schuch et al. 2009; Mouret et al. 2010). UV-A induced CPD formation has been recently demonstrated in cultured Chinese hamster ovary cells (Douki et al. 2003) and human skin keratinocytes and fibroblasts (Wondrak et al. 2006) suggesting that these photoproducts may be involved in the genotoxic effect of solar UV-A radiation (Courdavolt et al. 2004). From studies on effect of UV-A on DNA, Jiang et al. (2009) reported that the amount of CPDs produced in vitro was actually greater than the amount of CPDs generated by the same UV-A dose in vivo, suggesting that some, if not all, cellular CPDs are produced by UV-A directly. The fact that CPD formation in DNA may be caused by UV-A (as demonstrated in these experiments), confirms that UV-A is directly absorbed by DNA – a fact that was known, for long, to be true only for UV-B. The study of Hiraku et al. (2007) explained the mechanisms of UV-A-induced DNA damage in the presence of various photosensitizers and also proposed a new mechanism for its chemoprevention. UV-A radiation induces DNA damage at the 5′-G of 5′-GG-3′ sequence in double-stranded DNA through type I mechanism, which involves electron transfer from guanine to activated photosensitizers. Endogenous sensitizers such as riboflavin and pterin derivatives and an exogenous sensitizer nalidixic acid mediate DNA photodamage via this mechanism. The major type II mechanism involves the generation of singlet oxygen from photoactivated sensitizers, including hematoporphyrin and a fluoroquinolone antibacterial lomefloxacin, resulting in damage to guanines without preference for consecutive guanines. UV-A also produces superoxide anion radical by an electron transfer from photoexcited sensitizers to oxygen (minor type II mechanism), and DNA damage is induced by reactive species generated through the interaction of hydrogen peroxide with metal ions. UV-A exposure in the absence of photosensitizers induces phototransformation of 6-4PP into their Dewar valence isomers, another class of prominent DNA photolesions. Oxidative base modification can occur by UV-A photosensitization involving the triplet-state sensitizer (S*) with formation of base radical cations by electron transfer reactions

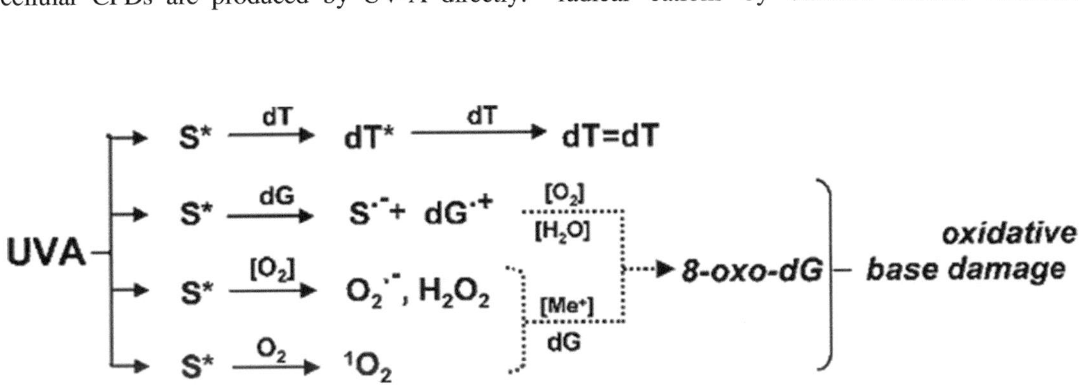

Fig. 3.7 UV-A-mediated damage of DNA including thymine dimmer and oxidative damage. (S* represents triplet-state sensitizer) (Taken from Wondrak et al. 2006)

followed by water incorporation, a process involved in riboflavin-mediated base oxidation (Kasai et al. 1992) (Fig. 3.7).

UV-A exerts its DNA-damaging effect through cellular photosensitizers that are photoexcited and involved in type I or type II photoreactions to produce reactive oxygen species (ROS), like $^{\cdot}OH$, O_2^{-}, H_2O_2, or $_1O^2$ (Fig. 3.8) (Cadet et al. 1997, 2003, 2009) in the presence or absence of metal ions. Singlet oxygen ($_1O^2$) is able to react with DNA molecule and may induce G to T transversions due to 8-oxodG generation. $_1O^2$ is also able to induce the expression of genes involved with the cellular responses to oxidative stress, such as NF-kB, c-fos, and c-jun, and genes involved with tissue damage and inflammation, as ICAM-1, interleukins 1 and 6 (Agnez-Lima et al. 2012).

HPLC coupled with electrochemical detection (HPLC-ED), and recently HPLC coupled with tandem mass spectrometry (HPLC-MS/MS), was used to measure 8-oxoguanine and all the different bipyrimidine photoproducts. In the same study, *Escherichia coli* Fpg/Nth protein was used to reveal oxidation products of purine (mainly 8-oxoG)/pyrimidine, respectively, and Nfo protein was used to reveal abasic sites (the sites without any base). The 8-oxoguanine (8-oxoG) was more specifically measured by HPLC-ED (Rastogi et al. 2010).

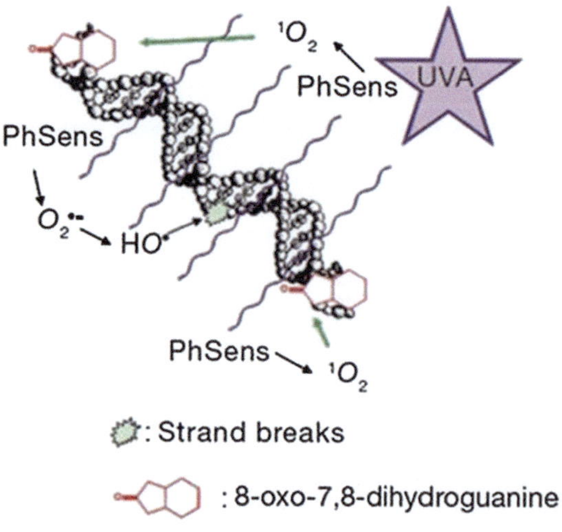

Fig. 3.8 UV-A-induced damage to cellular DNA by different kinds of ROS. *PhSens* photosensitizer (Taken from Cadet and Douki 2011)

3.4 Protein Damage by UV Radiation

Through the years of molecular evolution on earth, proteins have evolved spatial proximity between aromatic residues (Trp, Tyr, and Phe) and disulfide bridges (SS) for maintaining their structural stability (Petersen et al. 1999). After UV absorption, proteins undergo photomodification through photooxidation reactions mediated directly by UV-B or indirectly by UV-A mediated photosensitized production of active O_2 species and free radicals (Pattison and Davis 2006). Proteins are major cellular targets for photooxidation by UV radiation evidently due to their abundance in cells and also due to the fact that they carry chromophores. UV radiation is mainly absorbed by aromatic amino acids which act as nanosized antennas. After excitation by UV radiation, aromatic amino acids enter photochemical pathways causing distortion in protein structures. On the other hand, disulfide bridges in proteins act as quenchers to counteract the harmful effects caused by its aromatic amino acids, thereby restoring protein stability and activity.

3.4.1 Protein Damage by UV-B

Most proteins due to presence of the aromatic amino acids are direct targets of UV-B radiation. Aromatic amino acids such as tyrosine, phenylalanine, and tryptophan directly absorb ~280 nm or higher wavelengths of UV radiation (Khoroshilova et al. 1990) causing photomodification of proteins through photooxidation reactions. This excitation of the aromatic residues triggers electron ejection from their side chains (Bent and Hayon 1975a, b, c; Creed 1984a, b; Kerwin and Rammele 2007; Neves-Petersen et al. 2007). Cysteine is poor absorber in the UV-B region but undergoes UV-induced photolysis at high quantum efficiency (Creed

1984b). The disulfide group of cystine found to be splitted by UV-B radiation forming reactive sulfhydryl groups. The disulfide bridges are important for the maintenance of tertiary structure of many proteins, and the effect of UV-B on these bonds can strongly influence protein structure and function (Creed 1984b). The enzymatic antioxidants such as SOD and APX which protect the plants from other oxidative damages from ROS are also degraded by the UV-B radiation thereby confer no protection from UV-induced cell molecular damages.

3.4.2 Protein Damage by UV-A

UV-A like UV-B is also reported to cause various types of protein damage (Tyrrell and Sage 2011). Some proteins have bound prosthetic groups such as flavins and hemes which absorb in the UV-A range. As a result, direct photooxidation may occur that arises from absorption of radiation by the protein itself or by the bound prosthetic groups, thereby generating reactive oxygen species through photosensitizer type I processes. Indirect photooxidation of proteins via type II mechanism may also occur; this involves singlet oxygen which is generated by energy transfer from either protein-bound or other endogenous chromophores to oxygen (Wondrak et al. 2006). Vile and Tyrell (1995) have shown that UV-A radiation induces sulfhydryl oxidation in proteins that depends on iron, singlet oxygen, and hydrogen peroxide. UV-A also changes catalase charge in epidermis reconstituted with low phototype melanocytes, possibly due to oxidative modification of tryptophan and methionine residues. An example in this context is photoaging of skin that is associated with protein oxidation in the human skin, with accumulation of protein carbonyls, especially in the upper dermis. It was recently shown that the repair enzyme methionine-*S*-sulfoxide reductase (MSRA) was expressed in human epidermis and specifically upregulated by low doses of UV-A radiation (Ogawa et al. 2006). MSRA, by reducing the methionine sulfoxide to methionine, thus reverses the inactivation of proteins caused by oxidation of critical methionine residues. Another example is the base excision repair protein, DNA *N*-glycosylase hOGG1, which removes 8-oxoguanine and has been shown to be sensitive to oxidation, e.g., cysteine oxidation (Bravard et al. 2006). Girard et al. (2011) have recently shown that exposure of eucaryotic cells (studied in fission yeast *Saccharomyces pombe* and mammalian cells) to oxidative stress induced by UV-A radiation led to inhibition of DNA replication evidenced by impaired replication fork progression, as a consequence of UV-A-induced oxidative damage to proteins involved in replication of DNA. Singlet oxygen produced by photosensitization is responsible for PCNA modification.

3.5 Lipid Damage by UV Radiation

In the presence of oxygen (Kramer et al. 1991; Panagopoulos et al. 1990), unsaturated fatty acids of plant cell membrane (target of UV radiation) undergo peroxidation that leads to breakdown of membrane structure and function (Mishra and Singhal 1992; Murphy and Vu 1996). The major lipids of most cell membranes are phospholipids with phosphatidylcholine being the major component in contrast to the lipid composition of the chloroplast which is comprised largely of the galactolipids, monogalactosyldiacylglycerol (MGDG), and digalactosyldiacylglycerol (DGDG), accounting for approximately 70 % of the acyl lipid composition. These lipids of chloroplast (MGDGs, DGDGs) have been reported to be altered by the impact of UV radiation (Murphy and Vu 1996): the concentration of the MGDGs is reduced without affecting the overall DGDG concentration and the total phospholipid levels. This reduction of the MGDGs is important for maintaining the optimal membrane structure in chloroplasts for photosynthesis under UV-B radiation (Predieri et al. 1995).

The thylakoid membranes of chloroplast contain the sulfolipid, sulfoquinovosyldiacylglycerol, and unlike other membranes, the major phospholipid is phosphatidylglycerol containing the unique fatty acid – trans-A-3 hexadecanoic acid. These molecules are prime targets for UV-B-induced oxidative damage. Studies on barley, corn, bean, and radish showed a decrease in the galactolipids with the ratio of MGDG/DGDG decreasing as has been observed in senescing tissue (Tevini et al. 1981). In a study on cucumber cultivars, a decrease in unsaturated fatty acids was correlated with increased peroxidation (Kramer et al. 1991). The level of peroxidation caused by UV-B was also related to the UV-B sensitivity of the cultivars. In addition, increased saturation of membranes has recently been found to switch on desaturase genes (Vigh et al. 1993). Thus, UV-B-induced effect on membrane lipids can cause a cascade of widely different responses.

3.6 Membrane Damage by UV Radiation

UV-B can cause damage of cell membranes by photo absorption-mediated peroxidation of unsaturated fatty acids, thereby causing variation in membrane lipid composition (Jansen et al. 1998). UV-induced damage to membranes has been shown to mainly affect membrane transport (Murphy and Byczko 1990; Murphy et al. 1985; Murphy and Wilson 1982); high fluence UV-B causes more flow of molecules across membrane whereas low fluence UV-B inactivates membrane functions like transport systems, ATPase, and photosynthetic activity (Wright et al. 1981). UV-induced alteration in ion permeability has also been observed with experiments on plasma membranes (Doughty and Hope 1973) and cultured cells of higher plants (Murphy and Wilson 1982). UV-B-induced loss of K^+ from guard cells may be responsible for the observed loss of stomatal conductance in irradiated plants (Negash and Björn 1986).

3.7 Effect of UV Radiation on Pigments

UV-B exposure have been shown to induce both qualitative and quantitative changes in chlorophyll in many studies (Takeuchi et al. 2002; Strid et al. 1990; Deckmyn and Impens 1997) causing loss of photosynthetic capacity (Jordan et al. 1994). From their studies on annual desert plants, viz., *Malva parviflora* L., *Plantago major* L., *Rumex vesicarius* L., and *Sisymbrium erysimoides*, Salama et al. (2011) reported that the chlorophyll contents were affected by enhanced UV radiation. Experiments on pea seedlings have shown that the total chlorophyll decreased by 40 % over 8 days of UV-B treatment compared to control plants (Strid et al. 1990). *Chlb* is more affected than *Chla* (Marwood and Greenberg 1996), although the apparent relative stability of *chlb* may be due to its association with *chla* which makes them more stable than in isolation (Brown et al. 1991). *M. parviflora* showed highest constitutive levels of accumulated chlorophyll a, b, and total chlorophyll among the investigated plant species; *P. major* showed the lowest constitutive levels of chlorophyll a, b, and total chlorophyll (Salama et al. 2011). Chlorophylls (*Chl*) are more affected by UV-B radiation in comparison to carotenoids (Pfündel et al. 1992). In the enhanced UV radiation, carotenoid content was found to be increased compared with the control. Studies of Donkor and Hader (1996) on cyanobacteria *Anabaena variabilis*, *Oscillatoria tenuis*, *Nodularia baltica*, *Nodularia harveyana*, and *Phormidium uncinatum* (Baikal and Tubingen strain) indicated that pigmentation and energy transfer to the photosynthetic reaction centers were impaired under high intensity of UV radiation.

3.8 Effect of UV Radiation on Photosynthesis

UV radiation adversely affects the most important plant physiological process, viz., photosynthesis. Decrease in photosynthesis (3–90 %), particularly at higher UV-B doses, is due to both direct (effect on photosystem) and indirect (decrease in pigments and leaf area) effects (Kakani et al. 2003). Photosynthetic activity may be reduced by direct effects on photosynthetic enzymes, metabolic pathways, or indirectly through effects on photosynthetic pigments or stomatal function. In several plants, inhibitory effect of UV-B radiation has been reported both on net photosynthesis (Tevini 1993) and on the partial reactions (Jordan 1993). This inhibition appears to arise from disruptions at a number of points in the photosynthetic cycle including disruption of PSII reaction center, decrease in Rubisco activity, and damage to photosynthetic pigments (chlorophylls and carotenoids) (Strid et al. 1990). Photosynthetic structures are widely impaired by UV-B at several levels, such as through the induction of tissue chlorosis and necrosis, disruption of leaf ultrastructure including damage of chloroplast (Bornman 1989) and leaf anatomy (e.g., in the thickness of epidermal and palisade mesophyll cell layers), as well as degradation of photosynthetic pigments and thylakoid electron transport carriers (Bornman and Teramura 1993; Jenkins 2009; Lidon and Ramalho 2011). In growth chamber or glasshouse experiments leaf photosynthesis generally exhibit exaggerated adverse effects compared to adverse effects noted in field grown plants. This is because in artificial chambers interaction among other components of natural solar radiation (viz. longer wavelengths such as PAR and UV-A) remain precluded (Hu et al. 2006).

Increases in atmospheric CO_2 may negate or reduce the detrimental effects of elevated UV-B, since there is ample documentation that these two environmental parameters, applied both in isolation and in combination, mostly have contrasting effects on plant photosynthesis and growth (Sullivan 1997). There are however inconsistencies; the magnitude, duration, and direction of both UV-B and CO_2 effects are often species-specific and modified by the availability of other resources such as light, water, and nutrients in the growing environment (Idso and Idso 1994; Teramura and Sullivan 1994). Plant species that have either C3 or C4 photosynthetic pathways have been studied, and in general C3 species were found to be more sensitive than C4 plants (Basiouny et al. 1978; Vu et al. 1982). Under certain conditions however, important C4 crop species, such as *Zea mays*, have been reported to be appreciably susceptible to inhibition by UV-B radiation (Vu et al. 1982). The resistance of C4 plants compared to C3 plants may be due to the different internal cellular organization (Kranz anatomy) and biochemistry. Extensive study of UV-B radiation effect on the partial reactions of photosynthesis has demonstrated multiple sites of inhibition (Bornman 1989; Strid et al. 1990). The diverse processes that are affected include the range of molecules that absorb UV-B radiation and also the energetically unstable nature of the photosynthetic apparatus. These changes subsequently lead to inhibition of photosynthetic function and decreased efficiency of photosynthesis. Numerous studies have shown that in photophosphorylation processes, PSII is the most UV sensitive component of the thylakoid membrane (Brandle et al. 1977; Savitch et al. 2001; Lidon 2012), being related to rapid degradation of the D1 and D2 proteins of PSII (Renger et al. 1989; Lidon 2012). UV-B radiation has a lesser impact on PSI (Kulandaivelu et al. 1991) and cytochrome b6/f complex (Cen and Bornman 1990; Lidon et al. 2012), although strong UV-B-mediated effects on PSI linear electron transport (Fig. 3.9) (Lidon and Ramalho 2011) and on cyclic phosphorylation have also been reported under certain conditions (Kataria et al. 2014).

In a comprehensive study, Strid et al. (1990) showed a differential sensitivity of photosynthetic function to UV-B radiation. Photoinhibition of photosystem II is also caused by high levels of visible radiation. However, significant differences exist between UV-B damage and photoinhibition, such as in the recovery kinetics (Chow et al. 1992). The water oxidizing complex of PS II is an important target of ultraviolet B radiation (Skórska 2000; Szilárd et al. 2007; van Rensen et al. 2007). It has been shown that the damage by UV-B radiation occurs first at the acceptor site of photosystem II and later at the donor site. In addition to the direct effect of UV-B radiation, photosynthesis may also be indirectly affected. Stomatal function, and thus leaf gas exchange, is also commonly impaired (Teramura 1983; Tevini and Teramura 1989). Induction of stomata closure may reduce the efficiency of gas exchange; changes in leaf thickness and anatomy may alter

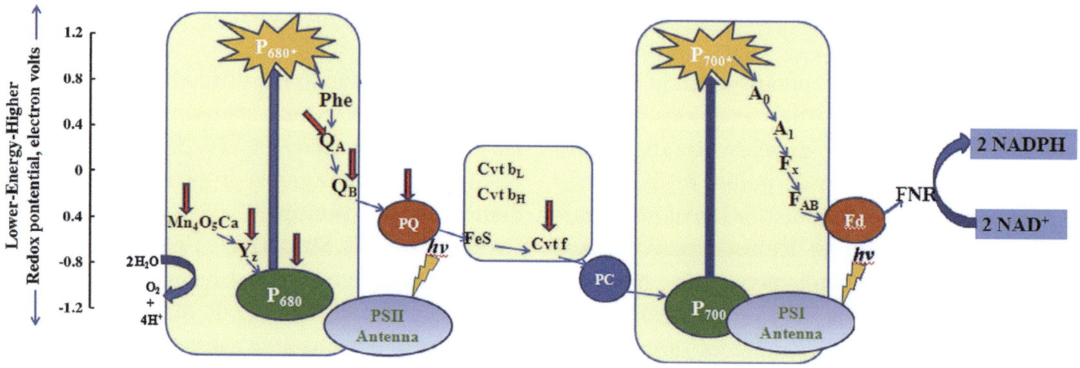

Fig. 3.9 Effects of UV-B on electron transport chain in chloroplast. Major UV-B targets are shown by *red arrows* (Taken from Chen and Han 2014)

light environment within the leaf, and modification in canopy morphology may also indirectly affect whole plant photosynthesis (Nogués et al. 1999; Hollosy 2002; Fagerberg and Bornman 2005). Furthermore, the integrity of the thylakoid membrane and structure seems to be even more sensitive than the activities of the photosynthetic apparatus bound within (Lidon and Ramalho 2011; Lidon 2012). Ribulose 1,5-bisphosphate carboxylase (Rubisco) is the primary enzyme of CO_2 fixation in C3 plants (Huffaker1982; Miziorko and Lorimer 1983; Spreitzer 1993). Rubisco exhibits reduced activity, being inhibited by UV-B exposure (Vu et al. 1982, 1984; Strid et al. 1990; Jordan 1993); this inhibition is likely to be caused by protein degradation and or inactivation of the enzyme. Ribulose-1,5-bisphosphate carboxylase/oxygenase (Rubisco) content and activity are also strongly affected by UV-B radiation in many field crops (Vu et al. 1982, 1984; Correia et al. 1998; Savitch et al. 2001). Degradation of Rubisco by UV-B causes decreased carbon dioxide fixation and oxygen evolution. Other important photosynthetic functions adversely affected by UV-B are photophosphorylation and CF, F, and –ATPase activity (Strid et al. 1990; Zhang et al. 1994).

3.9 Effect of UV on Growth and Productivity

UV radiation adversely affects plants through alteration of growth, development, and morphology associated gene activity (Mpoloka 2008) and also nonspecifically by DNA damage caused by formation of CPD that stalls DNA replication, thereby preventing cell division. This results in decreased growth and productivity (Teramura 1983). UV radiation–augmentation studies have identified many UV radiation resistant cultivars of higher plants that are of economic importance, including rice (Kumagai and Sato 1992; Bornman and Teramura 1993; Hidema et al. 1996; Correia et al. 1998). In rice, UV-sensitive rice cultivar Norin 1, a progenitor of many Japanese commercial rice strains (Kumagai and Sato 1992; Sato and Kumagai 1993; Hidema et al. 1996), is deficient in photorepair of CPDs (Hidema et al. 1997; Hidema and Kumagai 1998) and exhibits high sensitivity to UV-induced damage (Kumagai and Sato 1992; Hidema et al. 1996). On the other hand, UV-resistant rice cultivar Sasanishiki showed no such damaging effect under UV radiation. This observation demonstrates that there is a varietal difference, i.e., genetic control, in rice toward UV sensitivity. Figure 3.10 shows the effect of supplementary UV-B radiation on the growth of Norin 1 (Fig. 3.10a) and its close relative Sasanishiki (Fig. 3.10b).

In addition, UV has also been reported to induce changes in plant growth regulators, thereby producing changes in growth, general development, and flowering. UV-induced photolytic degradation of indole-3-acetic acid (IAA) has been demonstrated in sunflower seedling (Ros and Tevini 1995), causing inhibition of

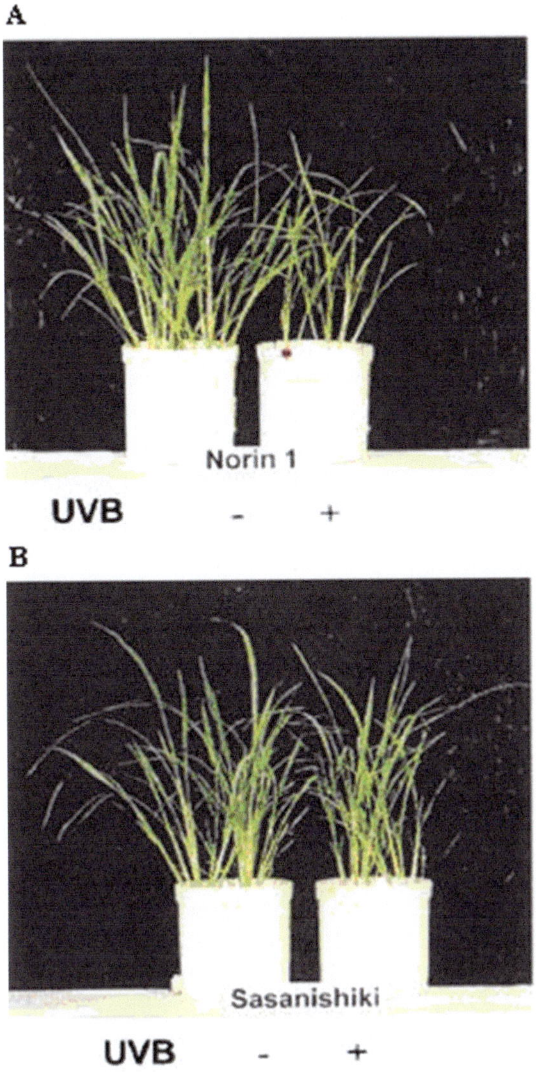

Fig. 3.10 UV radiation damage and photorepair in UV radiation-sensitive and UV radiation-resistant rice cultivars (Taken from Kumagai and Sato 1992). (**a**) UV-B radiation induces growth stunting and browning of UV radiation-sensitive Norin 1. (**b**) UV-B radiation has little effect on UV radiation-resistant Sasanishiki seedlings

hypocotyl growth. 2,4-Dichloro-phenoxyacetic acid (2,4-D), a synthetic auxin, is converted to 2,4-dichlorophenol after UV exposure leading to death of the cell. The phytohormone abscisic acid (ABA), gibberellins, and ethylene also absorb strongly in the UV-B region and can be inactivated by photolysis (Lindo et al. 1978, Ballare et al. 1991, Ros and Tevini 1995).

3.10 UV-Induced Programmed Cell Death

Programmed cell death (PCD) is a genetically controlled self destruction mechanism in all eukaryotic multicellular organisms and can be induced either as part of normal development or in response to stress such as high-dose UV. PCD or apoptosis is well studied in animal systems. Apoptosis is characterized by specific hallmarks such as cell shrinkage, nuclear condensation and fragmentation, and eventually the breakup of the cell into "apoptotic bodies" that are eventually engulfed by phagocytes (Adrain and Martin 2001; Reape and McCabe 2008). The progress of plant PCD is much slower but interestingly shares many conserved components with animal apoptosis. Despite the similarities between cell death pathways in plants and animals, there is no evidence for apoptotic bodies, as well as classical caspases in plant systems. Plant cell death is described as an apoptotic-like PCD (AL-PCD) (Danon et al. 2000). AL-PCD is now accepted as a fundamental cellular process since it plays an essential role during development, under stress conditions, in the senescence process and in response to pathogen infection in plants (Mittler and Lam 1996; Lam et al. 1999). Apart from developmental stimuli, biotic stresses, such as the pathogen-induced hypersensitivity response (HR), and abiotic stresses, such as heat stress and high fluence UV radiation, have been shown to induce AL-PCD (Lam and Pozo 2000; Wang et al. 1996; McCabe et al. 1997).

Enhanced UV-B radiation brought about by long-term depletion of the stratospheric ozone layer has been shown to induce instability of genome in plant populations (Ries et al. 2000) often causing cell cycle arrest and apoptosis (Kunz et al. 2005) in plants (studied in maize, Casati and Walbot 2003) and also in mammals (Kunz et al. 2005). Recently, it has been shown that UV-B radiation modulates cell cycle regulatory genes involved in G1-to-S transition in root tips of *Arabidopsis* (Jiang et al. 2011). The G1-to-S arrest in root tips, hypocotyls (Kubasek et al. 1992), evidently a consequence of UV-B induced DNA damage, allows time for repairing

DNA damage before start of replication/synthesis of cell protective compounds. The physiological manifestation of this could be the photomorphogenic effect of delayed seedling emergence (Biever et al. 2014) (see Case Study I in Chap. 4). Excessive DNA damage unattended by appropriate DNA repair machinery due to different genetic makeup (i.e., varietal difference) could be the possible reason behind observed plant growth inhibition under enhanced UV radiation and its effect on terrestrial ecosystems (Bjorn 1996).

The higher doses of UV-B radiation can induce oligonucleosomal DNA fragmentation resulting in nucleosomal fragments in multiples of 180 bp which is seen as typical apoptotic DNA ladder (Danon and Gallois 1998; Danon et al. 2004; Lytvyn et al. 2010). As this DNA laddering is closely associated with programmed cell death (PCD) in plant systems (Stein and Hansen 1999; Houot et al. 2001) albeit at a slower rate in comparison to animal system (Danon et al. 2000), such studies confirm the role of UV-B radiation in induction of programmed cell death (PCD). DNA damage is quantified by using terminal deoxynucleotidyl transferase-mediated dUTP nick-end labeling reaction (TUNEL), which detects in situ free 3′-OH DNA breaks (Gorczyca et al. 1994; Danon and Gallois 1998).

3.11 Effect of UV on Human

3.11.1 Damage of the Eye

Eye tissues are damaged by prolonged exposure to high-intensity UV (e.g., in sunbeds), causing a burning sensation called as "snow blindness" or photokeratitis. Ultraviolet light exposure to the eye has been associated with cataract formation and retinal degeneration. In both cases, it is hypothesized that ultraviolet light can initiate formation of free radicals, which can cause protein modification and lipid peroxidation. It has been estimated that 1 % depletion of ozone layer causes 0.5 % increase in cataract (UNEP 1994; Tevini 1993). WHO has estimated that up to 20 % of cataracts or three million per year could be due to UV exposure to the eye (http://www.who.int/uv/health/en/).

3.11.2 Damage of the Skin

The ubiquitous presence of solar UV radiation in human life leads to various adverse effects – the most commonly observed effect in this regard is that of pigmentation seen as melanin formation in skin. Sunburn cells, i.e., keratinocytes, which fail to efficiently repair damaged DNA undergo apoptosis; this helps to escape the risk of conversion of normal cells into malignant cells (Boccalandro et al. 2004). Although molecular mechanism leading to pigmentation in human skin is not completely understood till date, studies undertaken in this regard suggest that UV-induced DNA damage and/or the repair process initiate signals for development of melanogenesis (Brenner and Hearing 2008). One percent decrease in stratospheric ozone and associated UV enhancement is estimated to cause an increase of approximately 2.3% non-melanoma skin cancer (UNEP 1994; Tevini 1993). This exemplifies the predicament of the post "ozone hole" era. Till recovery of ozone thinning through strict adherence to regulations of Montreal Protocol, the threat for cell molecular damages will probably remain. In this context, it is interesting to note that for cellular protection, plant cells have evolved inherent cell molecular mechanisms for repair of UV-induced DNA damage. In vivo studies have shown that UV-B radiation affects epidermal morphology including increasing the mean stratum corneum (SC) thickness (Pearse et al. 1987) and disrupts the permeability barrier, causes morphological changes in SC lipids, increases trans epidermal water loss, and decreases SC hydration (Kambayashi et al. 2003; Meguro et al. 1999). The United Nations Environment Programme has estimated that over two million non-melanoma skin cancers and 200,000 malignant melanomas occur globally each year. In the event of a 10 % decrease in stratospheric ozone, with current trends and behavior, an additional 300,000 non-melanoma and 4500 melanoma skin cancers could be expected worldwide.

3.11.3 Damage of Immune System

Overexposure to UV radiation has a deleterious suppressing effect on the immune system. Scientists believe that sunburn can change the distribution and function of white blood cells in humans for up to 24 h after exposure to the sun. Repeated overexposure to UV radiation adversely affects the body's immune system. On a positive note, it is possible that UV-induced immunosuppression could provide positive advantages. It is mainly helpful in case of autoimmune disease such as asthma, Th1-mediated autoimmune diseases where immune system becomes impaired by UV-B and therefore the symptoms of the disease get diminished (Hart et al. 2011).

Bibliography

Adrain C, Martin SJ (2001) The mitochondrial apoptosome: a killer unleashed by the cytochrome seas. Trends Biochem Sci 26:390–397

Agnez-Lima LF, Melo JT, Silva AE, Oliveira AH, Timoteo AR, Lima-Bessa KM, Martinez GR, Medeiros MH, Di Mascio P, Galhardo RS, Menck CF (2012) DNA damage by singlet oxygen and cellular protective mechanisms. Mutat Res/Rev Mutat Res 751(1):15–28

Agrawal SB, Rathore D (2007) Changes in oxidative stress defense in wheat (Triticum aestivum L.) and mung bean (Vigna radiata L.) cultivars grown with or without mineral nutrients and irradiated by supplemental ultraviolet-B. Environ Exp Bot 59:21–22

A-H-Mackerness S, John CF, Jordan B, Thomas B (2001) Early signaling components in ultraviolet-B responses:distinct roles for different reactive oxygen species and nitric oxide. FEBS Lett 489:237–242

Ahmad M, Jarillo JA, Klimczak LJ, Landry LG, Peng T, Last RL, Cashmore AR (1997) An enzyme similar to animal type II photolyases mediates Photoreactivation in *Arabidopsis*. Plant Cell 9:199–207

Ajavon AN, Albritton DL, Watson RT (2007) Scientific assessment of ozone depletion: 2006, Rep. 50, Global Ozone Res. and Monitoring Project, World Meteorological Organisation, Geneva

Alscher RG, Donahue JL, Cramer CL (1997) Reactive oxygen species and antioxidants: relationships in green cells. Physiol Plant 100(2):224–233

Apel K, Hirt H (2004) Reactive oxygen species: metabolism, oxidative stress, and signal transduction. Annu Rev Plant Biol 55:373–399

Asada K, Takahashi M (1987) Production and scavenging of active oxygen in photosynthesis. In: Kyle DJ, Osmund CB, Arntzen CJ (eds) Photoinhibition. Elsevier, Amsterdam, pp 227–287

Baier MC, Barsch A, Kuster H, Hohnjec N (2007) Antisense repression of the Medicago truncatula nodule-enhanced sucrose synthase leads to a handicapped nitrogen fixation mirrored by specific alterations in the symbiotic transcriptome and metabolome. Plant Physiol 145:1600–1618

Balakrishnan V, Ravindran KC, Venkatesan K, Karuppusamy S (2005) Effect of UV-B supplemental radiation on growth and biochemical characteristics in Crotalaria juncea L. seedlings. Electr J Environ Agric Food Chem 4:1125–1131

Ballaré CL, Barnes WP, Kandrick RE (1991) Photomorphogenetic effect of UV-B radiation on hypocotyls elongation in wild type and stable-phytochrome-deficient mutant seedlings of cucumber. Physiol Plant 83:651–658

Banath JP, Olive PL (2003) Expression of phosphorylated histone H2AX as a surrogate of cell killing by drugs that create DNA double-strand breaks. Cancer Res 63:4347–4350

Bartholic JF, Halsey LH, Garrard LA (1975) Field trials with filters to test for effects of UV radiation on agricultural productivity. In: Nachtwey DS, Caldwell MM, Biggs RH (eds) Climatic Impact Assessment Program (CLAP), Monograph 5, Part 1 – Ultraviolet Radiation Effects, US Dept Transp., Report No. DOT-TST-75-55. (PB-247-725), National Technical lnformation Service. Springfield, Virginia, pp 4–61, 4–71 (Appendix A fluence is higher)

Basiouny FM, Van TK, Biggs RH (1978) Some morphological and biochemical characteristics of C_3 and C_4 plants irradiated with UV-B. Physiol Plant 42(1):29–32

Basu-Modak S, Tyrrell RM (1993) Singlet oxygen: a primary effector in the ultraviolet A/near-visible light induction of the human heme oxygenase gene. Cancer Res 53:4505–4510

Battag PR, Brennan TM (2000) Differential effects of short-term exposure to UV-B radiation upon photosynthesis in cotyledons of a resistant and a susceptible species. Int J Plant Sci 161:771–778

Beak SM, Lee YS, Kim JA (2004) NADPH oxidase and cyclooxygenase mediate the ultraviolet B-induced generation of reactive oxygen species and activation of nuclear factor-kappaB in HaCaT human keratinocytes. Biochimie 86:425–429

Beggs C, Wellmann E, Grisebach H (1986) Photocontrol of flavonoid synthesis. In: Kendsrick RE, Kronenberg G (eds) Photomorphogenesis in plants. Nijhoff/W. Junk Publishers, Dordrecht, pp 467–499

Bent DV, Hayon E (1975a) Excited-state chemistry of aromatic amino-acids and related peptides. I. tyrosine. J Am Chem Soc 97(10):2599–2606, ISSN (Print): 0002-7863

Bent DV, Hayon E (1975b) Excited-state chemistry of aromatic amino-acids and related peptides. II. phenylalanine. J Am Chem Soc 97 (10):2606–2612, ISSN(Print): 0002-7863

Bent DV, Hayon E (1975c) Excited-state chemistry of aromatic amino-acids and related peptides. III. tryptophan. J Am Chem Soc 97(10):2612–2619

Berkelsar EJ, Ormrod DP, Hale BA (1996) The influence of photosynthetically active radiation on the effects of ultraviolet-B radiation on Arabidopsis thaliana. Photochem Photobiol 64(1):110–116

Besaratinia, Kim SI, Pfeifer GP (2008) Rapid repair of UV-A-induced oxidized purines and persistence of UV-B-induced dipyrimidine lesions determine the mutagenicity of sunlight in mouse cells. FASEB J 22:2379–2392

Biever JJ, Brinkman D, Gardner G (2014) UV-B inhibition of hypocotyl growth in etiolated *Arabidopsis thaliana* seedlings is a consequence of cell cycle arrest initiated by photodimer accumulation. J Exp Bot 65:2949–2961

Biggs RH, Basiouny FM (1975) Plant growth responses to elevated UV-B irradiation under growth chamber, greenhouse and solarium conditions. In: Nachtwey DS, Caldwell MM, Biggs RH (eds) Climatic Impact Assessment Program (CLAP), Monograph 5, Part 1 – Ultraviolet Radiation Effects, US Dept Transp., Report No. DOT-TST-75-55. (PB-247-725), National Technical Information Service. Springfield, Virginia, pp 4–197, 4–249 (Part 1, Chpt. 4, Appendix L)

Bjorn LO (1996) Effect of ozone depletion and increased UV-B on terrestrial ecosystems. Int J Environ Stud 51:217–243

Blokhina O, Virolainen E, Fagerstedt KV (2003) Antioxidants, oxidative damage and oxygen deprivation stress: a review. Ann Bot (Lond) 91:179–194

Blumthaler M, Amback W (1990) Indication of increasing solar UV-B radiation flux in alpine regions. Science 248:206–208

Boccalandro F, Muench A, Sdringola S, Rosales OR (2004) Wireless laser-assisted angioplasty of the superficial femoral artery in patients with critical limb ischemia who have failed conventional percutaneous revascularization. Catheter Cardiovasc Interv 63:7e12

Bonura T, Smith KC (1975) Quantitative evidence for enzymatically-induced DNA double-strand breaks as lethal lesions in UV-irradiated pol+ and polA1 strains of *E. coli* K-12. Photochem Photobiol 22:243–248

Bornman JF (1989) New trends in photobiology: target sites of UV-B radiation in photosynthesis of higher plants. J Photochem Photobiol B:Biol 4:145–158

Bornman JF, Teramura AH (1993) Effect of ultraviolet-B radiation on terrestrial plants. In: Young AR, Bjorn LO, Moan J, Nultsch W (eds) Environmental UV photobiology. Plenum Press, New York, pp 427–471

Bornman JF, Barnes PW, Robinson SA, Ballaré CL, Flint SD, Caldwell MM (2015) Solar ultraviolet radiation and ozone depletion-driven climate change: effects on terrestrial ecosystems. Photochem Photobiol Sci 14:88–107

Box HC, Dawidzik JB, Budzinski EE (2001) Free radical-induced double lesions in DNA. Free Radic Biol Med 31:856–868

Brandle JR, Campbell WF, Sisson WB, Caldwell MM (1977) Net photosynthesis, electron transport capacity, and ultrastructure of *Pisum sativum* L. exposed to ultraviolet-B radiation. Plant Physiol 60:165–169

Bravard A, Vacher M, Gouget B, Coutant A, de Boisferon FH, Marsin S, Chevillard S, Radicella JP (2006) Redox regulation of human OGG1 activity in response to cellular oxidative stress. Mol Cell Biol 26 (20):7430–7436

Brenner M, Hearing VJ (2008) The protective role of melanin against UV damage in human skin. Photochem Photobiol 84(3):539–549

Britt AB, Chen J-J, Wykoff D, Mitchell D (1993) A UV sensitive mutant of *Arabidopsis* defective in the repair of pyrimidine- pyrimidone(6-4) dimers. Science 261:1571–1574

Brown SB, Houghton JD, Henry GAF (1991) Chlorophyll breakdown. In: Scheer H (ed) Chlorophylls. CRC Press, Boca Raton, pp 465–489

Buma AGJ, Helbling EW, Karin De Boer MK, Villafañe VE (2001) Patterns of DNA damage and photoinhibition in temperate South-Atlantic picophytoplankton exposed to solar ultraviolet radiation. J Photochem Photobiol B: Biol 62(1-2):9–18

Cadet J, Douki T (2011) Oxidatively generated damage to DNA by UV-A radiation in cells and human skin. J Invest Dermatol 131:1005–1007

Cadet J, Berger M, Douki T, Morin B, Raoul S, Ravanat JL, Spinelli S (1997) Effect of UV and visible radiation on DNA-final base damage. Biol Chem 378:1275–1286

Cadet J, Douki T, Gasparutto D, Ravanat J-L (2003) Oxidative damage to DNA: formation, measurement and biochemical features. Mutat Res 531:5–23

Cadet J, Douki T, Ravanat JL et al (2009) Sensitized formation of oxidatively generated damage to cellular DNA by UV-A radiation. Photochem Photobiol Sci 8:903–911

Caldwell MM, Teramura AT, Tevini M (1978) The changing solar ultraviolet climate and the ecological consequences for higher plants. Trends in Ecol Evol 4:363–367

Carletti P, Masi A, Wonisch A, Grill D, Tausz M, Ferretti M (2003) Changes in antioxidant and pigment pool dimensions in UV-B radiation maize seedlings. Environ Exp Bot 50:149–157

Casati P, Walbot V (2003) Gene expression profiling in response to UV radiation in maize genotypes with varying flavonoid content. Plant Physiol 132:1739–1754

Casati P, Walbot V (2004) Crosslinking of ribosomal proteins to RNA in maize ribosomes by UV-B and its effect on translation. Plant Physiol 136:3319–3332

Cen YP, Bornmann JF (1990) The response of bean plants to UV-B radiation under different irradiances of background visible light. J Exp Bot 41:1489–1495

Chatgilialoglu C, D'Angelantonio M, Kciuk G, Bobrowski K (2011) New insights into the reaction paths of hydroxyl radicals with 2′-deoxyguanosine. Chem Res Toxicol 24(12):2200–2206

Chen H, Han R (2014) He-Ne laser treatment improves the photosynthetic efficiency of wheat exposed to enhanced UV-B radiation. Laser Phys 24:105602–105608

Chow WS, Strid A, Anderson JM (1992) Short term treatment of pea plants with supplementary ultraviolet-B radiation: recovery time-courses of some photosynthetic functions and components. In: Murata N (ed) Research in photosynthesis, vol 4. Kluwer Academic Publishers, Dordrecht, pp 361–364

Corlett JE, Stephen J, Jones HG, Woodfin R, Mepsted R, Paul ND (1997) Assessing the impact of UV-B radiation on the growth and yield of field crops. In: Lumsden PJ (ed) Plants and UV-B: responses to environmental change society for experimental biology, vol 64. Cambridge University Press, Cambridge

Correia CM, Areal ELV, Torres-Pereira MS, Torres-Pereira JMG (1998) Intraspecific variation in sensitivity to ultraviolet-B radiation in maize grown under field conditions. I. Growth and morphological aspects. Field Crops Res 59:81–89

Courdavault S, Baudouin C, Charveron M et al (2004) Larger yield of cyclobutane dimers than 8 oxo-7,8-dihydroguanine in the DNA of UV-A-irradiated human skin cells. Mutat Res 556:135–142

Creed D (1984a) The photophysics and photochemistry of the near-UV absorbing amino acids .2. Tyrosine and its simple derivatives. Photochem Photobiol 39 (4):563–575

Creed D (1984b) The photophysics and photochemistry of the near-UV absorbing amino-acids.3. Cystine and its simple derivatives. Photochem Photobiol 39: 577–583

Culligan K, Tissier A, Britt A (2004) ATR regulates a G2-Phase cell-cycle checkpoint in *Arabidopsis thaliana*. Plant Cell 16:1091–1104

Danon A, Gallois P (1998) UV-C radiation induces apoptotic-like changes in *Arabidopsis thaliana*. FEBS Lett 437:131–136

Danon A, Delorme V, Mailhac N, Gallois P (2000) Plant programmed cell death: a common way to die. Plant Physiol Biochem 38:647–655

Danon A, Rotari VI, Gordon A, Mailhac N, Gallois P (2004) Ultraviolet-C overexposure induces programmed cell death in *Arabidopsis*, which is mediated by caspase-like activities and which can be suppressed by caspase inhibitors, p35 and defender against apoptotic death. J Biol Chem 279:779–787

Dat J, Vandenabeele S, Vranova E, Van Montagu M, Inze D, Van Breusegem (2000) Dual action of the active oxygen species during plant stress responses. Cell Mol Life Sci 57:779–795

Davies MJ (2004) Reactive species formed on proteins exposed to singlet oxygen. Photochem Photobiol Sci 3 (1):17–25

Deckmyn G, Impens I (1997) The ratio UV-B photosynthetically active radiation (PAR) determines the sensitivity of rye to increased UV-B radiation. Environ Exp Bot 37:3–12

Del Rio LA, Sandalio LM, Palma JM, Bueno P, Cospus FJ (1992) Metabolism of oxygen radicals in peroxisomes and cellular implications. J Free Radic Biol Med 13:557–580

Donkor VA, Hader DP (1996) Effects of ultraviolet radiation on photosynthetic pigments in some filamentous cyanobacteria. Aquat Microb Ecol 11(143–149):19

Doughty JC, Hope AB (1973) Effect of ultraviolet radiation on the membranes of Chara coralline. J Membr Biol 13:185–198

Douki T, Cadet J (2003) Formation of the spore photoproduct and other dimeric lesions between adjacent pyrimidines in UVC-irradiated dry DNA. Photochem Photobiol Sci 2:433–436

Douki T, Perdiz D, Moustacchi E, Cadet J, Sage E (1999) Oxidation of guanine in cellular DNA by solar UV radiation: biological role. Photochem Photobiol 70:184–190

Douki T, Reynaud-Angelin A, Cadet J, Sage E (2003) Bipyrimidine photoproducts rather than oxidative lesions are the main type DNA damage involved in the genotoxic effect of solar UV-A radiation. Biochemistry 42:9221–9226

Du H, Liang Y, Pei K, Ma K (2011) UV radiation-responsive proteins in rice leaves: a proteomic analysis. Plant Cell Physiol 52(2):306–316

Dubest S, Gallego ME, White CI (2002) Role of the AtRad1p endonuclease in homologous recombination in plants. EMBO Rep 3:1049–1054

El Ghissassi F, Baan R, Straif K et al (2009) A review of human carcinogens—part D: radiation. Lancet Oncol 10:51–52

Elstner EF (1987) Metabolism of activated oxygen species. In: Davies DD (ed) Biochemistry of plants. Academic Press, London, pp 253–315

Fagerberg WR, Bornman JF (2005) Modification of leaf cytology and anatomy in *Brassica napus* grown under above ambient levels of supplemental UV-B radiation. Photochem Photobiol Sci 4(3):275–279

Fedina I, Hidema J, Velitchkova M, Georgieva K, Nedeva D (2010) UV-B induced stress responses in three rice cultivars. Biol Plant 54(3):571–574

Fell LJ, Paul ND, McMillan TJ (2002) Role for non-homologous end-joining in the repair of UV-A-induced DNA damage. Int J Radiat Biol 78:1023–1027

Frohnmeyer H, Staiger D (2003) Ultraviolet-B radiation-mediated responses in plants. Balancing damage and protection. Plant Physiol 133:1420–1428

Gao W, Zheng Y, Slusser JR, Heisler GM, Grant RH, Xu J, He D (2004) Effects of supplementary ultraviolet-B irradiance on maize yield and qualities: a field experiment. Photochem Photobiol 80:127–131

Garcia V, Bruchet H, Camescasse D, Granier F, Bouchez D, Tissier A (2003) AtATM is essential for meiosis and the somatic response to DNA damage in plants. Plant Cell 15:119–132

Girard PM, Francesconi S, Pozzebon M, Graindorge D, Rochette P, Drouin R, Sage E (2011) UV-A-induced damage to DNA and proteins: direct versus indirect photochemical processes. J Phys: Con Ser 261:1–11

Girard M, Lariviere R, Parfitt DA et al (2012) Mitochondrial dysfunction and Purkinje cell loss in autosomal

recessive spastic ataxia of Charlevoix-Saguenay (ARSACS). Proc Natl Acad Sci U S A 109:1661–1666
Girotti AW (2001) Photosensitized oxidation of membrane lipids: reaction pathways, cytotoxic effects and cytoprotective mechanisms. J Photochem Photobiol B 63:103–113
Gorczyca W, Tuziak T, Kram A, Melamed MR, Darzynkiewicz Z (1994) Detection of apoptosis-associated DNA strand breaks in fine-needle aspiration biopsies by in situ end labeling of fragmented DNA. Cytometry 15(2):169–175
Greinert R, Volkmer B, Henning S, Breitbart EW, Greulich KO, Cardoso MC, Rapp A (2012) UV-A-induced DNA double-strand breaks result from the repair of clustered oxidative DNA damage. Nucleic Acids Res 40(20):10263–10273
Grether-Beck S, Bonizzi G, Schmitt-Brenden H, Felsner I, Timmer S, Sies H, Johnson JP, Piette J, Krutmann J (2000) Nonenzymatic triggering of the ceramide signalling cascade by solar UV-A radiation. EMBO J 19(21):5793–5800
Harper JV, Anderson JA, O'Neill P (2010) Radiation induced DNA DSB: contribution from stalled replication forks? DNA Repair 9:907–913
Hart PH, Gorman S, Finlay-Jones JJ (2011) Modulation of the immune system by UV radiation: more than just the effects of vitamin D? Nat Rev Immunol 11:584–596
He YY, Pi J, Huang JL, Diwan BA, Waalkes MP, Chignell CF (2006) Chronic UV-A radiation of human HaCaT keratinocytes induces malignant transformation associated with acquired apoptotic resistance. Oncogene 25:3680–3688
Heck DE, Vetrano AM, Mariano TM et al (2003) UV-B light stimulates production of reactive oxygen species: unexpected role for catalase. J Biol Chem 278 (25):22432–22436
Hidema J, Kumagai T (1998) UVB-induced cyclobutyl pyrimidine dimer and photorepair with progress of growth and leaf age in rice. J Photochem Photobiol B Biol 43:121–127
Hidema J, Makino A, Mae T, Ojima T (1991) Photosynthetic characteristics of rice leaves under different irradiances from full expansion through senescence. Plant Physiol 97:1287–1293
Hidema J, Kang HS, Kumagai T (1996) Differences in the sensitivity to UV-B radiation of two cultivars of rice (*Oryza sativa* L.). Plant Cell Physiol 37:742–747
Hidema J, Kumagai T, Sutherland JC, Sutherland BM (1997) Ultraviolet B–sensitive rice cultivar deficient in cyclobutyl pyrimidine dimer repair. Plant Physiol 113:39–44
Hidema J, Taguchi T, Ono T, Teranishi M, Yamamoto K, Kumagai T (2007) UV-B resistance and CPD photolyase activity in rice. Plant J 50:70–79
Hiraku Y, Ito K, Hirakawa K, Kawanishi S (2007) Photosensitized DNA damage and its protection via a novel mechanism. Photochem Photobiol 83:205–212
Holló sy F (2002) Effect of ultraviolet radiation on plant cells. Micron 33:179–197
Houot V, Etienne P, Petitot AS, Barbier S, Blein JP, Suty L (2001) Hydrogen peroxide induces programmed cell death features in cultured tobacco B4-2 cells in a dose-dependent manner. J Exp Bot 52(361):1721–1730
http://www.who.int/uv/health/en/
Hu Z, Li H, Chen S, Yang Y (2006) Interaction of UV radiation with other components of solar radiation. Physiol Plant 58:401–407
Huffaker RC (1982) Biochemistry and physiology of leaf proteins. In: Parthier B, Bolter D (eds) Encyclopedia of plant physiology. SpringerVerlag, Berlin, pp 370–400
Hunter JR, Kaupp SE, Taylor JH (1981) Effects of solar and artificial ultraviolet-B radiation on larval northern anchovy, engraulis mordax. Photochem Photobiol 34:477–486
Hutchinson F (1987) A review of some topics concerning mutagenesis by ultraviolet light. Photochem Photobiol 45:897–903
Hutchinson F, Yamamoto K, Stein J, Wood RD (1987) Effect of photoreactivation on mutagenesis of lambda phage by ultraviolet light. J Mol Biol 202:593–601
Idso K, Idso SB (1994) Plant responses to atmospheric CO_2 enrichment in the face of environmental constraints: a review of the past 10 year's research. Agric Forste meteorol 69:153–203
Ilwanzik W, Tevini M, Dohnt G, Vos M, Weiss W, Gra¨ber P, Renger G (1983) Action of UV-B radiation on photosynthetic primary reaction in spinach chloroplasts. Physiol Plant 58:401–407
Jackson AL, Loeb LA (2001) The contribution of endogenous sources of DNA damage to the multiple mutations in cancer. Mutat Res 477:7–21
Jansen MAK, Gaba V, Greenberg BM (1998) Higher plants and UV-B radiation: balancing damage, repair and acclimation. Trends Plant Sci 3:131–135
Jenkins GI (2009) Signal transduction in responses to UV-B radiation. Annu Rev Plant Biol 60:407–431
Jiang Y, Rabbi M, Kim M, Ke C, Lee W, Clark RL, Mieczkowski PA, Marszalek PE (2009) UV-A generates pyrimidine dimers in DNA directly. Biophys J 96(3):1151–1158
Jiang L, Wang Y, Bjorn LO, Li S (2011) UV-B induced DNA damage mediates expression changes of cell cycle regulatory genes in *Arabidopsis* root tips. Planta 233:831–841
Jordan BR (1993) The molecular biology of plants exposed to ultraviolet B radiation and the interaction with other stresses. In: Jackson MB, Black CR (eds) Interacting stresses on plants in a changing climate, vol 16, NATO ASI Series. Springer, Berlin, pp 153–170
Jordan BR, James PE, Strid A, Anthony RG (1994) The effect of UV-B radiation on gene expression and pigment composition in etiolated and green pea leaf tissue: UV-B induced changes in gene expression are gene specific and dependent upon tissue development. Plant Cell Environ 17:45–54
Kakani VG, Reddy KR, Zhao D, Mohammed AR (2003) Effects of ultraviolet-B radiation on cotton (*Gossypium hirsutum* L.) morphology and anatomy. Ann Bot 91:817–826

Kambayashi H, Odake Y, Takada K, Funasaka Y, Ichihashi M (2003) Involvement of changes in stratum corneum keratin in wrinkle formation by chronic ultraviolet radiation in hairless mice. Exp Dermatol 12(Suppl 2):22–27

Karin M (1998) Mitogen-activated protein kinase cascades as regulators of stress responses. Ann N Y Acad Sci 851:139–146

Kasai H, Yamaizumi Z, Berger M, Cadet J (1992) Photosensitized formation of 7,8-dihydro-8-oxo-2-′-deoxyguanosine (8-hydroxy-2′-deoxyguanosine) in DNA by riboflavin: a non singlet oxygen mediated reaction. J Am Chem Soc 114:9692–9694

Kataria S, Jajoo A, Guruprasad KN (2014) Impact of increasing Ultraviolet-B (UV-B) radiation on photosynthetic processes. J Photochem Photobiol B Biol 137:55–66

Kerwin BA, Remmele RL Jr (2007) Protect from light: photodegradation and protein biologics. J Pharmac Sci 96(6):468–1479

Khoroshilova EV, Repeyev YA, Nikogosyan DN (1990) UV photolysis of aromatic amino acids and related dipeptides and tripeptides. J Photochem Photobiol 7:159–172

Kielbassa C, Roza L, Epe B (1997) Wavelength dependence of oxidative DNA damage induced by UV and visible light. Carcinogenesis 18:811–816

Klem K, Alexander AC, Holuba P, Kovca D, Spundaa V, Robsonb M, Urbana O (2012) Interactive effect of PAR and UV radiation on the physiology, morphology and leaf optical properties of two barley varieties. Environ Exp Bot 75:52–64

Kolb CA, Laser MA, Kopecky J, Zotz G, Riederer M, Pfundel EE (2001) Effects of natutral intensities of visible and UV radiation on epidermal UV-screening and photosynthesis in grape leaves (vitis vinifera cv. Silvaner). Plant Physiol 127:863–875

Kramer GF, Norman HA, Krizek DT, Mirecki RM (1991) Influence of UV-B radiation on polyamines, lipid peroxidation and membrane lipids in cucumber. Phytochemistry 30:2101–2108

Kubasek WL, Shirley BW, Mckillop A, Goodman HM, Briggs W, Ausubel FM (1992) Regulation of flavonoid biosynthetic genes in germinating Arabidopsis seedlings. Plant Cell 4(10):1229–1236

Kulandaivelu G, Nedunchezhian N, Annamalainathan K (1991) Ultraviolet-B (280–320 nm) radiation induced changes in photochemical activities and polypeptide components of C3 and C4 chloroplasts. Photosynthetica 25:333–339

Kuluncsics Z, Perdiz D, Brulay E et al (1999) Wavelengths dependence of ultraviolet-induced DNA damage distribution: involvement of direct or indirect mechanisms and possible artefacts. J Photochem Photobiol B 49:71–80

Kumagai T, Sato T (1992) Inhibitory effect of increase in near-UV radiation on the growth of Japanese rice cultivars (*Oryza sativa* L.) in a phytotron and recovery by exposure to visible radiation. Jpn J Breed 42:545–552

Kunz BA, Anderson HJ, Osmond MJ, Vonarx E (2005) Components of nucleotide excision repair and DNA damage tolerance in *Arabidopsis thaliana*. Environ Mol Mutagen 45:115–127

Kvam E, Tyrrell RM (1997) Induction of oxidative DNA base damage in human skin cells by UV and near visible radiation. Carcinogenesis 18:2379–2384

Lam E, Pozo O (2000) Caspase-like protease involvement in the control of plant cell death. Plant Mol Biol 44:417–428

Lam E, Pontier D, del Pozo O (1999) Die and let live: programmed cell death in plants. Curr Opin Plant Biol 2:502–507

Landry LG, Stapleton AE, Lim J, Hoffman P, Hays J, Walbot V, Last R (1997) An Arabidopsis photolyase mutant is hypersensitive to ultraviolet-B radiation. Proc Natl Acad Sci U S A 94:328–332

Lee HC, Wei YH (2000) Mitochondrial role in life and death of the cell. J Biomed Sci 7:2–15

Lee YJ, Jeong SY, Karbowski M, Smith CL, Youle RJ (2004) Roles of the mammalian mitochondrial fission and fusion mediators Fis1, Drp1, and Opa1 in apoptosis. Mol Biol Cell 15:5001–5011

Lidon FC (2012) UV-B radiation in rice:interaction between micronutrients accumulation and the photosynthetic performance. J Plant Inter 7(1):19–28

Lidon FC, Ramalho JC (2011) Impact of UV-B radiation on photosynthetic performance and chloroplast membrane components in *Oryza sativa* L. J Photochem Photobiol B Biol 104:457–466

Lima-Bessa KM, Armelini MG, Chigançjas V, Jacysyn JF, Amarante-Mendes GP, Sarasin A, Menck CF (2008) CPDs and 6-4PPs play different roles in UV induced cell death in normal and NER-deficient human cells. DNA Repair 7(2):303–312

Lindo SJ, Seeley SD, Caldwell MM (1978) Effect of ultraviolet-B radiation on the abscisic acid status of Rumex patientia leaves. Physiol Plant 45:67–72

Liu Z, Hossain GS, Islas-Osuna MA, Mitchell DL, Mount DW (2000) Repair of UV damage in plants by nucleotide excision repair: *Arabidopsis* UVH1 DNA repair gene is a homolog of *Saccharomyces cerevisiae* Rad1. Plant J 21:519–552

Loft S, Poulsen HE (1996) Cancer risk and oxidative DNA damage in man. J Molec Med 74:297–312

Lu C, Zhu F, Cho YY, Tang F, Zykova T, Ma WY, Bode AM, Dong Z (2006) Cell apoptosis: requirement of H2AX in DNA ladder formation, but not for the activation of caspase-3. Mol Cell 23:121–132

Lytvyn DI, Yemets AI, Blume YB (2010) UV-B overexposure induces programmed cell death in a BY-2 tobacco cell line. Environ Exp Bot 68:51–57

Mackerness S, John CF, Jordan B, Thomas B (2001) Early signaling components in ultraviolet-B responses: distinct roles for different reactive oxygen species and nitric oxide. FEBS Lett 489:237–242

Marwood CA, Greenberg BM (1996) Effect of supplementary UV-B radiation on chlorophyll synthesis and accumulation of photosynthesis during chloroplast development in *Spirodela oligorrhiza*. Photochem Photobiol 64:664–670

Masaki H, Izutsu Y, Yahagi S, Okano Y (2009) Reactive oxygen species in HaCaT keratinocytes after UV-B radiation are triggered by intracellular Ca^{2+} levels. J Invest Dermatol Sym Proc 14:50–52

Mazza CA, Boccalandro HE, Giordano CV, Battista D, Zima AM, Scopel AL, Ballare CL (2000) Functional significance and induction by solar radiation of ultraviolet-absorbing sunscreens in field-grown soybean crops. Plant Physiol 122:117–125

McCabe PF, Levine A, Meijer PJ, Tapon NA, Penell RI (1997) A programmed cell death pathway activated in carrot cells cultured at low cell density. Plant J 12:267–280

Meguro S, Arai Y, Masukawa K, Uie K, Tokimitsu I (1999) Stratum corneum lipid abnormalities in UV-B-irradiated skin. Photochem Photobiol 69:317–321

Mirecki RM, Teramura AH (1984) Effects of ultraviolet-B irradiance on soybean. The dependence of plant sensitivity on the photosynthetic photon flux density during and after leaf expansion. Plant Physiol 74:475–480

Mishra RK, Singhal GS (1992) Function of photosynthetic apparatus of intact wheat leaves under high light and heat stress and its relationship with peroxidation of thylakoid lipids. Plant Physiol 98:1–6

Mittler R (2002) Oxidative stress, antioxidants and stress tolerance. Trends Plant Sci 7(9):405–410

Mittler R, Lam E (1996) Sacrifice in the face of foes: pathogen-induced programmed cell death in plants. Trends Microbiol 4:10–15

Mittler R, Vanderauwera S, Gollery M, Van Breusegem F (2004) The reactive oxygen gene network in plants. Trends Plant Sci 9:490–498

Miziorko HM, Lorimer GH (1983) Ribulose-1,5-bisphosphate carboxylaseoxygenase. Annu Rev Biochem 52:507–535

Mohammed AR, Tarpley L (2009) Effects of elevated ultraviolet-B radiation on productive tillers, spikelet sterility and grain characteristics of southern US rice (Oryza sativa L.) cultivars. J Agron Crop Sci 195:292–300

Moldau H (1999) Ozone detoxification in the mesophyll cell wall during a simulated oxidative burst. Free Radic Res 31:S19–S44

Molinier J, Lechner E, Dumbliauskas E, Genschik P (2008) Regulation and role of *Arabidopsis* CUL4-DDB1A-DDB2 in maintaining genome integrity upon UV stress. PLoS Genet 4:e1000093

Mouret S, Baudouin C, Charveron M, Favier A, Cadet J, Douki T (2006) Cyclobutane pyrimidine dimers are predominant DNA lesions in whole human skin exposed to UV-A radiation. Proc Natl Acad Sci U S A 103:13765–13770

Mouret S, Philippe C, Gracia-Chantegrel J, Banyasz A, Karpati S, Markovitsi D, Douki T (2010) UV-A-induced cyclobutane pyrimidine dimers in DNA: a direct photochemical mechanism ? Org Biomol Chem 8(7):1706–1711

Mpoloka SW (2008) Effect of prolonged UV-B exposure in plants. Afr J Biotech 7(25):4874–4883

Murphy TM (1990) Effect of broad-band ultraviolet and visible radiation on hydrogen peroxide formation by cultured rose cells. Physiol Plant 80:63–68

Murphy MG, Byczko Z (1990) Effects of polyunsaturated fatty acids on adenosine receptor function in N 1 Ell5 neuroblastoma cells. Can J Biochem Cell Biol 34:56–67

Murphy TM, Vu H (1996) Photoactivation of superoxide synthases of the plasma membrane from rose (*Rosa damascene Mill.*) cells. Photochem Photobiol 64: 106–109

Murphy TM, Wilson C (1982) The UV-stimulated K+ efflux from rose cells. Plant Physiol 70:709–713

Murphy TM, Hurrel HC, Sasaki TL (1985) Wavelength dependence of ultraviolet radiation-induced mortality and K+ efflux in cultured cells of *Rosa damascena*. Photochem Photobiol 42:281–286

Musil CF, Chimphango SBM, Dakora FD (2002) Effects of elevated ultraviolet-B radiation on native and cultivated plants of southern Africa. Annu Bot 90:127–137

Nakajima S, Sugiyama M, Iwai S, Hitomi K, Otoshi E, Kim S, Jiang CZ, Todo T, Britt AB, Yamamoto K (1998) Cloning and characterization of a gene (UVR3) required for photorepair of 6–4 photoproducts in *Arabidopsis thaliana*. Nucleic Acids Res 26:638–644

Nawkar GM, Maibam P, Park JH, Sahi VP, Lee SY, Kang CH (2013) UV-Induced cell death in plants. Int J Mol Sci 14:1608–1628

Negash L, Björn LO (1986) Stomatal closure by ultraviolet radiation. Physiol Plant 66:360–364

Neves-Peterson MT, Gajula GP, Peterson SB (2007) UV light effects on proteins: from photochemistry to nanomedicine. Mol Phytochem 186:128–163

Nishiyama Y, Yamamoto H, Allakhverdiev SI, Inaba M, Yokota A, Murata N (2001) Oxidative stress inhibits the repair of photodamage to the photosynthetic machinery. EMBO J 20:5587–5594

Nogue's S, Allen DJ, Morison JIL, Baker NR (1999) Characterization of stomatal closure caused by Ultraviolet-B radiation. Plant Physiol 121:489–496

Ogawa F, Sander CS, Hansel A, Oehrl W, Kasperczyk H, Elsner P, Shimizu K, Stefan H, Heinemann SH, Thiele JJ (2006) The repair enzyme peptide Methionine-*S*-Sulfoxide reductase is expressed in human epidermis and upregulated by UV-A radiation. J Invest Dermatol 126:1128–1134

Ohnishi T, Takahashi A, Nagamatsu A, Omori K, Suzuki H, Shimazu T, Ishioka N (2009) Detection of space radiation-induced double strand breaks as a track in cell nucleus. Biochem Biophys Res Commun 390(3):485–488

Osipov AN, Smetanina NM, Pustovalova MV, Arkhangelskaya E, Klokov D (2014) The formation of DNA single-strand breaks and alkali-labile sites in human blood lymphocytes exposed to 365-nm UV-A radiation. Free Radic Biol Med 73:34–40

Pakker H, Beekman CAC, Breeman AM (2000) Efficient photoreactivation of UV-BR-induced DNA damage in the sublittoral macroalga *Rhodymenia pseudopalmata* (Rhodophyta). Eur J Phycol 35:109–114

Panagopoulos I, Bornmann JF, Björn LO (1990) Effect of ultraviolet radiation and visible light on growth, fluorescence induction, ultraweak luminiscence and peroxidase activity in sugar beet plants. Photochem Photobiol 8:73–87

Pattison DI, Davies MJ (2006) Action of ultraviolet light on cellular structures. In: Bignold LP (ed) Cancer: Cell structures, carcinogens and genomic instability. Birkhäuser, Virlag

Peak MJ, Peak JG (1982) Single-strand breaks induced in *Bacillus subtilis* DNA by ultraviolet light: action spectrum and properties. Photochem Photobiol 35:675–680

Peak MJ, Peak JG (1990) Hydroxyl radical quenching agents protect against DNA breakage caused by both 365-nm UV-A and by gamma radiation. Photochem Photobiol 51(6):649–652

Peak MJ, Peak JG, Moehring MP, Webb RB (1984) Ultraviolet action spectrum for DNA dimer induction, lethality, and mutagenesis in *Escherichia coli* with emphasis on the UV-B region. Photochem Photobiol 40:613–620

Pearse AD, Gaskell SA, Marks R (1987) Epidermal changes in human skin following radiation with either UV-B or UV-A. J Invest Dermatol 88:83–87

Petersen MTN, Jonson PH, Petersen SB (1999) Amino acid neighbours and detailed conformational analysis of cysteins in proteins. Protein Eng Des Sel 12 (7):535–548

Pfundel EE, Pan RS, Dilley RA (1992) Inhibition of violaxanthin deep oxidation by ultraviolet-B radiation in isolated chloroplasts and intact leaves. Plant Physiol 98:1372–1380

Phillipson RP, Tobi SE, Morris JA, McMillan TJ (2002) UV-A induces persistent genomic instability in human keratinocytes through an oxidative stress mechanism. Free Radic Biol Med 32:474–480

Pouget J-P, Douki T, Richard M-J, Cadet J (2000) DNA damage induced in cells by and UV-A radiation as measured by HPLC/GC-MS and HPLC-EC and comet assay. Chem Res Toxicol 13(7):541–549

Predieri S, Norman HA, Krizek DT, Pillai P, Mirecki RM, Zimmerman RH (1995) Influence of UV-B radiation on membrane lipid composition and ethylene evolution in (Doyenne D' Hiver) pear shoots grown in vitro under different photosynthetic photon fluxes. Environ Exp Bot 35:151–160

Quaite FE, Sutherland BM, Sutherland JC (1992) Action spectrum for DNA damage in alfalfa lowers predicted impact of ozone depletion. Nature 358:576–578

Rao MV, Paliyath G, Ormrod DP (1996) Ultraviolet-B- and ozone-induced biochemical changes in antioxidant enzymes of Arabidopsis thaliana. Plant Physiol 110:125–136

Rapp A, Greulich KO (2004) After double-strand break induction by UV-A, homologous recombination and nonhomologous end joining cooperate at the same DSB if both systems are available. J Cell Sci 117:4935–4945

Rastogi RP, Richa KA, Tyagi MB, Sinha RP (2010) Molecular mechanisms of ultraviolet radiation-induced DNA damage and repair. J Nucleic Acid 56:1–32

Ravanat JL, Douki T, Cadet J (2001) Direct and indirect effects of UV radiation on DNA and its components. J Photochem Photobiol B Biol 63:88–102

Reape TJ, McCabe PF (2008) Apoptotic-like programmed cell death in plants. New Phytol 180:13–26

Rebeiz CA, Montazerzouhoor A, Mayasich JM, Tripathy BC, Wu SM, Rebeiz CC (1988) Photodynamic herbicides – recent developments and molecular basis of selectivity. Crit Rev Plant Sci 6:385–436

Regensburger J, Knak A, Maisch T, Landthaler M, Baumler W (2012) Fatty acids and vitamins generate singlet oxygen under UV-B radiation. Exp Dermatol 21:135–139

Renger G, Völker M, Eckert HJ, Fromme R, Hohm-Veit S, Gräber P (1989) On the mechanism of photosystem II deterioration by UV-B radiation. Photochem Photobiol 49:97–105

Ries G, Heller W, Puchta H, Sandermann H, Seidlitz HK, Hohn B (2000) Elevated UV-B radiation reduces genome stability in plants. Nature 406:98–101

Rizzo JL, Dunn J, Rees A, Runger TM (2011) No formation of DNA double-strand breaks and no activation of recombination repair with UV-A. J Invest Dermatol 131:1139–1148

Roleda MY, Hanelt D, Wiencke C (2006a) Exposure to ultraviolet radiation delays photosynthetic recovery in Arctic kelp zoospores. Photosynth Res 88:311–322

Roleda MY, Hanelt D, Wiencke C (2006b) Growth and DNA damage in young *Laminaria* sporophytes exposed to ultraviolet radiation: implication for depth zonation of kelps in Helgoland (North Sea). Mar Biol 148:1201–1211

Ros J, Tevini M (1995) Interaction of UV radiation and IAA during growth of seedlings and hypocotyls segments of sunflower. J Plant Physiol 146:295–302

Rousseaux MC, Ballare CL, Giordano CV, Scopel AL, Zima AM, Swarcberg-Brachitta M, Searles PS (1999) Ozone depletion and UV-B radiation: impact on plant DNA damage in southern South America. Proc Natl Acad Sci U S A 96:15310–15315

Rozema J, van de Staaij J, Bjorn LO, Caldwell M (1997) UV-B as an environmental factor in plant life: stress and regulation. Trends Ecol Evol 12:22–28

Saewan N, Jimtaisong A (2013) Photoprotection of natural flavonoids. J Appl Pharm Sci 3(09):129–141

Sage E (1993) Distribution and repair of photolesions in DNA: genetic consequences and the role of sequence context. Photochem Photobiol 57:163–174

Salama HMH, Al Watban AA, Al-Fughom AT (2011) Effect of ultraviolet radiation on chlorophyll, carotenoid, protein and praline contents of some annual desert plants. Saudi J Biol Sci 18:79–86

Sato T, Kumagai T (1993) Cultivar differences in resistance to the inhibitory effect of near-UV radiation among Asian ecotype and Japanese lowland and upland cultivars of rice. Jpn J Breed 43:61–68

Savitch LV, Pocock T, Krol M, Wilson KE, Greenberg BM, Huner NPA (2001) Effect of growth under UV-A radiation on CO2 assimilation, carbon partitioning, PSII photochemistry and resistance to UV-B radiation in Brassica napus cv Topas. Aust J Plant Physiol 28:203–212

Schuch AP, da Silva Galhardo R, de Lima-Bessa KM, Schuch NJ, Menck CF (2009) Development of a DNA-dosimeter system for monitoring the effect of solar-ultraviolet radiation. Photochem Photobiol Sci 8:111–120

Schuch AP, Lago JC, Yagura T, Carlos Frederico Martins Menck CFM (2012) DNA dosimetry assessment for sunscreen genotoxic photoprotection. PlosOne 7(6): e40344

Shorrocks J, Paul ND, McMillan TJ (2008) The dose rate of UV-A treatment influences the cellular response of HaCaT keratinocytes. J Invest Dermatol 128:685–693

Sileman TA, Nicholson WL (2000) Artificial and solar UV radiation induces strand breaks and cyclobutane pyrimidine dimers in *Bacillus subtilis* spore DNA. Appl Environ Microbiol 66:199–205

Singh R, Singh S, Tripathi R, Agrawal SB (2011) Supplemental UV-B radiation induced changes in growth, pigments and antioxidant pool of bean (*Dolichos lablab*) under field conditions. J Environ Biol 32:139–145

Sinha RP, Häder DP (2002) UV-induced DNA damage and repair: a review. Photochem Photobiol Sci 1 (4):225–236

Sinha RP, Margit Dautz M, Häder D-P (2001) A simple and efficient method for the quantitative analysis of thymine dimers in cyanobacteria, phytoplankton and macroalgae. Acta Protozool 40:187–195

Skórska E (2000) Responses of some crop plants on UV-B radiation (in Polish), Rozprawy, 192. Agricultural University Press, Szczecin

Spreitzer RJ (1993) Genetic dissection of Rubisco structure and function. Annu Rev Plant Physiol Plant Mol Biol 44:411–434

Staehelin J, Dütsch HU (1989) Zeigen die Schweizerischen Messungen eine Gefährdung der Ozonschicht? Chimia 43:338–348

Staehelin J, Blumthaler M, Ambach W, Torhorst J (1990) Skin cancer and the ozone shield. Lancet 336:502

Stein JC, Hansen G (1999) Mannose induces an endonuclease responsible for DNA laddering in plant cells. Plant Physiol 121:71–79

Stein B, Rahmsdorf HJ, Steffen A, Litfin M, Herrlich P (1989) UV-induced DNA damage is an intermediate step in UV-induced expression of human immunodeficiency virus type 1, collagenase, c-fos, and metallothionein. Mol Cell Biol 9:5169–5181

Strid A, Chow WS, Anderson JM (1990) UV-B damage and protection at the molecular level in plants. Photosynth Res 39:475–489

Sullivan JH (1997) Effects of increasing UV-B radiation and atmospheric CO2 on photosynthesis and growth implications for terrestrial ecosystems. Plant Ecol 128:194–206

Sullivan JH, Teramura AH (1989) The effects of ultravioletB radiation on loblolly pine. I. Growth, photosynthesis and pigment production in greenhouse-grown seedlings. Physiol Plantarum 77:202–207

Sutherland JC, Griffin KP (1981) Absorption spectrum of DNA for wavelengths greater than 300 nm. Radiat Res 86:399–409

Szilárd A, Sass L, Deák Z, Vass I (2007) The sensitivity of Photosystem II to damage by UV-B radiation depends on the oxidation state of the water-splitting complex. Biochim Biophys Acta 1767:876–882

Takeuchi A, Yamaguchi T, Hidema J, Strid A, Kumagai T (2002) Changes in synthesis and degradation of Rubisco and LHCII with leaf age in rice (*Oriza sativa* L.) growing under supplementary UV-B radiation. Plant Cell Environ 25:695–706

Teramura AH (1983) Effect of ultraviolet-B radiation on the growth and yield of crop plants. Physiol Plant 58:415–427

Teramura AH, Murali NS (1986) Intraspecific differences in growth and yield of soybean exposed to ultraviolet-B radiation under greenhouse and field conditions. Environ Exp Bot 26:89–95

Teramura AH, Sullivan JH (1991) Potential effects of increased solar UV-B on global plant productivity. In: Riklis E (ed) Photobiology. Plenum Press, New York, pp 625–634

Teramura AH, Sullivan JH (1994) Effects of UV-B radiation on photosynthesis and growth of terrestrial plants. Photosynth Res 39:463–473

Tevini M (1993) Effect of enhanced UV-B radiation on terrestrial plants. In: Tevini M (ed) UV-B radiation and ozone depletion: effect on humans, animals, plants, microorganisms, and materials. Lewis Publishers, Boca Raton, pp 125–153

Tevini M, Teramura AH (1989) UV-B effect on terrestrial plants. Photochem Photobiol 50:479–487

Tevini M, Iwanzik W, Thoma U (1981) Some effect of enhanced ultraviolet radiation on the growth and composition of plants: barley, maize, kidney beans, radishes. Planta 153:388–394

Tosserams M, Visser A, Groen M, Kalis G, Magendans E, Rozema J (2001) Combined effects of CO2 concentration and enhanced UV-B radiation on faba bean. Plant Ecol 154:195–210

Tsukaguchi T, Iida Y (2008) Effects of assimilate supply and high temperature during grain-filling period on the occurrence of various types of chalky kernels in rice plants (Oryza sativa L.). Plant Prod Sci 11:203–210

Tyrrell RM, Pidoux M (1989) Singlet oxygen involvement in the inactivation of cultured human fibroblasts by UV-A (334 nm, 365 nm) and near-visible (405 nm) radiations. Photochem Photobiol 49(4):407–412

Tyrrell RM, Sage E (2011) Handbook of photochemistry and photobiology, 3rd edn. CRC Press. ISBN 9781439899335–CAT#K14351

Tyrrell RM, Ley RD, Webb RB (1974) Induction of single-strand breaks (alkali-labile bonds) in bacterial and phage DNA by near UV (365 nm) radiation. Photochem Photobiol 20:395–398

UNEP (1994) Environmental effect of ozone depletion: 1994 assessment. In: Madronich S, Mckenzie RL, Caldwell MM, Björn LO (eds) Changes in ultraviolet radiation reaching the Earth's surface. Stratospheric Ozone and Human Health Project, United Nations Environment Programme (UNEP)

Van Breusegem F, Vranova E, Dat JF, Inze D (2001) The role of active oxygen species in plant signal transduction. Plant Sci 161:404–414

Van Rensen JJS, Vredenberg WJ, Rodrigues GC (2007) Time sequence of the damage to the acceptor and donor sides of photosystem II by UV-B radiation as evaluated by chlorophyll a fluorescence. Photosynth Res 94:291–29

Vigh L, Los DA, Horvath I, Murata N (1993) The primary signal in the biological perception of temperature: Pd-catalyzed hydrogenation of membrane lipids stimulated the expression of the desA gene in *Synechocystis* PCC6803. Proc Natl Acad Sci U S A 90:9090–9094

Vile GF, Tyrrell RM (1995) UV-A radiation-induced oxidative damage to lipids and proteins in vitro and in human skin fibroblasts is dependent on iron and singlet oxygen. Free Radic Biol Med 18 (4):721–730

von Sonntag C (2006) Free-radical-induced DNA damage and its repair. Springer, Berlin

Vu CV, Allen LH Jr, Garrard LA (1982) Effect of supplemental UV-B radiation on primary photosynthetic carboxylating enzymes and soluble proteins in leaves of C3 and C_4 crop plants. Physiol Plant 55(1):11–16

Vu CV, Allen LH, Gaward LH (1984) Effects of enhanced UV-B radiation (280–320 nm) on ribulose-1.5-bisphosphate carboxylase in pea and soybean. Environ Exp Bot 24:131–144

Vu CV, Allen LH, Gaward LH (1984) Effects of enhanced UV-B radiation (280–320 nm) on ribulose-1.5-bisphosphate carboxylase in pea and soybean. Environ. Exp. Bot. 24:131–144

Wang Q-W, Hidema J, Hikosaka K (2014) Is UV-induced DNA damage greater at higher elevation? Am J Bot 101(5):796–802

Warner CW, Caldwell MM (1983) Influence of photon flux density in the 400-700 nm waveband of inhibition of photosynthesis by UV-B (280-320 nm) radiation in soybean leaves: separation of indirect and immediate effects. Photochem Photobiol 38:341–346

Whitmore SE, Potten CS, Chadwick CA, Strickland PT, Morison WL (2001) Effect of photoreactivating light on UV radiation-induced alterations in human skin. Photodermatol. Photoimmunol Photomed 17(5): 213–217

Wondrak GT, Jacobson MK, Jacobson EL (2006) Endogenous UV-A-photosensitizers: mediators of skin photodamage and novel targets for skin photoprotection. Photochem Photobiol Sci 5:215–237

Wright LA, Murphy TM, Travis RL (1981) The effect of ultraviolet radiation on wheat root vesicles enriched in plasma membrane. Photochem Photobiol 33:343–348

Yagura T, Makita K, Yamamoto H, Menck C, Schuch AP (2011) Biological sensors for solar ultraviolet radiation. Sensors 11:4277–4294

Zhang J, Hu X, Henkow L, Jordan BR, Strid A (1994) The effect of ultraviolet-B radiation on the CF_0F_1-ATPase. Biochim Biophys Acta 1185:295–302

Zhang XS, Rosenstein BS, Wang Y et al (1997) Induction of 8-oxo-7,8-dihydro-2′-deoxyguanosine by ultraviolet radiation in calf thymus DNA and HeLa cells. Photochem Photobiol 65:119–124

UV Acclimation Strategy in Plants

4

4.1 Introduction

The predicament of solar UV radiation on life-forms is of particular concern with respect to plants. Despite holding a pivotal role (as primary producers) for life survival, plants are subject to deleterious UV radiation; spontaneously fluencing UV radiation at high intensity (by direct impact/ROS-mediated induction) causes massive cell molecular damage to plants that are stationarily posed in open field. Notwithstanding, plants still survive and flourish exhibiting UV acclimation under high UV ambience. This is enabled by the unique feature of cellular plasticity that allows development of UV-protective strategy specifically operative in plants. Such UV acclimation measures help plants to "outwit"/reduce harmful effects of UV radiation.

Oscillatory UV dose variation (under fluctuations of weather/aerosol/cloud cover, Frederick and Snell 1990) manifested as spells of low-dose UV (through signal transduction, Ulm et al. 2004) induces epigenetic upregulation of enzymes for synthesis of UV-protective (phenolic) compounds, viz., flavonoids (Brown et al. 2005), that provide prophylactic protection by serving as internal UV screen (Li et al. 1993). This mode of UV protection may be seen as personalized UV screening in plants that function in addition to the nonspecific screening of the stratospheric ozone layer that is enjoyed by all living forms on earth. In addition to serving as internal UV screen, flavonoids, due to their structural specialization, also serve as (nonenzymatic) antioxidants inherently (in plants) and through dietary source (in animals) for preventing oxidative damage (Larson 1988) thereby alleviating UV-B stress and adding to acclimation strategy in plants. Another group of phenolic compounds, viz., carotenoids, also serve as nonenzymatic antioxidants (White and Jhanke 2002). Another means of UV acclimation, induced by low-dose UV radiation, constitutes synthesis of DNA repair enzymes in plants (Brosche and Strid 2003). This represents another means of recovery from/resilience to UV-induced cellular damage. All such processes collectively confer UV acclimation via genetically controlled molecular rearrangements enabled by cellular plasticity uniquely operative in plants. Such protective mechanisms are manifested as trait variation (at the morphological/physiological/genetic level) thereby avoiding stress symptoms (Hectors et al. 2007).

In the "post ozone hole" era, ozone thinning caused increase in high-intensity UV flux on earth and raised alarm over the world. This prompted urgent research (Ajavon et al. 2007). In this context a comprehensive understanding on varietally regulated UV-protective molecular mechanisms in plants at different locations under variation of UV intensity have been reported by Jansen et al. (1998). Ballare et al. (2011)

S. Sen Mandi, *Natural UV Radiation in Enhancing Survival Value and Quality of Plants*,
DOI 10.1007/978-81-322-2767-0_4

documented that difference in UV-B acclimation contributes to variation in growth responses observed in plants on open field.

Reports that ambient UV radiation in the post ozone hole era is taking its toll often in the way of varietally regulated loss of biodiversity are sometimes discussed in the literature (Bornman et al. 2015). Cause for such loss of genotypes/biodiversity loss commonly relates to inefficient acclimation process. This may be observed in loss of viability of embryonic axis during postharvest dry storage of (orthodox) seeds where direct UV-caused damage in dehydrated cells cannot be countered since due to low cell moisture content, enzymatic processes for development of UV acclimation remain precluded (Flood and Sinclair 1981). If cellular damage incurred during postharvest storage exceeds varietally regulated UV acclimation potential/DNA repair activity, the postharvest (dehydrated) seed becomes nonviable. Such events pose a threat of genotype extinction/biodiversity loss. If, before the point of no return, the seed gets access to germination condition, repair of cell molecular damages (incurred in dehydrated embryonic cells of postharvest seed) and resumption of UV screening compounds may restore UV acclimation processes and prepare for germination-associated emergence, albeit delayed, of the embryonic axis (Bievers et al. 2014; Ballare et al. 2011). Studies on plant growth and development in closed chamber and under artificial regime of UV radiation are usually discounted due to lack of PAR in the plants' ambience. Studies of Berkelsar et al. (1996) establish the importance of controlling and comparing PAR levels when undertaking studies on UV-B sensitivity of plants.

High levels of white light as well as UV-A/blue light have been reported to mediate photorepair mechanisms that ameliorate UV-B-induced damaging effects (studied in soybean – Warner and Caldwell 1983; Mirecki and Teramura 1984). Such studies, demonstrating enhancement in sensitivity to UV-B radiation under lower than normal fluence of PAR, establish the importance of conducting experiments in open field under natural (open air) level of PAR. From field experiments Adamse and Britz (1992) have demonstrated that ambient PAR (1000–1600 μmol m^{-2} s^{-1}) is able to ameliorate the effects of UV radiation as high as 18 kJ m^{-2} per day. Data from experiments conducted in closed chamber, where experimental conditions lack natural interactions between wave bands of solar UV radiation and photosynthetically active radiation (PAR), exhibit exaggerated cell-damaging effect of UV (Deckmyn and Impens 1997; Corlett et al. 1997). Such data establish the importance of PAR in reducing damaging effects of UV. Resultant between UV-caused damage under high UV radiation and acclimation strategy developed under low UV dose (that induces cell molecular processes for UV acclimation) determines plant survival on earth. Studies focusing on this aspect have become more important in the "post ozone hole" era. Interesting information in this regard comes from studies on the plant *Gunnera magellanica*, a creeping perennial herb, growing at 82^0S at a region over which the ozone hole occurs and at 41^0S (at a region that although is in the Southern Hemisphere is much removed from the Antarctica where the ozone hole occurs); these regions represent sites with variation in UV fluence. A comparison of data from these studies show higher level of DNA damage in plants at the Antarctic region (Rousseaux et al. 1998) compared to the damage level occurring at 41^0S (Giodanoi et al. 2003) corroborating the concept that in the post ozone hole era, cell molecular damages incurred are higher. Such studies reiterate the importance of studying low-dose UV-induced UV acclimation strategy in plants.

Such studies indicate a genetic basis for variability in sensitivity of cultivars to UV – the variability in sensitivity reflecting varietal difference in UV protective acclimation process. Rozema et al. (1997) have studied leaf thickness and UV-B pigments (as parameters of UV acclimation) in four plant species growing under elevational gradient related high UV fluence along the Blue Mountains, Jamaica. Using a Robertson-Berger detector for measuring solar erythemal radiation at the Jungfraujoch High Mountain Station (Swiss Alps, 47^0N, 3576 m above sea level), Blumthaler and Ambach (1990) demonstrated that there has been a slight increase of about 1 % per year in the flux of solar ultraviolet-B radiation (290–330 nm) since 1981 – the increase being related to a long-term ozone depletion. This calls for focused studies on effect of UV on plants, manifested as a resultant of UV fluence countered by genetically controlled variation in synthesis of UV-protective compounds/pigments, viz., flavonoids, under variation of the (UV) inducer. Shyam Chowdhury and Sen Mandi (2011) have reported higher flavonoid content in tea clones growing at high altitude compared to flavonoid content of tea clones of the same genotype growing at lower altitude (both places within the same latitude) indicating a relation between flavonoid accumulation and UV fluence on site. Other studies on UV acclimation processes at different altitude also relate the data to high/low UV fluence and proportional variation in low-dose UV radiation resulting from attenuation of UV passing through cloud cover (Rozema et al. 1997), a frequent phenomenon at high altitude regions; the variation is evidently due to variation of flavonoid biosynthesis under variation of the (UV) inducer at the different altitude cultivation sites. Studies on field-grown plants at various locations have also established that absence of significant damaging effect of elevated UV radiation in plants under field conditions is correlated with a general efficiency of UV acclimation process in plants (Caldwell et al. 1994; Björn et al. 1997; Allen et al. 1999) in the open field.

Thus studies on relationship between UV and crop yield and quality conducted in open field would provide useful information not only on conservation of biodiversity but also on selecting cultivation sites that would be suitable for development of enhanced resilience to UV in plants, thereby ensuring high productivity, subject to genetic potential of the crop plant studied. UV-B response in crop plants such as maize (Casati and Walbot 2004; Falcone Ferreyra et al. 2010; Casati et al. 2011), barley (Kravets et al. 2012), cucumber (Shinkle et al. 2010), grapevine (Pontin et al. 2010; Martinez-Luscher et al. 2013), and some commercially important crops, viz., Chinese woad (Chen 2009) and *Populus* (Jia et al. 2009) have been explored and reported in the literature.

Authors have reported significant DNA damage due to enhanced solar UV radiation in this plant. Giordanoi et al. (2003) have studied the same species in the Southern Hemisphere – at 410S. DNA damage recorded in the plants in this region is lower than the data recorded on plants in the Antarctic region. Such variation in data is evidently because DNA repair and accumulation of UV absorbing sunscreens together exhibit higher efficiency of acclimation strategy in plants of this region. Studies on plants at various other locations have also established that absence of significant damaging effect of elevated UV radiation in plants under field conditions is correlated with a general efficiency of UV acclimation process in plants (Caldwee et al. 1994; Bjorn et al. 1997; Allen et al. 1999) in the open field.

4.2 UV-Induced Damage

Though some fraction of UV radiation is reflected back from leaf epidermal surface, and some is absorbed by phenolic compounds that, being deposited albeit ununiformly along the inner surface of epidermal cells, serve as internal UV screen, a considerable amount of UV radiation enters the inner tissue. High-intensity UV radiation entering into inner tissue causes damage, particularly to DNA and proteins with aromatic amino acids that constitute direct targets for UV radiation. Both UV-A and UV-B cause DNA double-strand and single-strand breaks as well as photodimer formation such as CPD and 6-4PP (Grienert et al. 2012). UV-B that directly impinges on DNA, proteins, and phospholipids may also cause oxidative damages to macromolecules. UV-A, while causing direct damage particularly to DNA, also causes oxidative damage to other macromolecules. UV-induced oxidative damage constitutes indirect damaging effects on nucleic acids, membrane lipid proteins, and enzymes (Roleda et al. 2006a, b) through generation of a range of ROS, viz., H_2O_2, OH, O^{2-} including ${}_1O^2$ (Moldau 1999). Such phenomena have been studied in several plant species (Agrawal and Rathore 2007; Du et al. 2011; Singh et al. 2011). UV-A (lower in energy of impact compared to UV-B) also generates ROS, particularly singlet oxygen (${}_1O^2$) that can damage DNA via indirect photosensitizing reactions (Alscher et al. 1997). In addition to causing damage to DNA, singlet oxygen also damages proteins and other macromolecules through oxidative damage (Wondrak et al. 2006).

4.3 UV Acclimation Strategy in Plants

High-intensity UV radiation fluencing through clear sky into earth's atmosphere induces cell molecular damage, posing a threat to survival of life on earth; yet, life continues evidently through variable weather-related intermittent spell of low-dose UV (resulting from attenuation when passing through cloud cover)-induced development of inherent UV acclimation processes. UV-induced damaging effect evident in field-grown plants represents manifestation of the resultant manifestation of potential damaging effect of UV countered by genetically regulated UV acclimation strategy. For prophylactic as well as retaliatory means of protection from UV radiation plants, enabled by the unique feature of cellular plasticity (for interconversion between primary and secondary pathway), undertake upregulation in synthesis of phenolics, viz., flavonoids, for surviving under the brunt of UV radiation in open field. While variation in UV-induced cellular protection is commonly studied as a manifestation of varietal/species difference, some studies report on variation in UV protection as a manifestation of difference in growth phase-related upregulation in synthesis of UV-protective compounds (see Fig. 4.5). In this context studies on resilience to UV in plants at various levels of growth (Hidema and Kumagai 1998), development (Schmitz-Hoerner and Weissenböck 2003), and yield (Teramura 1983) of crop plants in open field are available in the literature.

Dose of fluencing UV radiation varies with time of day/season/weather conditions including aerosol/cloud cover that determine extent of sunshine and thus UV radiation fluencing on earth. UV dose correlates with extent of cellular damage and, interestingly, also with damage restraining cell molecular rearrangement that confers UV acclimation strategy in plants: while high-dose UV radiation causes macromolecular damage, UV radiation in low dose induces (UV signaling – see Chap. 5) based upregulation of genes for synthesis of UV screening compounds (Klein 1978; Jordan 1996; Rozema et al. 1997; Ryan et al. 2001) that in turn restrict penetration of damaging UV radiation into leaf cells (Alvero-Bascos and Ungson 2012). Deposited along the inner wall of epidermal cells, such UV screening phenolics, viz., flavonoids, serve as internal UV screen thereby providing prophylactic protection; this represents the first line of defense conferring a UV acclimation strategy in plants. Several studies have reported varietal difference in extent of UV screening phenolic compounds, viz., flavonoids, accumulated in plants (leaf) (Kolb et al. 2003; Kakani et al. 2003; Caldwell et al. 2007). The flexibility by which primary metabolic pathway is diverted to the secondary metabolic pathway (enabled by cellular plasticity) for synthesis of flavonoids (Alvero-Bascos and Ungson 2012; Frohnemeyer and Staiger 2003) determines plant's efficiency for survival at locations of high UV intensity.

As an additional attribute, structural specialization of flavonoid allows dual functioning and thus serves (in addition to UV screening) as nonenzymatic antioxidants providing retaliatory protection from (UV-induced) oxidative damage of macromolecules in cells. This widens the ambit of flavonoid-mediated cell molecular protection under UV acclimation processes in plants. UV at low dose via signal transduction also upregulates synthesis of antioxidative enzymes, viz., SOD and APX (Nawkar et al. 2013). Under cellular conditions when enzymatic antioxidants are rendered ineffective due to loss of macromolecular integrity under UV radiation particularly in dehydrated cells of embryonic axis of orthodox seed (Shyam Chowdhury and Sen Mandi 2012), flavonoids representing non-enzymatic antioxidants take to protection from oxidative damage in cells. Low-dose UV-B radiation can also induce alterations in antioxidant status through regulation of glutathione pathways, cinnamate, phenylpropanoid pathway (of which flavonoid is one of the end products), and pyridoxine biosynthesis pathways (Hideg et al. 2013) and also by some alkaloids (Lyddon et al. 1987). UV at low dose via signal transduction upregulates synthesis of antioxidative enzymes, viz., SOD and APX (Nawkar et al. 2013). Mild and ecologically relevant UV-B radiation dose may trigger early adaptation, so that when conditions worsen (under high UV intensity), plants are protected from distress situations (Muller Xing et al. 2014).

To cope with cell molecular damages inflicted by the extent of UV that escape the internal UV screening arrangement, UV acclimation strategy to ensure resilience to UV-induced damage have to be adapted. Of the UV-inflicted cell molecular damages, damage of DNA presents the most hazardous effect, relating to persistence of life since this causes disruption in cell cycle process (Kunz et al. 2005) and if not repaired in time would lead to programmed cell death/apoptosis. Thus, a very important aspect of UV acclimation is through macromolecular recovery. This is affected by low-dose UV-induced upregulation of DNA repair enzyme, viz., photolyase (Hutchinson 1987), for repair of DNA damage.

The epigenetic route for upregulation in synthesis of flavonoids and DNA repair enzymes allows effective UV acclimation under variation of UV fluence. Weather-related intermittent spell of low-dose UV induces development of inherent UV acclimation processes as a manifestation of varietal difference. Intraspecific variations in antioxidant defense responses and sensitivity of soybean varieties to ambient UV radiation have been reported by Baroniya et al. (2013). Changes in antioxidant and UV screening compounds, viz., flavonoids in UV-B irradiated maize seedlings under UV radiation have been reported by Carletti et al. (2003). Hidema et al. (1997) have reported variation in DNA repair activity due to alteration in CPD photolyase in different cultivar of rice as a part of UV acclimation strategy. Such studies establish a genetic basis for variability in sensitivity of cultivars to UV – the variability in sensitivity reflecting varietal difference (Sects. 4.3.1.3 and 4.3.4.1) in UV-protective acclimation process.

Macromolecular rearrangements conferring UV acclimation are generally associated with trait variation at the level of gene expression and associated variation in morphological features that allow avoidance of UV stress response (Hectors et al. 2007). In this context studies of Rozema et al. (1997) on leaf thickness and UV-B pigments (as parameters of UV acclimation) of four plant species growing under elevational gradient-related high UV fluence along the Blue Mountains, Jamaica provide interesting information.

Other commonly studied photomorphogenic modification, in this regard, are delayed emergence of hypocotyls (Ballare et al. 2011)/embryonic axis (Ganguli et al. 1992) and need based stem extension (Brown et al. 2005) for affecting UV avoidance till cell molecular events may ensure protection of the embryonic tissue from UV stress (Ballare et al. 2011; Biever et al. 2014). Requirement for synthesis of UV-protective compounds/repair enzymes prior to exposure of dividing cells to ambient UV environment represents avoidance strategy for UV acclimation in plants. Other UV-induced photomorphogenic effects include leaf thickening, shortened inflorescence stem, cotyledon curling, stem and leaf elongation, axillary branching, and shifts in the root-shoot ratio (Barnes et al. 1990; Cen and Bornman 1993; Boccalandro et al. 2001). In summary, a UV-specific signaling component orchestrates UV protection/acclimation in plants. Species-based studies on UV acclimation in plants establish genetically controlled UV protection/acclimation in plants involving synthesis of flavonoid (Jende-Strid 1993; Shirley 1996) and DNA repair enzymes (Britt 1999). UV upregulated synthesis of cell-protective compounds/enzymes that are specifically synthesized in plants as a means of acclimation under UV environment serves as plants' personalized additional protection in addition to the stratospheric ozone screening that is availed of, nonspecifically, by all life-forms.

4.3.1 Restraint of UV Penetration into internal tissue

For UV radiation to confer appropriate (damaging/protective) effect on internal processes of plants, the radiation should first have access to internal cell molecular entities. The first line of defense in restraining UV penetration into plant cells is leaf surface reflection that is provided by variation in leaf surface morphology, the most common being wax deposition on leaf surface/epicuticular waxes that reflect a small portion of the fluencing UV radiation (Jetter and Schäffer 2001; Perkins et al. 2005) back to the atmosphere (Fig. 4.1) while allowing the remaining fraction to enter into internal tissue. Other leaf characters for restraining UV radiation from penetrating through leaf surface is development of increased leaf thickness. When studying plants with different leaf surface texture, Roberecht et al. (1980) reported that while reflectance from glabrous leaf is less than 10%, pubescent or glaucous leaf surface reflects more than 40%. Due to such partial restriction of UV penetration through the epidermal surface (as observed in different leaf types), some UV radiation still enters the leaf surface barrier. The fraction of UV escaping the leaf surface barrier is then subject to absorption by UV screening compounds that appear deposited, albeit ununiformly, along the inner surface of epidermal cells (Tevini et al. 1991). Rozema et al. (1997) reported that leaf thickness and accumulation of UV-B absorbing pigments varies in relation to elevational gradient that represent locations of variation of UV fluence (Sect. 1.5.3).

Strategies developed in plants for survival under UV radiation thus include restraining penetration of UV radiation into internal tissue (Perkins et al. 2005) by either (i) reflection of

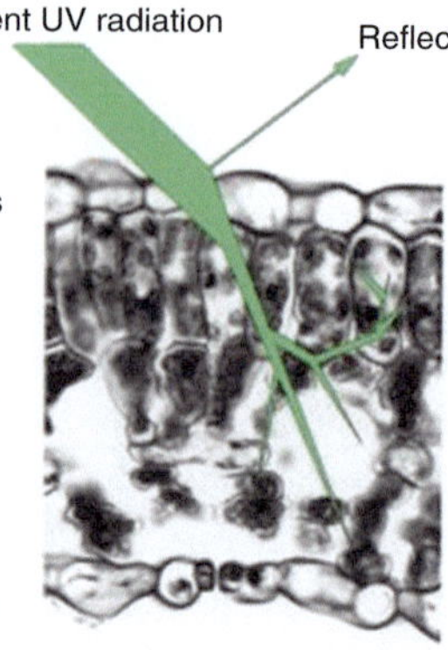

Fig. 4.1 Reflection, refraction, and absorption of incident UV-B radiation by a typical leaf. Bandwidths represent energy levels and show strong absorption by the upper epidermis in comparison to lower epidermis (Taken from Caldwell et al. 1983a)

UV radiation caused by physical features of leaf such as waxy leaf surface (ii) absorption of UV radiation by specific UV screening compounds that accumulate in epidermal cells (Sect. 4.3). Such modifications in morphology and chemical composition of leaf cells constitute UV avoidance/UV acclimation strategies operative in plants. Combination of field and laboratory studies have often been conducted to explore the nature of photomorphogenic effects of ultraviolet-B radiation on plant morphology and to evaluate the ecological consequences of these alterations in interspecific morphology for light competition (Barnes et al. 1996), UV radiation.

In addition to structural features adapted by leaf epidermis to restrain UV radiation, a more effective means of restraining UV radiation is provided by phenolic compounds, viz., flavonoids. For a long time, the main phenolic-dependent-protective mechanism in plants was thought to be associated with early defense against phytopathogens and herbivores (War et al. 2012); this paradigm changed with the development of a wider concept (Close and McArthur 2002) relating to the UV-absorbing feature of flavonoids that confers a role of internal UV screening (Sect. 4.3.2). Penetration of UV-B into internal tissue largely depends upon the extent of its attenuation by phenolic compounds that differentially (subject to varietal difference) accumulate in the epidermis. The concept that flavonoids may serve as UV is supported by a large body of experimental evidence as studied in grapes (Kolb et al. 2003), *Vicia faba* (Olsson et al. 1998), rice (Markham et al. 1998), and *Arabidopsis* (Sheahan 1996). In plants exposure to UV-B induces a rapid and species-related proportional increase in enzyme activities of the phenylpropanoid biosynthetic pathway for synthesis of the major protective pigments, viz., water-soluble, colorless flavonoids including flavones, flavonols, and isoflavonoids and also the colored compounds, viz., flavonoid-derived pigment, viz., anthocyanin. These compounds are deposited mainly in the vacuoles and cell walls (Sect. 4.3). Morales et al. (2010) have reported that temporal variation in epidermal flavonoids due to altered solar UV radiation is moderated by leaf position (studied in *Betula pendula*). Wilson et al. (2001) have shown report on effects of UV-A exposure on UV-B-induced accumulation of specific flavonoids in *Brassica napus*. The accumulated phenolic compounds, viz., flavonoids, form an internal UV screen that reduces epidermal penetration of UV (Bornman and Vogelmann 1991) without interfering with the part of solar radiation that relates to photosynthesis (Caldwell et al. 1983b). Such avoidance strategy for preventing UV entry into the mesophyll tissue of a leaf by the accumulation of UV-absorbing compounds in cell vacuoles and/or cell walls of the epidermis has been reported and studied by Caldwell et al. (1983a) and Hutzler et al. (1998). This UV acclimation strategy that relates to restriction of UV entry through leaf surface is affected by presence of flavonoids and hydroxycinnamic acids, the two main groups of phenolics that absorb high-energy (UV) radiation (Sect. 4.1) thereby restricting entry of this deleterious radiation into plant cells (Cerovic et al. 2002). Both these groups of phenolics, deposited in the leaf epidermis, act as UV-absorbing sunscreens for restraining the sun's UV radiation from entering into inner tissue (Jordan 2002; Bassman 2004), while transmitting the photosynthetically active radiation (PAR) into the mesophyll cells for the process of photosynthesis.

Penetration of UV-B largely depends upon the extent of attenuation by phenolic compounds differentially accumulated in the epidermis (subject to varietal difference) that serve as internal UV screens. Wilson et al. (2001) have shown effects of UV-A exposure on UV-B induced accumulation of specific flavonoids in *Brassica napus*.

The most abundant phenolics detected in leaf epidermis are derivatives of the flavonols, quercetin, and myricetin; kaempferol derivatives are found in smaller quantity. Epidermally located ultraviolet (UV)-absorbing phenolic compounds, viz., flavonoids and hydroxycinnamic acid esters (HCAs), serve as shield for protecting underlying tissues against harmful UV radiation (Burchard et al. 2000). A strong correlation between the content of the two phenolic groups of compounds, viz., soluble HCA esters and flavone glycosides, and the screening efficiency of epidermal tissue has been reported by Burchard et al. (2000). Other flavonoid compounds such as leucoanthocyanins are also found bound to the cell walls.

Spatially differential accumulation and spread of flavonoid compounds constitute a nonhomogeneous UV filter that causes differential UV penetration into leaf tissue (Fig. 4.2, Caldwell et al. 1983a). For studying the spatial pattern of UV-B penetration through the epidermis, Day (1993) developed a thin polymer film which fluoresced blue when irradiated with UV-B. Fresh epidermal leaf peels, when placed over the film and irradiated with UV-B and examined microscopically from below, showed that the film fluoresces below cell wall regions, but not in regions below epidermal and guard cell protoplasts. Results from the fluorescing film technique were substantiated by direct measurements of UV-B transmittance through epidermal peels using a fiber optic microprobe run paradermally along the bottom or inner side of irradiated peels. Such studies indicate that UV-B transmittance is much greater through anticlinal cell wall regions than through protoplasts. Extensive study also demonstrates that UV-B transmittance through large areas of epidermal cells could be affected by plasmolysis.

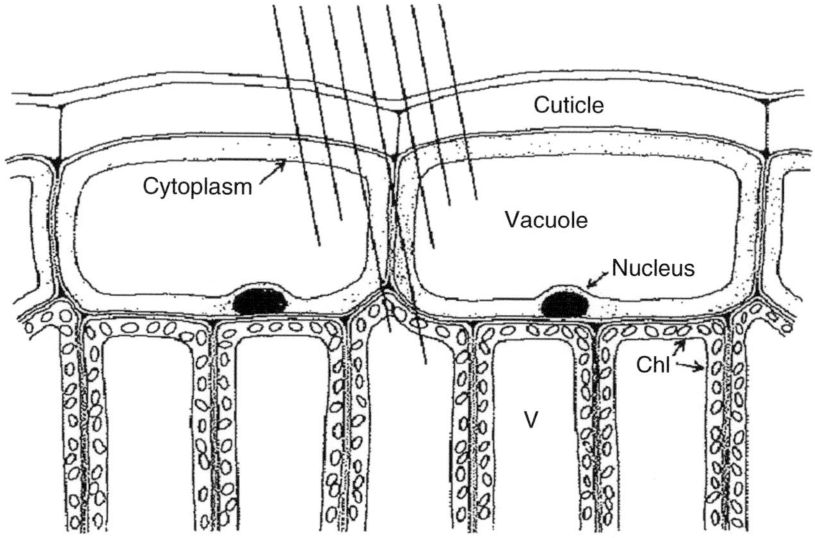

Fig. 4.2 Depiction of the epidermis as a nonuniform filter of UV-B radiation (*inclined parallel lines*). Ultraviolet-absorbing pigments should occur principally in the vacuoles; although some compounds such as leucoanthocyanins may be bound to the cell wall (Taken from Caldwell et al. 1983a)

Epidermal transmittance was also relatively high through stomatal pores (and what appear to be nuclei in *Smilacina*), but relatively low through stomatal guard cells.

The ununiformity in deposition of UV-absorbing phenolic compounds in epidermal cells and its implication in plant's sensitivity to UV radiation (Robberecht and Caldwell et al. (1982b), often, varies in an age- and irradiation-dependent manner (Burchard et al. 2000). Repeated measurements on identical leaves over a time course of 6 days demonstrated a strong age dependence in the capacity for the synthesis of UV-A screening compounds upon exposure to UV radiation. Based on the principle that accumulation of UV-absorbing compounds in the epidermis should reduce the fluorescence excited by UV beam, without affecting the fluorescence excited by the visible beam, Bilger et al. (2001) used a noninvasive method as a tool for estimating UV penetration without introducing any perturbation in the optical properties of the leaves and using natural UV target as a reporter of the UV climate within the mesophyll (Fig. 4.3). During the last decade, equipments have been developed which allow the nondestructive assessment of UV-A screening even under field conditions. It is concluded that the new method may provide a valuable tool for investigation of UV acclimation process in plants; accompanied by HPLC analysis of the reaction of phenolic metabolism correlated with environmental stimuli, this would provide utilizable information.

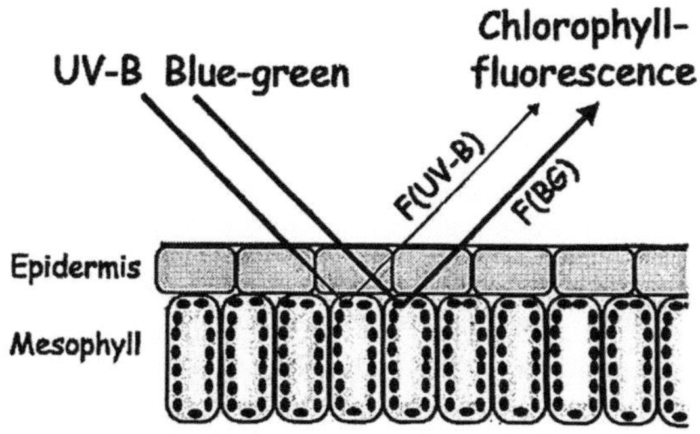

Fig. 4.3 Schematic drawing of the adaxial part of a leaf cross section illustrating the principle of the chlorophyll fluorescence method for assessment of epidermal transmission for UV-B radiation. The *thickness* of the beams indicates relative intensity (Taken from Bilger et al. 2001) [F(UV-B – UV-B fluorescence, F(BG) – *Blue green* fluorescence]

Upregulation of genetically conferred biosynthesis of different phenolic compounds (serving as internal UV screen) is a reflection of metabolic cross-talk between related pathways – the variation in time of selected pathways being a reflection of cellular requirement for survival under varying environmental UV radiation. Under low-dose UV radiation (fluencing through aerosol/cloud cover/seasonal/diurnal variation), flavonoid synthesis is upregulated by macromolecular rearrangements enabled by cellular plasticity, a unique feature present in plants; metabolic plasticity is a key contributor to adaptive strength in plants and is of paramount importance for developing adaptation to their appropriate terrestrial ecosystem. The most commonly found group of protective phenolic compounds found in terrestrial higher plants, viz., polyphenols/flavonoids, serve protective functions in plants. Polyphenols have been shown to serve multiple roles in plants in response to stressful (UV) environment. Studies of Burchard et al. (2000) also established that a correlation between epidermal UV-A and UV-B absorbances may be assessed by chlorophyll fluorescence measurements and contents of epidermal flavonoids.

Studies of Wilson et al. (2001) have reported that UV-A is able to penetrate deeper (than UV-B) into leaves where it produces active oxygen species that constitute the major means of UV-A-induced cell molecular damage. Wavelengths in the UV-A region have been reported to induce the accumulation of flavonols in *Mesembryanthemum crystallinum* leaves (Ibdah et al. 2002). Again, accumulation of specific flavonoids in *Brassica napus* (Wilson et al. 2001) has been reported to be induced by UV-B. In silver birch, Kotilainen et al. (2009) found that concentrations of two individual flavonoids and total phenolics decreased in fully expanded leaves with exclusion of UV-A. Burchard et al. (2000) reported that in early stages of development, the epidermal hydroxycinnamic acids (HCAs) are the dominant UV-B-protective compounds of the primary leaf; this function is increasingly taken over by the epidermal flavonoids during leaf development and acclimation.

Correlation between epidermal UV penetration and absorbing compounds has also been demonstrated in isolated leaf epidermis and mesophyll of *Urginea maritima* (Grammatikopoulos et al. 1999). Functional significance and induction by solar radiation of ultraviolet-absorbing sunscreens in field-grown soybean crops has also been reported by Mazza et al. (2000). In barley primary leaf hydroxycinnamic acid conjugates and also flavonoid glycosides have been found to serve as UV screen in leaf epidermis (Schmitz-Hoerner and Weissenböck 2003).

While studying variation in UV penetration in leaf of different plants at various locations (with variation in UV fluence), Day (1993) observed that UV penetration varies among different plant species, and this is correlated with sensitivity of the respective genotype. The author reported that penetration was particularly high in herbaceous dicotyledons (broad-leaved plants) and was progressively less in woody dicotyledons, grasses, and conifers (Day et al. 1992). UV screening compounds in leaf epidermis has also been shown to vary with age of the tissue.

UV screening compounds in plant cells

Dose of fluencing UV radiation varies with time of day/season/weather conditions including aerosol/cloud cover that determine extent of sunshine and thus UV radiation fluencing on earth. UV dose correlates with extent of cellular damage: while high-dose UV radiation causes macromolecular damage, UV radiation in low dose induces (UV signaling – see Chap. 5) upregulation of genes for synthesis of UV screening compounds (Klein 1978; Jordan 1996; Rozema et al. 1997; Ryan et al. 2001) that in turn restrict penetration of damaging UV radiation into leaf cells (Alvero-Bascos and Ungson 2012). Several studies have reported varietal difference in extent of UV screening phenolic compounds, viz., flavonoids, accumulated in plants (leaf) (Kolb et al. 2003; Kakani et al. 2003; Caldwell et al. 2007). Working on lettuce (green and red types), Tsormpatsidis et al. (2010) have shown that enhancement in natural ultraviolet (UV) radiation, while conferring cellular protection by increasing secondary metabolic products such as phenolics, significantly inhibits biomass accumulation (evidently due to diversion of primary metabolism to the secondary metabolic pathway; see Fig. 6.10). Such

UV-induced rearrangement of metabolic events, enabled by the unique feature of cellular plasticity, constitutes a natural strategy of plants for survival on earth.

Adverse (damaging) effect of the high-energy UV-B radiation is countered to different extent in plants – this effect (enabled by UV acclimation strategy developed in plants) exhibiting inter- and intraspecific variation often manifested as variation in plant growth efficiency, reflected as physiological, biochemical, and cell molecular events for restoring UV-induced macromolecular damages (Kramer et al. 1991; Mpoloka 2008; Fedina et al. 2010). This is induced by low-dose UV radiation through upregulation in synthesis of UV-absorbing compounds (Hahlbrock and Scheel 1989) mainly phenolics such as flavonoids and hydroxycinnamate esters. Accumulating in the vacuoles of epidermal cells in response to UV (both UV-B and UV-A) radiation, these phenolic compounds, by absorbing the fluencing UV radiation, attenuate penetration of UV radiation into deeper cell layers. Bieza and Lois (2001) have reported constitutively high accumulation of UV-absorbing compounds, viz., flavonoids and sinapate esters, in the *Arabidopsis* mutant uvt1 that is tolerant to lethal ultraviolet-B levels; the increased levels of flavonoids in the leaves block the entry of UV-B radiation into internal tissue and confers elevated resistance to this mutant to UV-B radiation-induced damage. Plant genotypes that are impaired in flavonoid biosynthesis are susceptible to damage by UV-B (Stapleton and Walbot 1994). *Arabidopsis* flavonoid mutants have been reported to be hypersensitive to UV-B radiation (Li et al. 1993; Ryan et al. 2001). Action spectrum analysis of UV radiation identified the action peak at 295 nm for stimulating synthesis of constitutive flavonoid compounds in parsley cell cultures (Beggs and Wellmann 1994).

Increase in flavonoids and also other phenolics, viz., sinapate esters, is a consistent and widespread (genetically regulated) plant response to enhanced UV radiation-associated low-dose UV-B (Jenkins 2009; Stracke et al. 2010; Kakani et al. 2003; Caldwell et al. 2007). Accumulating in the epidermal layer, UV screening compounds, viz., flavonoids, can increase to high concentrations between 1 and 10 mM (Vierstra et al. 1982), thereby providing the major means of attenuation of UV radiation in plants. Development of UV-restraining strategy in plants is observed both under controlled environments (Biggs et al. 1981, from studies on 19 cultivars of soybean) and also in natural plant communities (Wilson et al. 1998, from studies on *Brassica napus* L. cv Topas).

Chloroplasts have also been reported to contain flavonoids (Jansen et al. 2012) and appear to be capable of flavonoid biosynthesis (Pollastri and Tattini 2011; Hernández et al. 2009) thereby providing internal filters for restraining UV penetration into internal tissue (Caldwell et al. 1983a) at the epidermal layer as well as in chloroplasts (Jansen et al. 2012). UV screening compounds, generally known to be deposited along inner wall of leaf epidermal cells are also reported to be present in chloroplasts. Tevini et al. (1991) using UV-B pre irradiated plants with accumulated flavonoids in the epidermal layer have demonstrated flavonoid related protection to the photosynthetic apparatus. During cellular dehydration flavonoids within the chloroplast have been shown to preserve the integrity of the membrane through lipid remodeling (such as galactocylation) thereby preventing oxidative damage (Moellering et al. 2010; Inoue 2011).

Flavonoids, by virtue of their structure-activity relationship (SAR), perform a dual role, (a) as UV screens and (b) as active oxygen scavengers (Mackerness 2000; Ruhland et al. 2005; Jansen et al. 2008), serving to protect against oxidative damage including those caused by UV radiation (Sect. 4.3.2). This two-pronged protective mechanism by flavonoids thus provides holistic means for supporting plant for acclimation and survival under varying UV fluence. Studies of Mewis et al. (2012) on UV-B-induced overall secondary metabolite biosynthesis and associated gene expression in broccoli sprouts show that UV-B exposure leads to elevated levels of antioxidative flavonoids in the sprouts. Their studies demonstrated pronounced accumulation of antioxidative flavonoids, viz., kaempferol.

The level of ambient UV-B radiation varies with latitude and is expected to be relatively higher in tropical regions than in temperate regions. In the tropics, due to small solar zenith angle and the thin stratospheric ozone layer, terrestrial plants encounter much higher level of UV-B radiation than at higher latitudes (Caldwell et al. 1989; Madronich et al. 1995). However, due to higher ozone depletion over mid- and higher latitude areas in the last half of the twentieth century, life-forms of these regions are now known to suffer from high UV-B radiation. Since plants in this region are not preadapted (through the generally prevalent means of preexposures; see Chap. 7) to high UV, plants in this region exhibit pronounced damaging effect (Madronich and de Gruijl 1994). DNA damage effects are also evident in humans developing nonmelanoma skin cancer.

Evolutionary History of Protective Compounds in Plants

With the gradual development of the stratospheric ozone layer formed from a gradual build up of O_2 that was released by early photosynthetic organisms, earth's atmosphere gradually became suitable, albeit poorly, to start with, for supporting life forms on terrestrial habitat; migration of early life forms from underwater protection to land thus continued. The most significant of these were the oxygen-producing life forms, viz., the plants, that provided oxygen for building up the stratospheric ozone layer for an all pervasive UV screen over earth's surface. Of all the life forms, plants with their stationary lifestyle had to develop particularly efficient (internal) UV screens for self-protection from UV fluence in the open field. Presumably, the first enzymes capable of synthesizing flavonoids (the UV-screening compounds) were not as plentiful nor as efficient as present day forms, so that large amounts of flavonoids did not accumulate initially (Stafford 1991).

UV-screening compounds in early land plants consist of mycosporine-like amino acids (MAAs) and scytonemins (Rozema et al. 2002); these simple UV-absorbing compounds in aquatic plants are as efficient as the more complex flavonoids in terrestrial higher plants. The scytonemins and MAAs have also been reported to have some absorption in UV-C range; this may have conferred an advantage for survival of life under early atmosphere that at that time was devoid of the stratospheric ozone layer with consequent presence of UV-C on earth' surface. Wolf et al. (2010) studied function and regulation of UV-B response in early plant, viz., *Physcomitrella patens*, with an aim to follow the evolutionary aspects of UV-B tolerance and acclimation that laid the ground for plants to conquer land. In particular, it is evident that key genes regulating the response to UV-B (viz., those encoding AP2/EREBP transcription factors) for mediating molecular protection genes (involving PAL and CHS) have been conserved since the water-to-land transition of plants. In the course of evolution, polyphenolics/flavonoids have gradually evolved and provided the internal UV screens thereby providing acclimation strategies for establishing cellular protection under high UV radiation. The UV-B screening as well as nonenzymatic antioxidative functions of flavonoids enabled by structural specialization of these compounds (during the colonization of land by plants) are supposed to have followed several branch pathways of phenylpropanoid metabolism (Cockell and Knowland 1999). The "early" steps in the pathway are found even in the bryophytes (mosses), and it has been suggested that synthesis of flavones, flavanones, and flavonols may have evolved first to provide chemical messengers (for inducing effect at a distance from the point of perception) for effecting gene expression for the production of UV sunscreens (Stafford 1991). Different classes of phenolic compounds providing UV screen are described below:

4.3.1.1 Hydroxycinnamic Acid

Other phenolic compounds acting as UV-B attenuators include hydroxycinnamic acid derivatives (e.g., p-coumaric, ferulic, and caffeic acid), with emax in the 310 to 325 nm waveband (Harborne and Williams 2000; Tattini et al. 2004). These groups of phenolics exhibit a C6–C3 carbon skeleton which differs from flavonoids which have a C6–C3–C6 backbone (Cockell and Knowland 1999). However, the ratio of flavonoids to hydroxycinnamates increases steeply upon exposure to UV-B or strong sunlight (Burchard et al. 2000; Tattini et al. 2000; Agati et al. 2002; Kotilainen et al. 2008). Flavonoids seem to "replace" hydroxycinnamic acid derivatives as a leaf develops under UV irradiance (Burchard et al. 2000), and soluble hydroxycinnamates are usually confined to tissues receiving the lowest doses of UV radiation (Olsson et al. 1999; Tattini et al. 2004). Consequently, the UV-B-induced preferential biosynthesis of flavonoids, which have emax >335 nm (with the exception of some acylated forms), suggests that UV screening is just one of the multiple roles played by flavonoids in photoprotection (Markham et al. 1998; Harborne and Williams 2000). This suggestion is strongly supported by the deactivation of hydroxycinnamate in favor of the flavonoid branch pathway in the glandular trichomes of *Phillyrea latifolia* leaves following severe UV stress (Tattini et al. 2000; Agati et al. 2002), a decrease in the concentrations of the highly effective UV attenuators p-coumaric and chlorogenic acid derivatives coupled with a steep increase in the concentration of quercetin 3-O-glycosides (emax >350 nm) in plants growing under enhanced UV-B or UV-B + UV-A irradiance (Kotilainen et al. 2008, and flavonoid accumulation in full sunlight-treated leaves in the absence of UV irradiance (Kolb et al. 2001; Agati et al. 2009) or in plants treated with excess copper ions growing under photosynthetically active radiation (Babu et al. 2003). Flavonoids and hydroxycinnamic acid conjugates are the most efficient UV-screening compounds because of their strong absorbance in this wavelength region (Caldwell et al. 1995; Cockell and Knowland 1999). Studies of Burchard et al. (2000) on rye primary leaves using chlorophyll fluorescence measurements showed that hydroxycinnamic acids (representing one of the two phenolic epidermal sunscreens) are important as constitutive protecting compounds; the importance of this compound in UV screening depends on the developmental stage of the leaf. Since epidermal HCAs are present from the beginning, they apparently represent an important constitutive factor of protection against UV-B radiation. When epidermal flavonoids are absent, HCAs can contribute up to 100 % of the epidermal screening. The authors also reported an excellent correlation between epidermal UV-A and UV-B absorbances and contents of epidermal flavonoids; in this study HCAs showed an additional contribution to UV-B shielding. The degree to which HCAs support flavonoid screening differs markedly between species; these phenolic compounds were found to contribute minimally to UV-B screening in leaves of *Vicia faba* (Marksta¨dter et al. 2001) but considerably in *Arabidopsis thaliana* and grape vine (Landry et al. 1995; Sheahan 1996; Kolb et al. 2001).

Epidermally located ultraviolet (UV)-absorbing phenolic compounds, flavonoids, and hydroxycinnamic acid esters (HCAs) can shield the underlying tissues in plants against harmful UV radiation (Burchard et al. 2000).

4.3.1.2 Sinapate Esters

While flavonoids and anthocyanins are most commonly known as UV-screening compounds in plants, sinapate esters provide greater UV attenuation than flavonoids as studied in some plants (*Arabidopsis* Sheahan 1996). It is pertinent to mention that *Arabidopsis* has been reported also to contain flavonoids as UV-screening compounds. *Arabidopsis* mutants that do not accumulate flavonoids and/or synaptic esters are highly UV sensitive (Landry et al. 1995). Flavonoids and sinapate esters protect by specifically absorbing in the wavelength region from 280 to 340 nm (but not in the PAR waveband, which would diminish photosynthetic yields).

4.3.1.3 Flavonoids

Flavonoids as well as other phenolics, viz., hydroxycinnamic acids, anthocyanins, and

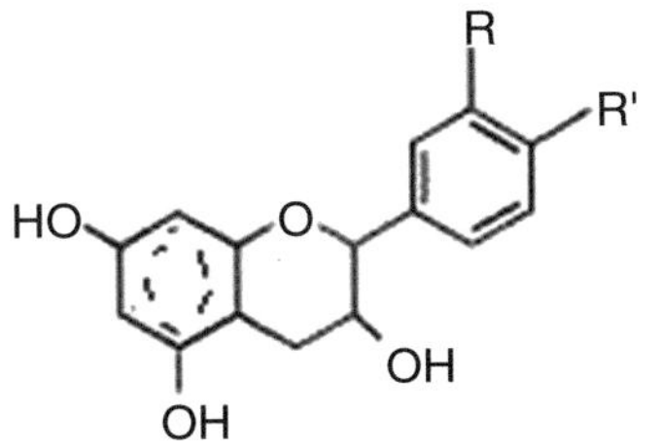

Fig. 4.4 π ↔ π* transition in the core nucleus enable flavonoids to act as internal (cellular) UV screens

sinapate esters, act as internal UV screen in plants, serving to restrain fluencing UV radiation from entering leaf internal tissue. Of these, flavonoids are particularly relevant for developing UV acclimation strategy. This is because, due to structural specialization, flavonoids in addition to serving as internal UV screens are also capable of functioning in dual role as (nonenzymatic) antioxidants. During the process of evolution of terrestrial plant from marine and freshwater plant life, flavonoids developed structural specialization for absorption of UV through π ↔ π* reversible transition in the core nucleus (Fig. 4.4) that allows stability of the compound under high energy of UV radiation thereby enabling flavonoids to restrain/screen UV penetration into cells. A comparison between absorption spectrum of flavonoids (Fig. 4.9) and the wavelengths of UV radiation entering into earth's atmosphere (through the stratospheric ozone layer) provides evidence to establish the role of flavonoids as UV screen in plant cells as a manifestation of structure-function relationship of flavonoids.

The UV-B screening functions of flavonoids during the plant colonization of land are likely to have originated from other ancestral primary roles. One such function is the role of flavonoids as antioxidants. Both the UV screening and antioxidant functions of flavonoids are enabled by structure-activity relationship (SAR) of the compound. While π → π* transition in core nucleus allows screening function, phenol quinone tautomerism in the side chain of dihydroxy B-ring of flavonoid allows antioxidative function. Concentrations of flavonoids in early terrestrial plants (estimated at μM range) are evidently lower (Raushar 2006; Stafford 1991) than the concentration required for functioning as effective shield against fluencing UV radiation for which flavonoid concentrations in the mM range are required (Agati and Tattini 2010; Cockell and Knowland 1999). That flavonoids in ancient plants served particularly as antioxidants is evident from the fact that genes for biosynthesis of dihydroxy B-ring-substituted flavonoids are detectable in liverworts and mosses (Raushar 2006).

Flavonoids are synthesized epigenetically as an outcome of molecular rearrangements involving diversion of primary metabolism to secondary metabolism enabled by the unique feature of cellular plasticity particularly operative in plants (Heijde and Ulm 2013) under temporal recurrence of transitory spells of weather change-related aerosol/cloud cover caused low-dose UV fluence signals (via UVR-8; see Chap. 5).

Internal UV screening/DNA repair enzymes provide prophylactic protection in plant cells by (a) UV screening function (Li et al. 1993; Hahlbrock and Scheel 1989) and (b) conferring antioxidative potential (Sichel et al. 1991; Yamasaki et al. 1997). Flavonoids by virtue of their structural specialization-associated SAR enables (a) π ↔ π* transition in core nucleus that allows high-energy UV absorption thereby serving as inherent UV screen for partially restraining UV entry into cells and (b) phenol ↔ quinone tautomerism in side chain also serve as antioxidants (Tarrant et al. 2014 to protect cells from oxidative damages (Roleda et al. 2006a). These cell molecular mechanisms together with low-dose UV-induced upregulation of DNA repair enzymes, viz., photolyase (Hutchinson 1987), for repairing UV-inflicted DNA damage constitute UV acclimation strategy in plants.

Chemistry and Types of Flavonoids

Flavonoids are a group of secondary metabolites in plants generally found to be upregulated under stress such as of enhanced UV radiation and pathogen attack. While UV-induced flavonoid accumulation occurs particularly in the epidermal layer and vacuoles thereby allowing some of the UV radiation to pass through into cells, flavonoids synthesized due to other stress such as pathogen attack may be accumulated in the mesophyll layer. Concentration of flavonoids in epidermal layers is significantly upregulated by low-dose UV-B radiation. Studies combining flavonoid quantification with measurements of epidermal transmission by UV radiation indicate that flavonoid levels increased by UV radiation result in improved UV screening (Burchard et al. 2000; Bilger et al. 2001; Krause et al. 2003). The major classes of flavonoids are anthocyanins (red to purple pigments), flavonols (colorless to pale yellow pigments), flavanols (colorless pigments that become brown after oxidation), and proanthocyanidins (PAs) or condensed tannins.

Anthocyanins are water-soluble pigments derived from flavonoids via the shikimic acid pathway, and similar to hydroxycinnamate and naringenin, are found to be developmentally transient, appearing only in juvenile or senescing tissues; in some tissues again, they may be permanent. In most cases, anthocyanins are found in or just below the upper epidermis of leaves (Krol et al. 1995; Burger and Edwards 1996; Tevini 1994). Shade plants accumulate anthocyanins in their lower epidermis, perhaps to assist in light capture (Lee et al. 1978, 1987); anthocyanins have also been reported to occur only in the mesophyll layers of some plant species including Mahonia, Viburnum and Rhododendron (Kaku et al. 1992) and the tropical woody genus Syzygium (Woodall and Stewart 1998). Study of Tsormpatsidis et al. (2007) demonstrated that both UV-B and UV-A are involved in anthocyanin photo-induction.

Stapleton and Walbot (1994) demonstrated that anthocyanins interposed between DNA and a UV source protects the DNA from CPD formation thereby establishing their role as UV screen. UV screening activity of anthocyanins has also been demonstrated in in vitro experiments where anthocyanins have been shown to form complex with DNA and protect it against oxidative damage. Anthocyanin formation has also been reported in suspension cell studies on Centuureu cyunus (Takahashi et al. 1991). Like several other phenolic compounds viz. hydroxycinnamic acid and naringenin, anthocyanins have also been observed to provide transitory protection from UV in leaf tissue as in juvenile leaves of two Syzygium species (S. luehmannii and *S. wilsonii*). These compounds however are notable exceptions (compared to other UV screening phenolic compounds) in that they have substantial absorption in the visible waveband and little in the UV-B (Caldwell et al. 1983b) in comparison to flavonoids (commonly known UV screen) which have substantial UV absorbance and very less absorption in the visible range. Unlike the flavonoids that due to their light absorption spectrum between 230 nm and 380 nm can directly act as UV screen in plants, the anthocyanin compounds that absorb maximally around 530 nm undergo esterification to cinnamic acid before these compounds may provide UV-B protection. A commonly cited example in agriculture is the occurrence of this group of coloured flavonoids (viz. anthocyanins) in a rice cultivar purple puttu that is often used as a marker variety in the agricultural field (Sharma 2001).

HPLC analysis for separating UV absorbing anthocyanins from other soluble UV absorbing phenolic compounds shows that the anthocyanin component, (predominantly malvidin-3, 5-diglucoside) contributes little to the total absorbance of UV-A and UV-B radiation. This is presumably because non-acylated anthocyanins only effectively absorb shortwave UV-B radiation.

Other flavonoid compounds viz. extremely bitter flavonoid glycosides naringin and neohesperidin accumulate in juvenile, rapidly growing grapefruit and pummelo (Citrus grundis) (Horowitz 1986). The pathway of biosynthesis of bitter flavanone-glycosides in Citrus has been elucidated (Fig. 4.5) and found to be similar to the well-documented pathway i.e. Phenyl Propanoid pathway found in other species (cf. Ebel and Hahlbrock 1982; Stafford 1990). The synthesis of naringin and prunin is diminished as the plants mature and the enzymes (viz. 7-Glucosyl transferase and 1–2 Rhamnosyl transferase) responsible for production of these compounds are no longer detectable during maturation (Bar-Peled et al. 1993). As the production of naringenin diminish, the other end products of PP pathway take over the role of internal UV screen compounds.

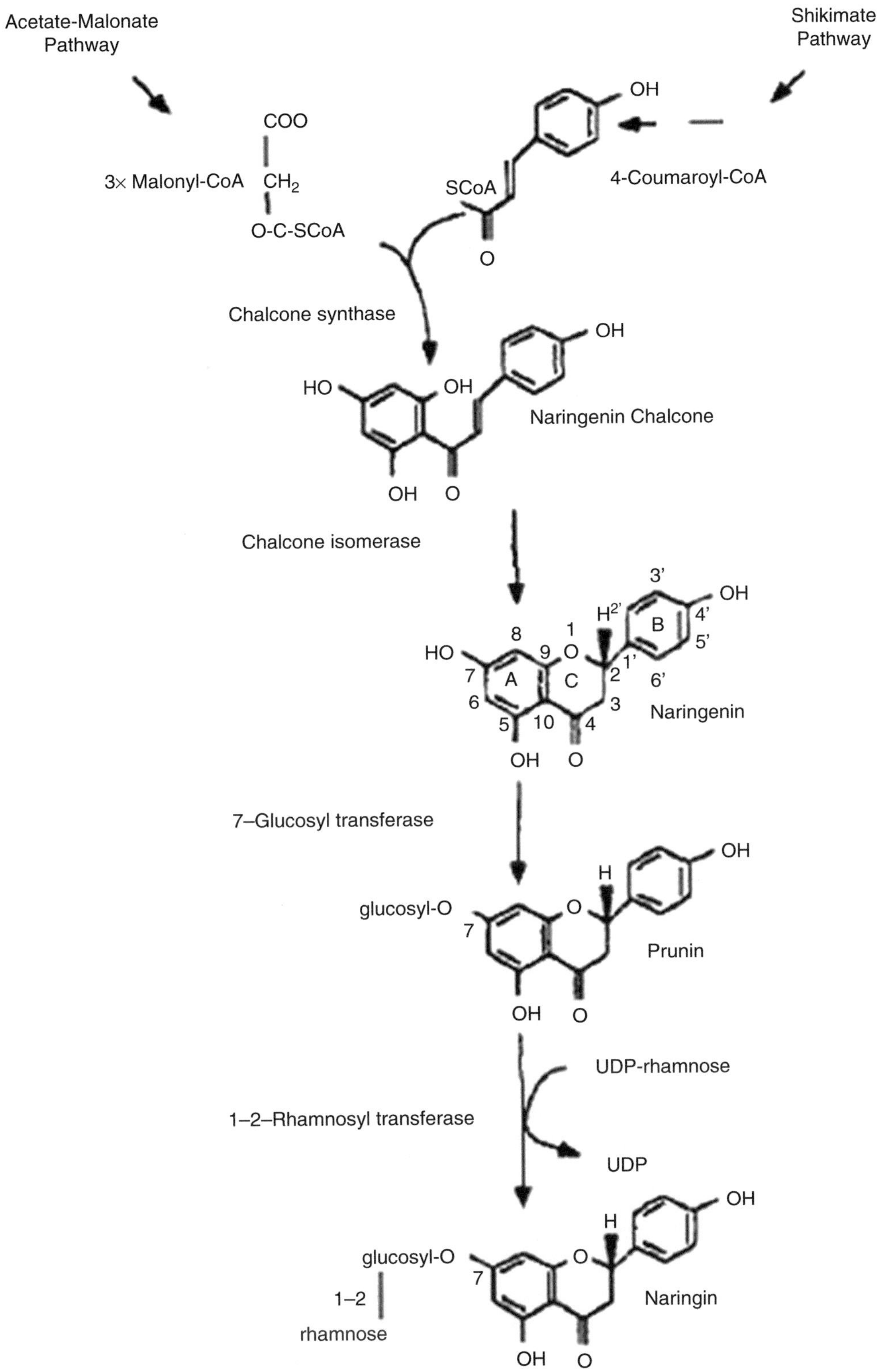

Fig. 4.5 Biosynthesis of naringin in *Citrus* (based on Lewinsohn et al. 1989; McIntosh et al. 1990; Bar-Peled et al. 1991) during early stage of plant maturation

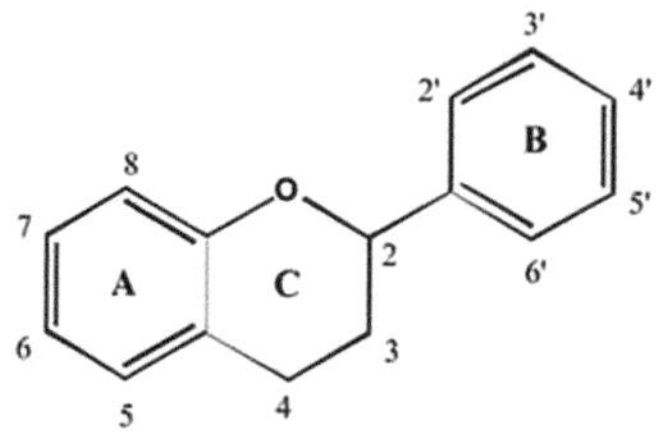

Fig. 4.6 Basic three-ring structure of flavonoid

Under stress of high intensity UV fluence as occurs naturally at high altitude locations (Sect. 1.5.3), primary metabolism in cells is diverted under intermittent low dose to the secondary metabolic pathway for production and accumulation of high content of flavonoids for providing UV screen and antioxidant potential to internal tissue as an acclimation strategy undertaken by plants for survival under high intensity UV radiation. Such macromolecular rearrangement is enabled as a unique feature viz. cellular plasticity particularly operative in plants. Such metabolic diversion occurs under low dose UV radiation, intermittently fluencing (in high UV regions) during seasonal/daily weather based fluctuation in aerosol formation/cloud covering earth's atmosphere.

Chemically, flavonoids are based upon a 15-carbon skeleton (C6-C3-C6) consisting of two benzene rings (A and B) linked via a heterocyclic pyran ring (C) (Fig. 4.6).

A number of studies have indicated superior stress protection of B ring dihydroxylated flavonoids over their monohydroxylated counterparts, e.g., under high light (Agati et al. 2009), low temperature (Albert et al. 2009), and elevated UV-B radiation (Hoffmann et al. 2003). They can be divided into a variety of classes such as flavones (e.g., flavone, apigenin, and luteolin), flavonols (e.g., quercetin, kaempferol, myricetin, and fisetin), flavanones (e.g., flavanone, hesperetin, and naringenin), and others (Fig. 4.7). The various classes of flavonoids differ in the level of oxidation and pattern of substitution of the C ring, while individual compounds within a class differ in the pattern of substitution of the A and B rings (Middleton 1998).

During the process of evolution of terrestrial plant from marine and freshwater plant life, flavonoids developed structural specialization for absorption of UV through $\pi \leftrightarrow \pi^*$ transition in the core nucleus (Fig. 4.4) that allows stability of the compound under high energy of UV radiation thereby enabling functioning to restrain/screen UV penetration into cells thereby exhibiting match between absorption and action spectrum of flavonoids in plant cells.

The UV-B screening functions of flavonoids during the plant colonization of land have likely to have originated from other ancestral primary roles. One such function is the role of flavonoids as antioxidants. Both the UV-screening and antioxidant functions of flavonoids are enabled by structure–activity relationship (SAR) of the compound. While $\pi \leftrightarrow \pi^*$ transition in core nucleus allows screening function, phenol quinone tautomerism in the side chain of dihydroxy B ring of flavonoid allows antioxidative function. Concentrations of flavonoids in early terrestrial plants (estimated at μM range) are evidently lower (Rausher 2006; Stafford 1991) than the concentration required for functioning as effective shield against fluencing UV radiation for which flavonoid concentrations in the mM range are required (Agati and Tattini 2010; Cockell and Knowland 1999). That flavonoids in ancient plants served particularly as antioxidants is evident from the fact that genes for biosynthesis of dihydroxy B ring-substituted flavonoids are detectable in liverworts and mosses (Rausher 2006).

The basic flavonoid structure is that of an aglycone (Fig. 4.8.b); this with substitutions gives rise to commonly occurring glycosides and methylated derivatives in plants through glycosidic linkage, normally located in positions 3 or 7 thereby forming flavonoid glycoside (Fig. 4.8.a). The carbohydrate linked by the glycosidic linkage can be L-rhamnose, D-glucose, glucorhamnose, galactose, or arabinose (Middleton 1984). Most flavonoids, except for the subclass of catechins, are present in plants bound to sugars as beta-glycosides. The aglycone moiety confers aroma in plant cells (Fig. 6.16).

Under stress of high UV fluence, primary metabolism in cells is diverted to the secondary metabolic pathway for production of flavonoids.

Difference in phenolic compounds produced in different phases of development has been

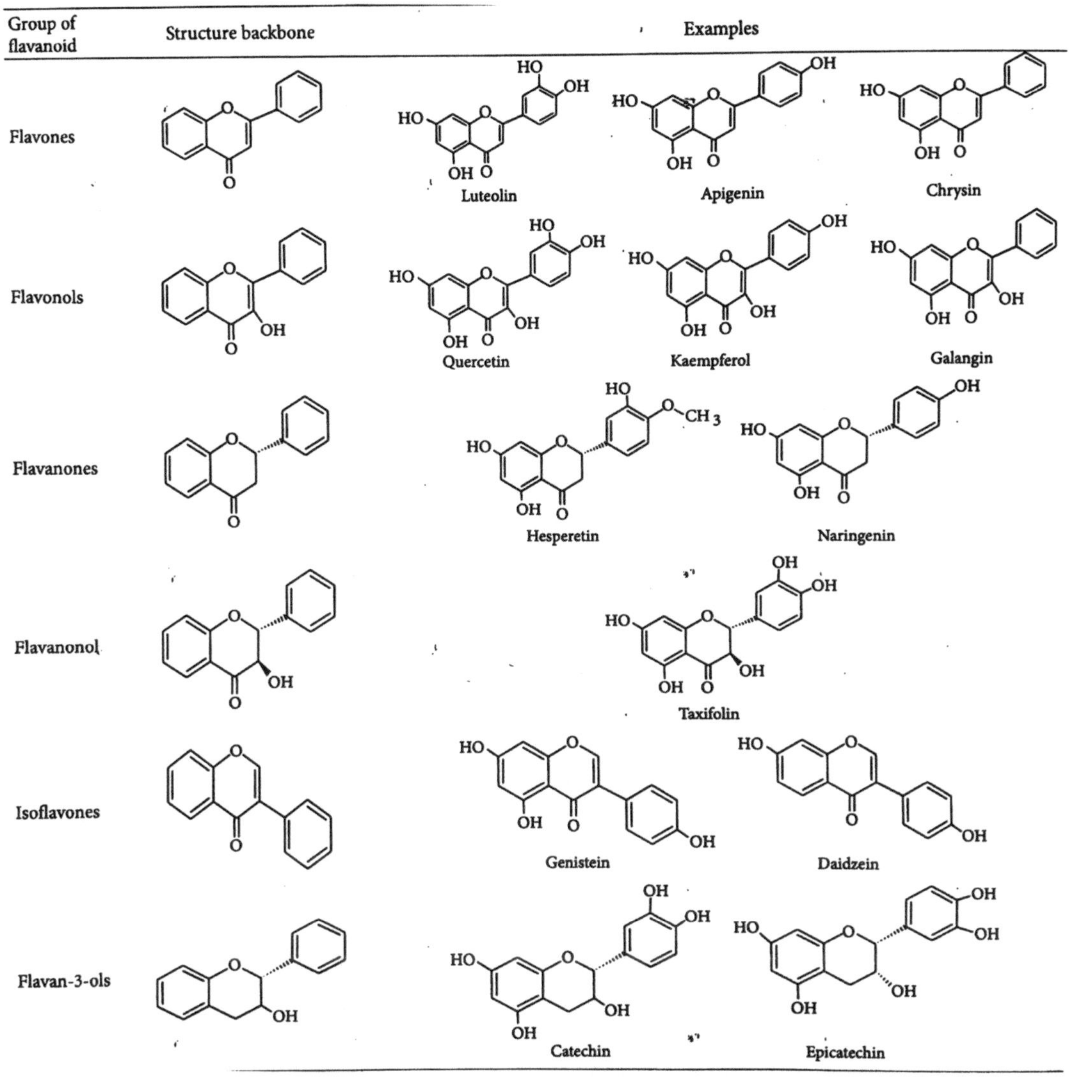

Fig. 4.7 Classes of flavonoids with their chemical structures

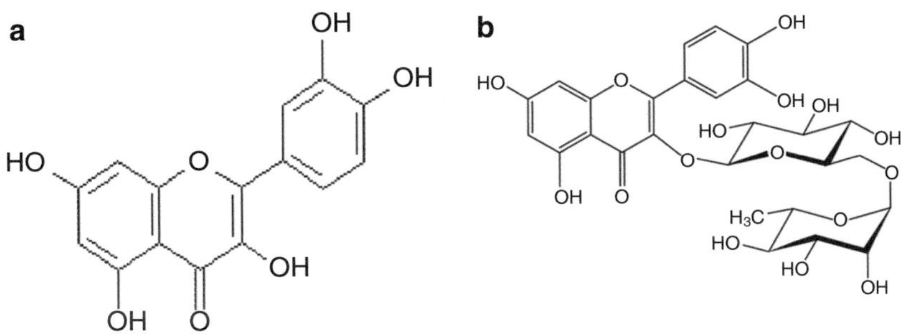

Fig. 4.8 (**a**, **b**) Structure of (**a**) flavonoid glycoside and (**b**) aglycone flavonoid

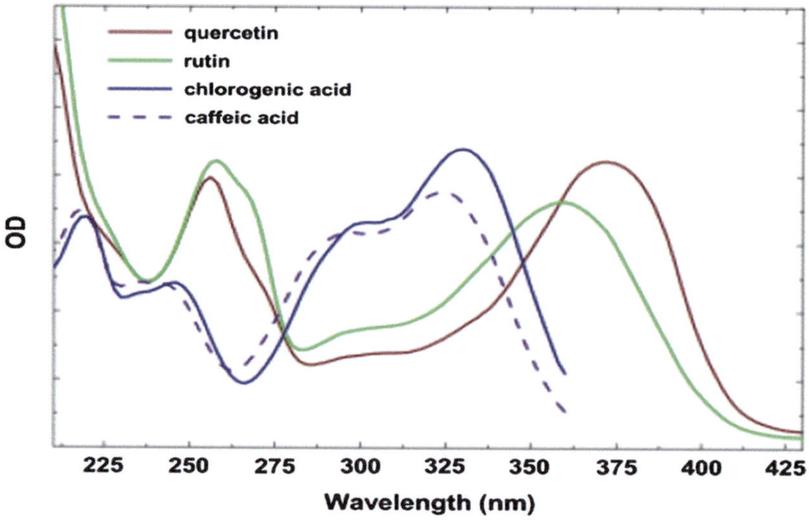

Fig. 4.9 Absorbance spectra of certain flavonols and phenolic acids in methanol (Taken from Solovchenko 2010)

reported (Strack et al. 1982) from primary leaves of rye (*Secale cereale* L.) which shows a close correlation between tissue differentiation and metabolism of flavones (luteolin derivatives), anthocyanins (cyanidin glycosides), and C-glycosylflavones (6-C-hexosyl-apigenin). The first two groups of phenolic compounds were found to be located exclusively in the mesophyll, whereas the third accumulated predominantly in the lower and upper epidermis. Flavonoids accumulate rapidly in the first 5 days of leaf development. In subsequent developmental stages, the amount of the mesophyll flavonoids, luteolin, and cyanidin derivatives drastically drops, reaching at day 8 half the concentration present at day 5. In contrast, the two major epidermal 6-C-hexosyl-apigenins increase slightly up to day 8. During vegetative growth total monoterpene (UV-upregulated group of antioxidants participating in plant acclimation (cf. alternative role of flavonoids) content of mature leaves was significantly higher at low PAR and ambient UV-B radiation compared to high PAR in the absence of UV-B radiation. Finally, at the stage of flowering, total monoterpene content was significantly enhanced at high PAR in the presence of UV-B radiation compared to the treatments without UV-B radiation.

Horowitz (1986) has reported that juvenile, rapidly growing grapefruit and pummelo (*Citrus grundis*) accumulate large amounts of the extremely bitter flavanone-glycosides naringin and neohesperidin. The pathway of biosynthesis of bitter flavanone-glycosides in *Citrus* has been elucidated (Fig. 4.8) and found to be similar to the well-documented pathway, i.e., phenylpropanoid pathway found in other species (cf. Ebel and Hahlbrock 1982; Stafford 1990). The synthesis of naringin and prunin is diminished as the plants mature and the enzymes (viz., 7-glucosyl transferase and 1–2 rhamnosyltransferase) responsible for production of these compounds are no longer detectable during maturation (Bar-Peled et al. 1993). As the production of naringenin diminish, the other end products of PP pathway take over the role of internal UV screen compounds.

Spectral Characteristics of Flavonoids

Spectroscopic study on absorption spectrum of flavonoids revealed variation in absorption pattern correlated with the different structural moiety of the molecule. Most flavones and flavonols exhibit two major absorption bands: Band I (320 nm–385 nm) represents the B-ring absorption, while Band II (250 nm–285 nm) corresponds to the A-ring absorption. Variations in absorption however may occur with difference in functional groups attached to the flavonoid skeleton: thus, a shift in absorption peak from 367 nm for kaempferol (3,5,7,4'-hydroxyl groups) to 371 nm for quercetin (3,5,7,3',4-'-hydroxyl groups) and to 374 nm in myricetin (3,5,7,3',4',5'-hydroxyl groups) has been observed (Yao et al. 2004). The absence of a 3-hydroxyl group in flavones distinguishes them from flavonols. Flavanones have a saturated heterocyclic C-ring, with no conjugation between the A- and B-rings, as determined by their UV spectral characteristics (Rice-Evans et al. 1996); these molecules exhibit a very strong Band II absorption maximum between 270 and 295 nm, with 288 nm for naringenin and 285 nm for taxifolin. Band I in these molecules is represented by only a shoulder at 326 and 327 nm. Band II appears as one peak (270 nm) in compounds with a monosubstituted B-ring. When a di-, tri-, or o-substituted B-ring is present, however, the molecule may exhibit two prominent peaks or one peak at 258 nm and a shoulder at 272 nm.

Characteristic absorption spectra of certain flavonols and phenolic acids in methanol are shown below.

Biosynthesis of Flavonoids

Flavonoids, a secondary metabolite, represent a constitutive trait specific to plants (Koes et al. 2005). This compound constitutively synthesized in plants and accumulating in leaf epidermis and mesophyll cells is enhanced under various stress conditions including insect/pathogen attack (Vogt 2010). A wider concept regarding effects of flavonoids in plants relate to upregulated synthesis of this compound that serve in UV-protective roles (Ryan et al. 2002; Berli et al. 2010; Agati et al. 2011). UV radiation at low dose (passing through aerosol/cloud cover) also induces upregulation of flavonoid for cell molecular protection in plants. In addition to the constitutive level of flavonoid biosynthesis, need (for survival under high-intensity UV radiation)-based upregulation of the constitutive level of flavonoid biosynthesis constitutes an important UV acclimation strategy (providing both UV screen and antioxidative potential) specifically evident in plants; the need for upregulation of this compound for sustenance of plant population/biodiversity has markedly increased in the post ozone hole era. The cell molecular features that enable plants to synthesize UV-protective compounds through diversion of primary to secondary metabolic pathway (Fig. 7.1), via malonyl CoA for synthesis of flavonoids through phenylpropanoid pathway (Fig. 4.10), were made possible due to the unique feature of cellular plasticity particularly operative in plants. This represents UV acclimation strategy in plants to confer UV-induced damaging effect so as to perform as primary producers for life sustenance on earth.

Constitutive biosynthesis of flavonoids occurs via the phenylpropanoid pathway, being initiated at the step of deamination of phenylalanine, the end product of shikimate pathway (Herrmann and Weaver 1999; Tzin and Galili 2010) by the enzyme phenylalanine ammonia lyase (PAL) to yield trans-cinnamic acid. The pathway is schematically represented in Fig. 4.10.

In the next step, cinnamate 4-hydroxylase (C4H) which is a cytochrome P450-dependent monooxygenase catalyzes the hydroxylation of cinnamate to yield 4-coumarate (also known as *p*-coumarate). Then, 4-coumarate CoA ligase (4CL) catalyzes the ATP-dependent formation of the CoA thioester 4-coumaroyl CoA (also known as *p*-coumaroyl CoA). This p-coumaroyl CoA then conjugates with three molecules of malonyl CoA to yield the true backbone of flavonoids, constituting a group of two phenyl rings containing compounds called chalcones. This reaction is catalyzed by chalcone synthase enzyme which is a rate-limiting enzyme of this phenylpropanoid pathway; flavonoids thus synthesized are readily converted to flavonoid glycosides (Fig. 6.17). Flavonoid glycosides on cleavage by beta-D glucosidase produce aglycon moiety that confer desirable traits such as aroma in plants. After synthesis of chalcone, flavanone is formed by the help of enzyme chalcone isomerase (CHI) which ultimately produces flavones and isoflavones. Flavanone also synthesizes

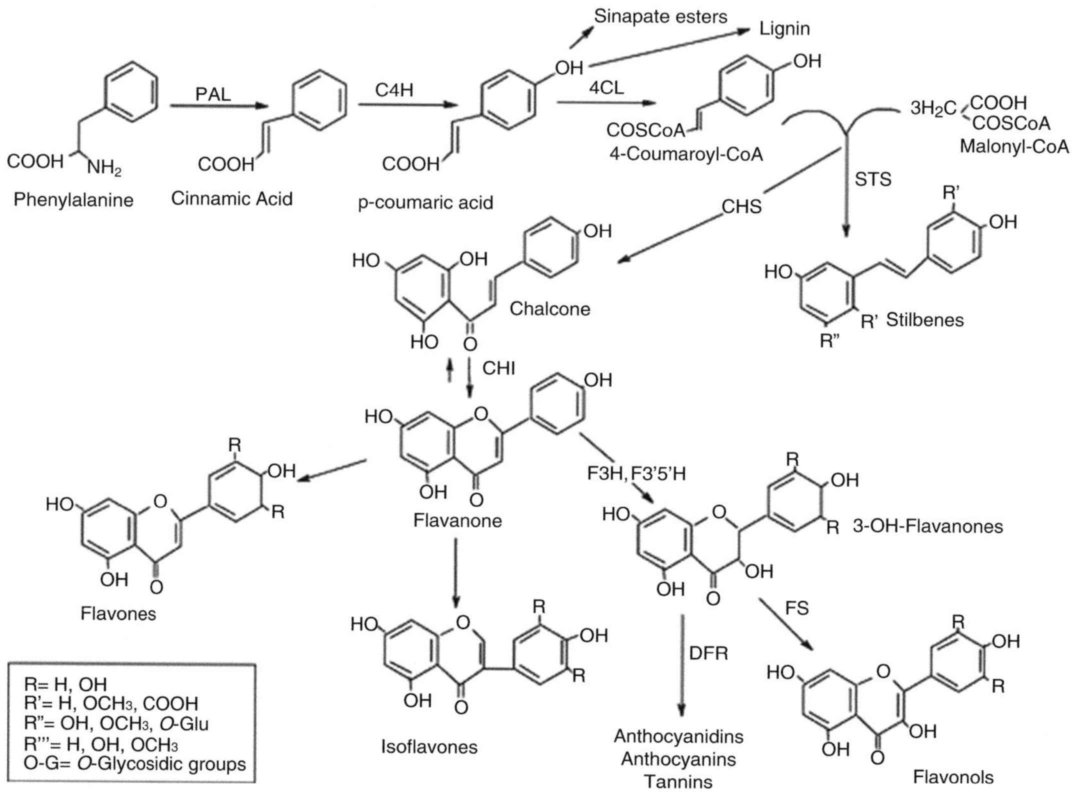

Fig. 4.10 Phenylpropanoid pathway showing biosynthesis of flavonoids and related compounds

anthocyanins, tannins, and anthocyanidins with the enzyme dihydrofolate reductase (DFR). Flavonols are synthesized from flavanone by the reaction catalyzed by flavonol synthase (FS). Stilbenes are produced by condensation of p-coumaroyl CoA and malonyl CoA with the help of stilbene synthase enzyme.

Biosynthesis of these compounds has been demonstrated to be specifically localized in the epidermal cells (studied in parsley leaves, Schmelzer et al. 1988) where CHS mRNA and related enzymes have been shown to increase (Brown et al. 2005) which have been correlated with flavonoid accumulation. UV-B directly induces CHS expression (rate-limiting enzyme of PP pathway) mainly in epidermal cells, where the flavonoids are localized (Fuglevand et al. 1996). Analysis of the CHS gene promoter revealed two UV-B responsive units that provide binding sites for transcription factors. UV-A/blue light however induces CHS gene by CRY1 photoreceptor (Fuglevand et al. 1996). Likewise, transacting factors have been identified that downregulate CHS transcription. The resulting modulation of CHS activity allows the plant to redirect the flow of intermediates of the phenylpropanoid pathway in response to environmental stimuli (Jenkins 2009). While high-intensity UV causes damaging effects in living cells, intermittent low-dose UV radiation (during temporal variation in weather/aerosol/cloud cover) signals epigenetic upregulation of gene expression for enhanced synthesis of flavonoids in plants; recurrence of low-dose UV under oscillatory function of weather allows accumulation of flavonoids that serve as internal UV screen and also as antioxidants to counter (UV-induced) oxidative damage – the dual role being enabled by the structural specialization of these compounds (Figs. 4.4 and 4.13). Synthesis of flavonoids thus confers a genetically controlled phenomenon (for developing UV acclimation) specific in plants. It is interesting to note however that although the genetic complement (upregulated by UV fluence at location) is important in determining the UV screening feature of a plant, it is important to note that composition and extent of UV screening compound is often found to vary with the developmental stage of leaves (Fig. 4.5).

Genetic control of flavonoid production

Genetic control of flavonoid production in plants is reported in studies on a single gene dominant mutation (uvt1) of *Arabidopsis* (Bieza and Lois 2001). These plants exhibit remarkable tolerance to UV-B radiation – conditions that would kill wild-type plants. Pigment analyses show a high (compared to wild type) content of UV-absorbing compounds in ultraviolet tolerant1 mutant (uvt1); this allows higher extent of UV screening in the uvt1 leaves. This allows higher absorption of UV-B radiation by the screening compounds thereby effectively restraining UV-B radiation from entering into cells. The high extent of blocking UV radiation from entering into the leaf is evidently due to higher flavonoid and sinapate accumulation in leaf epidermis. In these plants expression of chalcone synthase (CHS) mRNA was also shown to be constitutively elevated in uvt1 plants suggesting that the increases in absorption could be a consequence of changes in gene expression (compared to the wild-type plants). Further expression of CHS in uvt1 was shown to be also inducible by UV, indicating that the uvt1 lesion may not affect the cell molecular mechanism of UV-mediated regulation of CHS gene expression (Brown et al. 2005).

Some plant species synthesize specialized forms of flavonoids, such as the isoflavonoids that are found in legumes and a small number of nonlegume plants. Sorghum (*Sorghum bicolor*), maize (*Zea mays*), and gloxinia (*Sinningia cardinalis*) are among the few species known to synthesize 3-deoxyanthocyanins (or phlobaphenes in the polymerized form). The stilbenes, which are closely related to flavonoids, are synthesized by yet another group of unrelated species that includes grape (*Vitis vinifera*), peanut (*Arachis hypogaea*), and pine (*Pinus sylvestris*). Upregulation in synthesis of flavonoids and flavonoid glycoside conjugates in high-altitude leaf is observed where high UV radiation causes enhancement in flavonoid production (Shyam Choudhury and Sen Mandi 2012). Another cell molecular mechanism that correlates with high flavonoid synthesis in high altitude has also been found to be correlated with a single nucleotide polymorphism in the CHS gene that constitutes the rate-limiting enzyme in flavonoid biosynthetic pathway. Ghosh and Sen Mandi (2015) from studies on ginger landraces collected from high- and also low-altitude regions (of high and low UV fluence) have demonstrated that one SNP in chalcone synthase gene is correlated with a change in one amino acid located at the catalytic triad of the CHS enzyme. As this catalytic triad is responsible for substrate binding and subsequent product (6-gingerol- medicinally most important compound found in *Zingiber officinale*) formation, this change of amino acid is evidently the cause of variation in 6-gingerol observed in landraces of ginger.

Quantification of Flavonoids

The range of flavonoid occurring in plant materials is usually large, and every plant has an original and unique flavonoid profile. These compounds, known to function as cell protectants, are widely distributed in different amounts, according to the plant species, organ, developmental stage, and growth conditions (Debeaujon et al. 2001). Different spectrophotometric (such as UV-VIS, IR) and analytical techniques (such as HPLC, GC-MS) are usually used for proper quantification of different types of flavonoids. Piao et al. (2014) have demonstrated in eggplant by HPLC that four flavonoid aglycones differ in their content (Fig. 4.11).

Transport of Flavonoids

Flavonoids are synthesized via a well-characterized multienzyme complex localized in the cytoplasmic surface of endoplasmic reticulum. Efficient flavonoid transport systems then deliver these metabolites to appropriate cellular locations through membrane transporter-mediated transport as well as vesicle-mediated transport (Fig. 4.12) (Kitamura 2006; Grotewold 2004; Zhao and Dixon 2009) across different

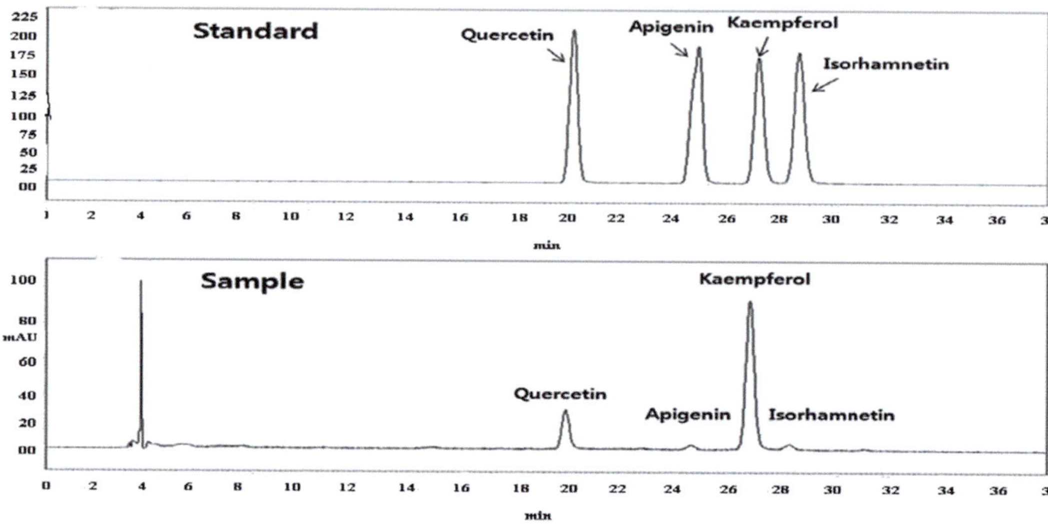

Fig. 4.11 Standard and sample chromatograms of flavonoid aglycones in eggplant (Taken from Piao et al. 2014)

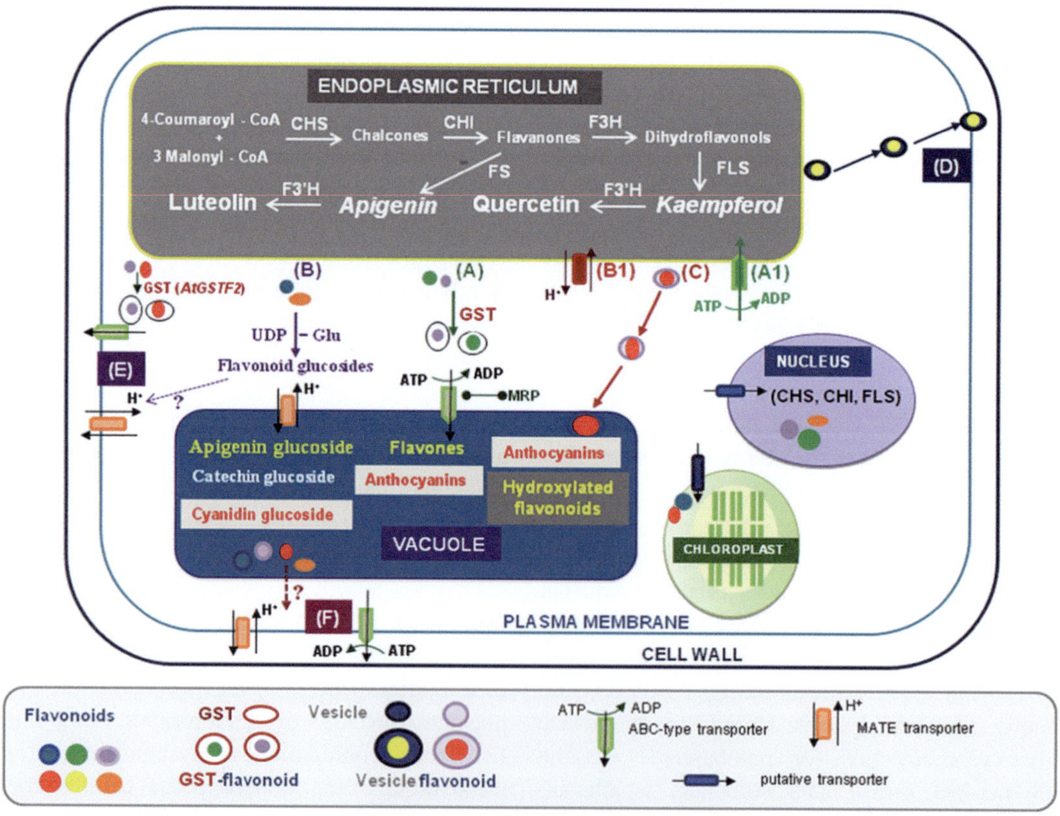

Fig. 4.12 A schematic representation of flavonoid transport system in plant cells (Taken from Agati et al. 2012)

membrane-limited compartments (Kitamura 2006). In internal cellular locations, flavonoids may exert antioxidative role in addition to its other role (i.e., UV screening) that specifically occurs in plants. The picture below (Fig. 4.12) describes the various transporters and channels which take significant role in flavonoid transport.

Flavonoids as internal UV screen

Flavonoids, structurally capable of remaining stable under UV radiation accumulating in leaf epidermis and mesophyll of herbaceous species are posed for serving as internal UV screen, thereby restricting UV penetration into inner tissue (Robberecht et al. 1980; Day et al. 1992). During subsequent studies using advanced techniques aimed at determination of precise location of flavonoids in the epidermis particularly of herbaceous species, it has revealed that flavonoids exclusively accumulate in the vacuoles of leaf epidermal cells (Hutzler et al. 1998). Due to such accumulation of flavonoids in leaf epidermal cells, UV penetration into deeper tissue is allowed only through channels located in the cytoplasm (Day 1993; Vogelmann 1993). In woody species flavonoids have been shown to accumulate in the wall and other compartments of epidermal layer (Hutzler et al. 1998; Strack et al. 1988; Schnitzler et al. 1996) thereby exhibiting superior UV-screening effectiveness with respect to that of herbaceous plants (Day 1993). Flavonoids have been shown to accumulate in the mesophyll of leaves exposed to UV-B radiation (Liu et al. 1995; Reuber et al. 1996), above all in the adaxial tissues (Alenius et al. 1995; Olsson et al. 1999). These phytochemicals constitute an important component of herbal medicines. While flavonoids have mostly been known and studied in plants, these compounds, through dietary source, also found access to animal cells (Agati et al. 2013).

Flavonoids as Developmental Regulators

It has long been documented that flavonoids control auxin movements (Jacobs and Rubery 1988; Brown et al. 2001; Peer and Murphy 2007), and a nuclear distribution of chalcone synthase (CHS) and chalcone isomerase (CHI) (Saslowsky et al. 2005) is consistent with control exerted by flavonoids on the transcription of genes required for growth and development, such as the auxin transport facilitator proteins (Peer et al. 2004). Stafford (1991) suggested that flavonoids probably played a primary role as internal regulators during the evolution of early land plants, with distinct functions carried out by the mono- and dihydroxy B ring structures in regulating auxin metabolism. These functions may be accomplished by relatively low flavonoid concentrations, as has probably occurred in early land plants, in contrast to the relatively high concentrations (in the mM range) actually needed to effectively attenuate UV radiation (Stafford 1991). Sunlight irradiance closely controls a plant's shape, with shade plants having long internodes and large leaves, whereas sunny plants usually display very short internodes and small, thick leaves. These morphological traits are under hormonal, particularly auxin control (Jansen 2002). Antioxidative flavonoids have been reported to be the most effective regulators of auxin transport in vivo (Jacobs and Rubery 1988; Brown et al. 2001; Taylor and Grotewold 2005), quercetin aglycone being much more effective than quercetin 3-O-rutinoside in inhibiting the basipetal transport of auxin (Jacobs and Rubery 1988). Therefore, the high light-induced preferential biosynthesis of "antioxidant" flavonoids may have a role in regulating whole plant and individual organ architecture (Lazar and Goodman 2006; Beveridge et al. 2007; Buer and Djordjevic 2009) and increase self-shading (Barnes et al. 1996; Jansen 2002).

Stafford (1991) suggested that flavonoids with distinct functions carried out by the mono- and dihydroxy B ring structures in regulating auxin metabolism probably played a primary role as internal regulators during the evolution of early land plants evidently through controlling auxin movements (Jacobs and Rubery 1988; Brown et al. 2001; Peer and Murphy 2007). Reports of a nuclear distribution of chalcone synthase (CHS) and chalcone isomerase (CHI) (Saslowsky et al. 2005) are consistent with the suggestion that flavonoids control transcription of genes required for growth and development, such as the auxin transport facilitator proteins (Peer et al. 2004).

4.3.2 Antioxidative Protection in plants

Ground level oxygen through energy transfer from UV radiation gives rise to ROS (Apel and Hirt 2004). For normal cell functioning ROS are scavenged by antioxidative enzymes/nonenzymatic antioxidants. Flavonoids by their dual role may function as nonenzymatic antioxidants (in addition to serving as internal UV screen – Sect. 4.3.1.3). This constitutes part of the acclimation strategy in plants (Tarrant et al. 2014; Brosche and Strid 2003). In plants, carotenoids, which are UV upregulated terpenoid compounds, though not effective as UV screen, serve specifically as antioxidants; these compounds are implicated in the direct protection of the photosystems against UV-B radiation (Middleton and Teramura 1993). *Arabidopsis* mutants with defects in the synthesis of such antioxidative compounds were tested for UV-B sensitivity: the transparent testa-4(tt4) mutant which has reduced flavonoids (capable of functioning as antioxidative compounds) and normal level of sinapate esters (serving as UV screens) was found to be more sensitive to UV-B than the wild type when grown under high UV-irradiance (Li et al. 1993). In addition to flavonoids, other groups of compounds such as terpenoids, carotenoids, alkaloids, and polyamines also act as nonenzymatic antioxidants which are upregulated by UV radiation (Gil et al. 2010; Schweikert et al. 2011; Lyddon et al. 1987).

4.3.2.1 Flavonoids

Flavonoids, in addition to the UV screening role, have been shown to quench effectively active oxygen species (Larson 1988; Rice-Evans et al. 1997). These compounds have drawn particular attention in studies on plant survival under UV fluence particularly because of their unique structural features that confer dual role in plants' UV acclimation strategy:

(a) $\pi \leftrightarrow \pi^*$ transition in the core nucleus (Sect. 4.3.1.3 and Fig. 4.4) conferring stability through absorption of high-energy UV radiation allows functioning of flavonoids as internal UV screen.
(b) Phenol ↔ Quinone tautomerism in the dihydroxy B-ring confers free radical scavenging activity, e.g., dihydroxy B-Ring-substituted flavonoids such as quercetin derivatives effectively scavenge ROS (thus preventing oxidative damage in various subcellular compartments, Agati et al. 2013) by donating hydrogen and an electron to hydroxyl, peroxyl, and peroxynitrite radicals, reflecting Phenol ↔ Quinone tautomerism, Fig. 4.13) thereby stabilizing them and giving rise to a relatively stable flavonoid radical (Cao et al. 1997).

Agati and Tattini (2010) argued that the main protective role of the phenolics is associated with their antioxidative capabilities more than its UV screening role. These pigments have long been known to accumulate mostly in the vacuoles of epidermal cells for protecting underlying tissues by absorbing UV-B photons, thereby serving the role of internal UV screen in plant cells (Stapleton and Walbot 1994; Baudry et al. 2004). Recent studies confirm that UV screening is one but probably not the most important

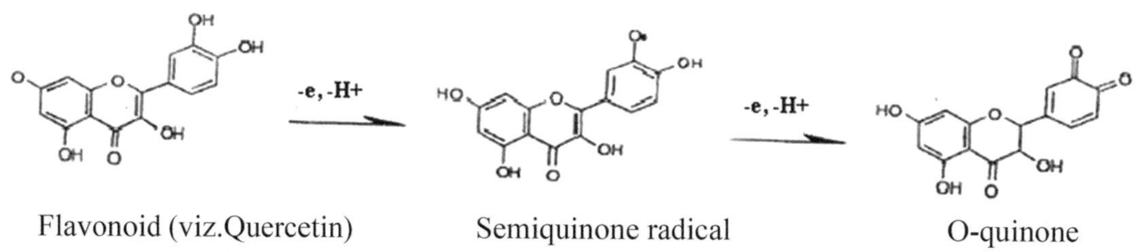

Fig. 4.13 Phenol ↔ Quinone tautomerism in the side chain of flavonoids enable these classes of compounds to act as antioxidants in plant cells (Korkina 2007)

protective function of flavonoids that due to its structural specialization also serve as nonenzymatic antioxidative compounds for countering oxidative damages (Lim et al. 2007); thus, flavonoids act in dual role providing an all pervasive protection of plants from UV radiation. This fits the observation that flavonoids can be found in tissues not directly exposed to UV-B and also in subcellular domains as far apart as chloroplasts, vacuoles, and nuclei (Jansen et al. 2012).

Chloroplasts appear to be capable of flavonoid biosynthesis (Pollastri and Tattini 2011; Hernández et al. 2009). Chloroplast-located flavonoids, serving as antioxidants, scavenge H_2O_2 and singlet oxygen generated under excess light stress and help in avoiding programmed cell death. Flavonoids on the outer surface of the chloroplast envelope might additionally quench ROS formed outside the chloroplast (Mullineaux and Karpinski 2002). Dihydroxy B-ring-substituted flavonoids, serving as antioxidants, also occur in nuclei (Polster et al. 2006; Saslowsky et al. 2005). Tournaire et al. (1993) have reported the results of a systematic study of the reactivity of 13 selected flavonoids (from the flavonol, flavone, flavanone, and flavane families) with singlet oxygen ($_1O^2$) in order to establish a structure-activity relationship. Various studies have suggested that certain flavonoid compounds serve not only as UV screening/filtering agents but also as antioxidants by dual functioning (Markham et al. 1998; Ryan et al. 1998); this has been further supported in subsequent studies (Tattini et al. 2004; Hofmann et al. 2003).

Functioning of flavonoids as nonenzymatic antioxidants is of particular importance since the commonly known enzymatic antioxidants, viz., SOD, POD, and APX, are degraded by UV radiation. The nonenzymatic antioxidants, viz., flavonoids, due to their structural specialization, remain stable under high UV radiation. It is pertinent to mention that the other "nonenzymatic" antioxidants, viz., ascorbic acid, get converted to dehydroascorbic acid under high heat and are thus not suitable for cell protection under prolonged exposure to ambient UV radiation. In plants under UV radiation thus the nonpeptide antioxidants, viz., flavonoids, are important – being more important in dehydrated cells where enzymatic turnover is precluded. The nonenzymatic antioxidative role of flavonoids constituting part of plant's acclimation strategy is particularly relevant in dehydrated system, viz., in orthodox seeds, where enzymatic antioxidant activity is precluded during postharvest ambient storage.

4.3.2.2 Carotenoids and other Terpenoids

Terpenoid group of compounds, upregulated by UV radiation, acts as nonenzymatic antioxidants for protecting plant cells from UV-induced oxidative damage. Terpenoids do not participate in UV screening in plants although some of these compounds may cause some reflection and scattering of incident radiation. Synthesis of terpenes under UV-B radiation has been studied in grape plants, the upregulation being a function of UV fluence rate (Gil et al. 2010). Low-intensity UV-B was found to promote the MVA (mevalonic acid) pathway, while high-intensity UV-B stimulated the MEP (2-c methyl – D erythritol 4 phosphate) pathway – both pathways relating to production of terpenoids, including carotenoids, that are synthesized by MEP pathway in plastid under low fluence of UV-B (Gil et al. 2010). White and Jhanke (2002) have reported that carotenoids have contrasting effects as these compounds can protect plants (*Dunaliella sp.*) from UV-A-induced damage, but it cannot protect the same species from UV-B-caused damage. UV-B promotion of terpenoids is observed particularly in members of the Lamiaceae family (Karousou et al. 1998; Maffei and Scannerini 2000; Behn et al. 2010). Johnson et al. (1999) reported accumulation of terpenoid compounds at seedling stage in basil leaves. Carotenoids, an important group of terpenoids represented mostly by carotenes and xanthophylls, are formed by condensation of eight isoprenoid monomers and are implicated in the direct protection of the photosystems against UV-B radiation (Middleton and Teramura 1993). This is because UV-B radiation is a potential oxidative factor (Panagopoulos et al. 1990) and carotenoids representing terpenoids as also flavonoids (Fig. 4.14)

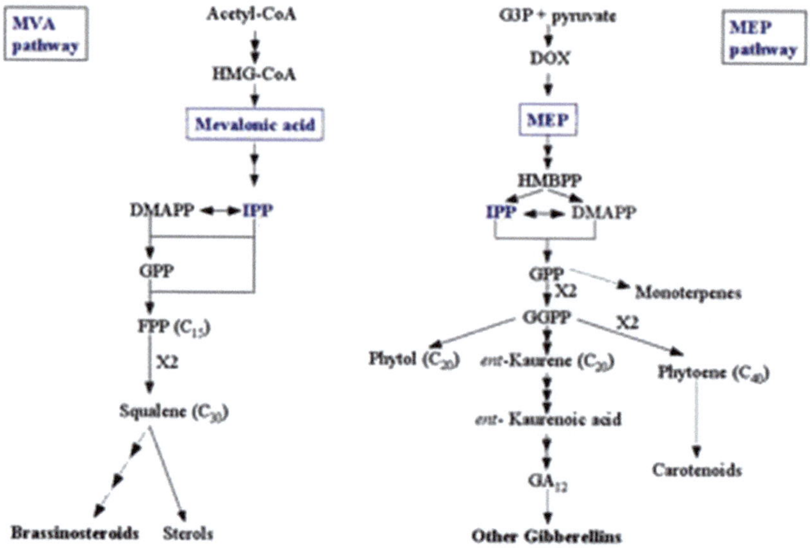

Fig. 4.14 Two pathways for production of terpenoids (including carotenoids) in plants (Taken from Gil et al. 2010)

Fig. 4.15 Developing *Arabidopsis* seedlings at 2, 3, 4, and 5 days showing disappearance of colored anthocyanins (Taken from Kubasek et al. 1992)

have been shown to quench effectively active oxygen species (Larson 1988; Rice-Evans et al. 1997). The carotenoids are implicated in the direct protection of the photosynthesis against UV-B radiation (Middleton and Teramura 1993). Carotenoids, a UV-upregulated group of terpenoid compounds, that act particularly as active oxygen scavengers have also been studied by Mewis et al. (2012) (Figs. 4.15, 4.16, and 4.17).

4.3.2.3 Other Antioxidants Conferring UV Protection

In addition to terpenoid and flavonoid group of compounds, polyamines, specific alkaloids and waxes (as deposition on leaf surface) contribute to UV tolerance. Polyamine synthesis has also been reported to be upregulated by UV radiation (Schweikert et al. 2011) in response to environmentally relevant doses of UV-B (Kramer et al.

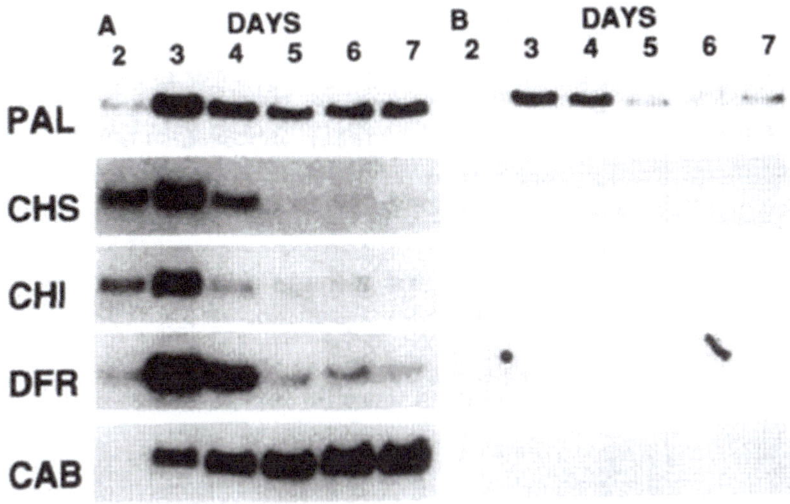

Fig. 4.16 Accumulation of flavonoid gene mRNAs in *Arabidopsis* seedlings. Steady-state transcript levels of PAL1, CHS, CHI, DFR, and CAB mRNAs were determined by RNA gel blot analysis as a function of seedling age (days 2–7) and light condition. (**a**) Grown in continuous *white light*. (**b**) Grown in continuous darkness (Taken from Kubasek et al. 1992) (PAL, CHS, CHI, and DFR represent enzymes of the flavonoid biosynthetic pathway; CAB represents chlorophyll a/b binding protein)

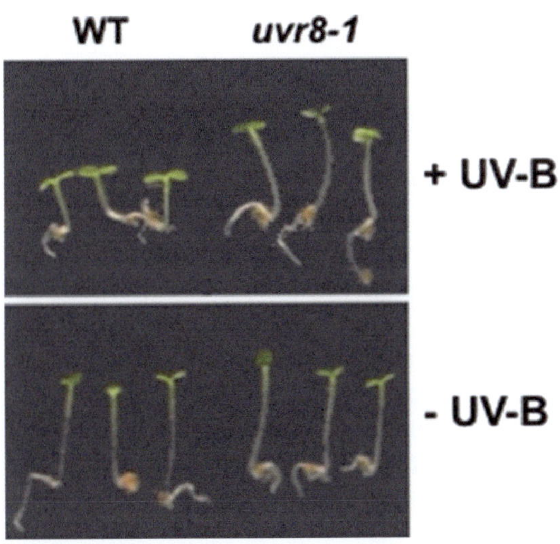

Fig. 4.17 Hypocotyl lengths of wild-type *Arabidopsis* (*LanDSBerg erecta*) and uvr8-1 seedlings grown for 4 days in 1.5 μmol $m^{-2}s^{-1}$ fluorescent *white light* supplemented (+), or not (−), with 1.5 μmol $m^{-2}s^{-1}$ narrowband UV-B (Taken from Hectors et al. 2007)

1992). In soybean, a correlation was found between levels of polyamines and tolerance to UV-B (Kramer et al. 1992). Possibly, the radical-scavenging activity of polyamines (and polyamine conjugates) moderates UV-B radiation stress, as was demonstrated for other free radical scavengers (Jansen et al. 1996).

Selected alkaloids that absorb UV-B wavelengths or possess free radical-scavenging activity also contribute to UV protection. Levels of UV-absorbing tetrahydrocannabiol increase linearly with UV-B dose in *Cannabis* (Lyddon et al. 1987). Similarly, cannabinoid content increases with altitude at which plants are grown evidently due to altitude-related enhanced UV fluence. This indicates a role of UV radiation in upregulating these compounds. In our preliminary studies on *Murraya koenigii*, we have found that one of the main active principles of this plant called mahanine (a carbazole alkaloid) is also upregulated by UV radiation. Plants from high-altitude areas of North East India contain higher amount of this compound in comparison to the amount of mahanine found in lower-altitude-grown plants in India.

4.3.2.4 Enzymatic antioxidants

Although macromolecules viz. proteins/enzymes are generally disrupted by UV through direct impact/indirectly by ROS mediated damage in field growing plants, low dose UV radiation through signal transduction upregulates synthesis of enzymatic antioxidants (Nawkar et al. 2013).

Enzymatic antioxidants provide major cellular redox buffers. From studies on algae, Rautenberger et al. (2013) and Janknegt et al. (2009) have demonstrated that enzymatic antioxidants such as superoxide dismutase (SOD), ascorbate peroxidase, and glutathione reductase (GSH) protect cells from UV radiation. GSH is oxidized by ROS forming oxidized glutathione (GSSG), whereas ascorbate is oxidized to monodehydroascorbate (MDA) and dehydroascorbate (DHA) – ascorbate peroxidase acting as scavenger of superoxide radicals (Surplus et al. 1998). It has been reported that mutant *Arabidopsis* that lacks ascorbate vtc1 gene is highly sensitive to UV-B (Conklin et al. 1996; Gao and Zhang 2008).

4.3.3 Photomorphogenesis

At the level of ecosystem photomorphogenic effect involving UV, avoiding structure and composition of growing stem is exhibited by plants. These features allow protection from fluencing UV radiation as a means of UV acclimation/amelioration in plants for survival and growth under ambient UV radiation. The time taken for repair processes such as UV-induced upregulation of photolyase for DNA repair (Biever et al. 2014) and development of internal UV screen, viz., flavonoids (Ballare et al. 2011), is often reflected as delayed growth of emerging stem embryonic axis that may be seen as photomorphogenic activity – one of the acclimation strategies in plants. Commonly observed phenotypic manifestation of UV adaptive trait (involving temporal cessation of primary metabolism to feed secondary metabolism) is temporal inhibition of hypocotyl/stem elongation (serving as means of avoiding UV exposure to dividing cells) until protective compounds through functioning of secondary metabolism become available (Barnes et al. 1990; Sullivan et al. 1992; Dixon et al. 2001). Photomorphogenesis in seedlings has been reported to be controlled by blue/UV-A-absorbing cryptochromes and also red/far red-absorbing phytochromes (phy A-E) (Batschauer 1999; Quail 2002). However, UV-B radiation functioning as a trait selective agent is the major driver of photomorphogenic effects in plants. Induced inhibition of stem elongation as a means of protection of cells from ravages of UV radiation is a typical example of photomorphogenesis (Kubasek et al. 1992; Gardner et al. 2009) which is effective in countering loss of biodiversity on earth. Delayed stem extension represents a temporal phenotypic trait (Gardner et al. 2009) exhibiting varietal difference (Muller-Xing et al. 2014). Other UV-induced photomorphogenic effects include leaf thickening, cotyledon/leaf curling, stem and leaf elongation, axillary branching, and shifts in the root–shoot ratio (Barnes et al. 1990; Cen and Bornman 1993; Boccalandro et al. 2001; Wilson and Greenberg 1993). These regulatory reactions, effective in restraining UV with potential for damaging effects, are evidently not stress responsive and can be considered as photomorphogenic in nature (Hectors et al. 2007). Cotyledon/leaf curling cause reflection of incident UV (Jetter and Schaffer 2001; Perkins et al. 2005) and thereby serve to restrain UV entry into cells. Leaf thickening is thought to diminish penetration of UV radiation to inner tissue due to reduced penetration of ambient UV-B; Cen and Bornman (1993) have reported effect of exposure to enhanced UV radiation on the penetration of monochromatic and polychromatic UV-B radiation in leaves of Brassica napus. Similarly, the often observed UV-related reduction of apical dominance (decreased shoot length and increased axillary branching), within a canopy, is expected to diminish exposure to direct sunlight (including its UV-B component) as leaves of short, bushy plants are more likely to be shaded (Barnes et al. 1996). Phototrophic curvature in etiolated Arabidopsis seedling, representing UV related photomorphogenic effect, has been studied by Vandenbussche et al. (2014). The authors report that early differential growth is mediated by phototropins (UV-A related photoreceptor – see Sect. 5.3.2) but clear phototropic bending to UV-B was observed in phot1 phot 2 double mutants. It was further observed that this phototropin-independent phototropic response to UV-B requires the UV-B photoreceptor UVR8. Broad UV-B-mediated repression of auxin-responsive genes suggests that UVR8 regulates directional bending by affecting auxin signaling. Kinetic analysis shows that UVR8-dependent directional bending occurs later than the phototropin response. We conclude that plants may use the full shortwavelength spectrum of sunlight to efficiently reorient photosynthetic tissue with incoming light.

Delay in emergence of hypocotyls allows time for the emerging tissue to upregulate synthesis of protective phenolics. Thus (low-dose) UV-B (in addition to inducing biosynthesis of protective phenolic compounds) may also provoke adaptive plant photomorphogenic responses (Tevini and Teramura 1989) exhibiting avoidance to UV radiation. During early germination when the emerging seedling is about to be exposed to UV radiation, the emerging hypocotyl takes to UV avoidance. This represents a short-term photomorphogenic modifications during early stage of embryonic growth (Lercari et al. 1990). The delayed germination may be a manifestation of inhibition of hypocotyl elongation that allows avoidance of UV exposure to seedlings under ambient UV radiation during early stage of embryonic growth (Lercari et al. 1990). The delayed germination may be due to (a) time required to complete repair of DNA damages (Biever et al. 2014) that have accumulated during post-harvest aging of the embryonic axis and (b) apparently allowing time for biosynthesis of flavonoids in the growing hypocotyl for development of protection from the UV screen in the growing hypocotyl (Ballare et al. 2011).

Although UV fluence rate determines UV-related molecular events in plants, such effects are primarily determined not only by the fluence rate of exposure but also dependent on whether the plants have been acclimated by prior exposure to UV-B (Muller-Xing et al. 2014). Other UV-induced morphological changes in acclimated plants include decreased rosette diameter, decreased inflorescence height, and increased numbers of flowering stems, indicating that long-term UV-B treatment causes a redistribution of molecular events involving primary and secondary metabolism rather than a cessation of growth thereby affecting photomorphogenic traits. *Arabidopsis thaliana* plants acclimated to low-dose rates of ultraviolet B radiation show specific changes in morphology and gene expression instead of showing stress symptoms (Hectors et al. 2007). The most commonly studied photomorphogenic phenomenon is temporal inhibition of hypocotyl elongation where the delay in hypocotyl elongation is observed to be a short-term event (Frohnmeyer and Staiger 2003). A time-course experiment on hypocotyl elongation inhibition reported by Gardner et al. (2009) showed hypocotyl growth inhibition that occurs within 6 h after UV radiation and lasted until 3–4 days later. Such temporal regulation is found to be correlated with NER-mediated DNA repair (Biever et al. 2014)/synthesis and accumulation of effective UV-screening compounds (Ballare et al. 2011; Kubasek et al. 1992) during early germination prior to emergence of the hypocotyl/embryonic axis. Kubasek et al. (1992) have demonstrated increasing synthesis of purple-colored anthocyanins around the rims of the cotyledons until 5 days of germination. Beyond this period, presumably when hypocotyl cells have accumulated appropriate amount of protective (phenolic) compounds, viz., anthocyanins, in the growing hypocotyl prior to emergence from the seed (protective cover), the hypocotyl resumes normal growth, and anthocyanin development decreases (Fig. 4.15).

During early hypocotyl growth, many plants exhibit a similar transient accumulation of flavonoid pigments (Beggs et al. 1986; Ehmann et al. 1991; Kubasek et al. 1992). The ubiquity of this phenomenon suggests it is of some fundamental importance to the plants. One possibility is that flavonoids are produced prophylactically to protect the emerging seedlings against UV light when exposed to ambient solar radiation (Ballare et al. 2011). This temporal regulation is found to be correlated with synthesis of cell-protective phenolic compounds (Fig. 4.16); this is indicative of a genetic control over photomorphogenic effects. From studies on *Arabidopsis* ecotypes, Torabinejad and Caldwell (2000) demonstrated that UV-induced flavonoid-based UV tolerance manifested as photomorphogenic effect in plants is subject to genetic variation.

Investigation of Kubasek et al. (1992) to explore the developmental dependence of flavonoid gene expression under continuous white light supplemented with UV-B light showed higher levels of mRNA accumulation for all the four genes (viz., PAL1, CHS, CHI, DFR) associated with flavonoid biosynthesis and CAB (chlorophyll a/b binding protein) associated with photosynthesis in 3-day-old seedlings (Fig. 4.15). Gel photographs representing a time course for synthesis of these enzymes is shown in Fig. 4.16.

In the presence of UV-B, the uvr8-1 mutant shows higher growth compared to the wild type where primary metabolism due to (inherently present) UVR-8 mediated signal transduction is diverted to secondary metabolism for upregulation of flavonoid synthesis for UV protection. In absence of UV-B, growth rate of both uvr8-1 mutant and WT are almost identical (Fig. 4.17) because of non interruption of primary metabolism by UV induced secondary metabolism for synthesis of UV protective (flavonoid) compounds.

Some of the reports on photomorphogenic effects include decrease of shoot height in *Triticum aestivum* and *Avena sativa* but increase in *Amaranthus retroflexus* and *Kochia scoparia* (Barnes et al. 1990); in all the four species, branching (tillering or axillary branching) increased in response to UV-B (Barnes et al. 1990; Biever et al. 2014).

Hypocotyl growth inhibition induced by UV-B in etiolated *Arabidopsis* seedlings, a classic photomorphogenic response, is influenced by signals originating from UV-B absorption by DNA and subsequent damage of DNA. Ultraviolet (UV) radiation thus induces photomorphogenic responses that is manifested as reduced hypocotyls / stem extension due to delayed cell cycle/cell cycle arrest (Biever et al. 2014). as well as enhanced fkllvonoid synthesis (Ballare et al. 2011) that represent acclimation strategy for avoiding UV exposure. Such changes generally cause variation in phenotypic traits but do not result in reduction of biomass (i.e., do not represent stress response).

4.3.4 DNA Repair

DNA, the most important biomolecule, is susceptible to UV radiation that causes DNA strand breaks directly or via the process of oxidative damage – the process being maximally affected by UV-B (at high dose) and to a lesser extent by UV-A. Constitutive DNA and RNA polymerases are incapable of proper functioning in the presence of UV-induced photoproducts involving pyrimidine dimers (viz., CPD and 6-4 PP); elimination of the photoproducts is essential for repair and proper DNA replication and transcription for cell survival to be ensured (Britt and May 2003). To avoid the deleterious effects of UV-induced DNA damage, most organisms have developed a complex set of repair mechanisms including photoreactivation (reinstating DNA damage by monomerization of photodimers in DNA by photolyase enzyme) induced by low dose of UV (Hutchinson 1987); this enzyme is upregulated via signal transduction (through UVR-8)-mediated gene expression as part of UV acclimation (cf. upregulation of flavonoid synthesis – see Chap. 5). One other pathway for DNA repair, viz., homologous recombination repair, is induced by high-dose UV radiation (Dubest et al. 2002). Studies of Ries et al. (2000b) indicate that although homologous recombination process for DNA repair is primarily induced by high dose of UV-B, this process may also require energy provided by photosynthetic activity. A nonspecific DNA repair process called nucleotide excision repair (NER) also operates in cells to repair DNA damage caused by UV radiation (Molinier et al. 2008; Liu et al. 2000). In addition to these different types of UV-induced repair (light repair), other repair mechanisms such as base excision repair (BER) and mismatch repair (MMR) are also involved in repair of damaged DNA. These processes do not require light energy for activation and are collectively referred to as “dark repair”; these are generally not involved in protection from UV-induced damage of DNA.

4.3.4.1 Photoreactivation

UV radiation induces dimerization of adjacent pyrimidine bases in DNA leading to production of photodimers such as cyclobutane pyrimidine dimers (CPDs) and pyrimidine (6-4) pyrimidinone dimers (6-4 PPs) (Fig. 3.5) (Frohnmeyer and Staiger 2003). Both these lesions (CPD and

6-4 PP) distort DNA structure, causing bends or kinks and thereby cause blockage in transcription and replication. Such damages caused by UV-B to the genetic machinery of the cell are potentially significant problems for plant growth and development. For resilience under damaging effects of UV radiation, it is imperative that plants have an effective system of DNA repair such as "photoreactivation" that refers to a pathway for reactivating/reinstating DNA structure by monomerization of UV-induced photodimers thereby reinstating the DNA structure. This constitutes the major DNA repair pathway (Brosche and Strid 2003) involving removal of photodimers; this process occurs under light and ensures acclimation of plants through their stationary habit in open field. This process of reinstating/repairing the DNA is mediated by a photoreactivating enzyme, viz., photolyase (Sancar 1994), which have sequence similarity with cryptochromes, whereas photolyases are DNA repair enzymes and cryptochromes act as photoreceptors (for UV-A) and also circadian clock proteins. Photolyase is an evolutionary stable protein found in almost all organisms. The light energy (>380 nm) is trapped by the antenna molecules of photolyase, and it transfers them to catalytic cofactor FADH – which becomes excited and transfers energy to the pyrimidine dimer in the form of free electron (e^-). After splitting the CPD into two monomeric units, the electron is transferred back to the flavin molecule. Figure 4.18 represents the various steps in DNA repair by photolyase after photodimer formation by UV radiation. Photoreactivation is also called "light repair" because of its dependence on light energy (Rastogi et al. 2010).

Photolyase enzymes exhibit substrate specificity. There are specific DNA photolyase enzyme which binds specifically either to (a) CPD (CPD photolyase) or (b) 6-4 PP (6-4 PP photolyase) as the structures of the lesions are very different. Both type of photolyase directly monomerizes the cyclobutane ring of the pyr <> pyr thereby protecting the genome from deleterious effects of UV radiation (Essen and Klar 2006). That photoreactivation greatly enhances the rate of removal of dimers that have been demonstrated from studies on *Arabidopsis*, rice, and alfalfa. Studies on *Arabidopsis* reveal that photolyases exhibit sub-

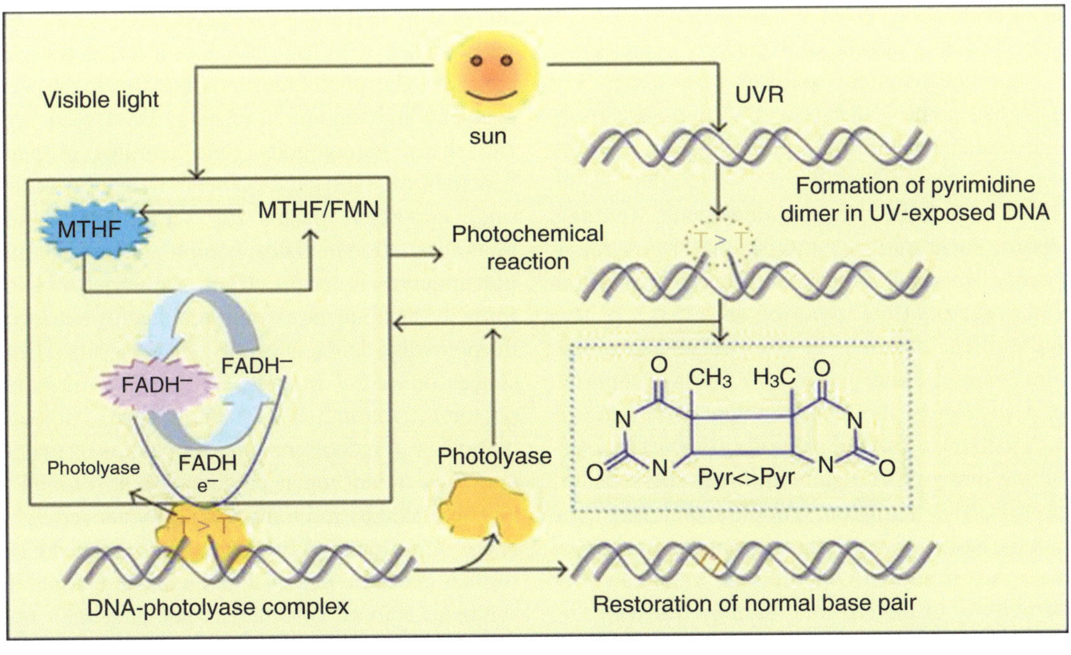

Fig. 4.18 Schematic diagram showing photoreactivation by photolyase enzyme (Taken from Rastogi et al. 2010)

strate specificity eliminating either CPDs or 6-4 PPs, respectively (Hoffman et al. 1996; Ahmad et al. 1997); the UVr2 gene encodes a photolyase that acts only on CPDs (Ahmad et al. 1997), whereas the UVr3 gene encodes a photolyase specific for 6-4 photoproducts (Nakajima et al. 1998). White light or UV-B radiation (at low dose) induces CPD photolyase expression in *Arabidopsis* tissues (through UVR-8-mediated signal transduction) which contrasts with the 6-4PP photolyase protein which is constitutively expressed and not regulated by either white or UV-B light (Waterworth et al. 2002). *Arabidopsis* CPD photolyase activity is only detected when plants have been exposed to visible light prior to, as well as during, the period of repair (Chen et al. 1994). This is because transcription of the gene is regulated by visible and UV-B light (Ahmad 1997; Waterworth et al. 2002). Studies of Britt (2004) have demonstrated that plants grown in the presence of photoreactivating radiation can eliminate the majority of both 6-4 PP and CPD lesion within hours, or in some cases minutes, of their induction. In absence of photoreactivating light also, dimers are eliminated from bulk DNA, albeit slowly (Chen et al. 1994; Kang et al. 1998). 6-4PPs are generally observed to be repaired more quickly than CPDs. This is due to the fact that photoreactivating enzymes particularly 6-4 PP photolyase may also be constitutively expressed.

Photolyase mutants exhibit variation in UV response, some exhibiting UV hyposensitivity and some exhibiting UV hypersensitivity (reported from *Arabidopsis* – Landry et al. (1997) and from rice – Hidema and Kumagai (2006); such studies establish that DNA repair mechanisms are subject to genetic variation. Hidema et al. (1997) reported on UV-B sensitive rice cultivar deficient in cyclobutyl pyrimidine dimer repair. Landry et al. (1997) have reported an *Arabidopsis* photolyase mutant (Uvr2-1) that is defective in photorepair of cyclobutane pyrimidine dimer (CPD) and exhibits hypersensitivity to UV-B radiation. Hidema and Kumagai (2006) have reported that variation in one or two bases in the CPD photolyase gene can alter the activity of the enzyme, thereby transforming those plants resistant to UV-B radiation. This would exhibit UV tolerance as adaptive advantage in enhanced UV environment such as at high altitude (see Chap. 7). Hidema and Kumagai (2006) have also proposed that by selective breeding or bioengineering of the genes, encoding CPD photolyase UV-resistant varieties may be produced. Giordanoi et al. (2003) reported that plants exposed to solar UV had higher CPD repair capacity (under optimal conditions of temperature 25 °C) compared to plants grown under artificially attenuated UV; there was no measurable repair at 8 °C. The rates of CPD repair in *G. magellanica* plant were modest (in comparison with other species) under similar conditions and exhibited lower repair rates – about 50 % lower than the repair rates of *Arabidopsis thaliana* (Ler ecotype). In contrast to photoreactivation which is induced by low dose of UV, other DNA repair processes, such as homologous recombination, are reported to be activated by high dose of UV radiation when cell undergoes severe UV damage.

4.3.4.2 Homologous Recombination

Repair of UV-B-induced DNA damage in plants is also brought about by high-dose UV through the process of homologous recombination between adjacent strands (Ries et al. 2000a) (cf. photoreactivation-associated DNA repair that is associated with low UV dose-associated repair enzymes, viz., photolyase activity). Due to the presence of a high amount of repeated DNA sequences and many homologous gene families in plant genomes, this molecular event is a frequently occurring important means of CPD elimination/monomerization in plants; homologous recombination frequency is correlated with the amount of CPD formed – its frequency being significantly enhanced in photolyase-deficient genotypes, as recorded from studies on *uvr2-1* mutants lacking CPD-mediated photoreactivation (Ries et al. 2000b). Although homologous recombination in plants is generally known as a dark repair process, it is also found to be stimulated by red light (but not by far-red light) exposure following UV-B treatment-induced CPD formation. These observations indicate that photosynthetic activity or other as yet undefined processes dependent on photosynthetic active radiation (PAR – 400–700 nm) may promote UV-

B-induced homologous recombination in plants (Ries et al. 2000b – PAR-related increased number of recombination events being evident only after UV-B radiation. By inducing different levels of CPD, Ries et al. (2000b) demonstrated that the amount of CPDs in the genome of irradiated plants is proportional to the activity of repair-related recombination after high-level UV-B radiation indicating that homologous recombination processes may be involved in repairing DNA lesions directly at damaged loci (Fig. 4.19).

In *Arabidopsis* and tobacco plants, it has been reported that elevated solar UV-B dose increases frequency of somatic homologous DNA recombination. Increases in recombination are associated with a strong induction of photolyase and other related gene expression putatively involved in major DNA repair pathways, viz., photoreactivation and recombination repair (Britt 1999; Vispe et al. 1998). In mutant *Arabidopsis* plants that are deficient in photoreactivating ultraviolet-induced cyclobutane pyrimidine dimers, recombination under elevated UV-B regimes greatly exceeds wild-type levels. Studies of Ries et al. (2000a) indicate that homologous recombination repair pathways might be involved in eliminating UV-B-induced DNA lesions in plants indicating that increases in terrestrial solar UV-B radiation as projected for the early twenty-first century may affect genome stability in plants (Ries et al. 2000a). As such studies on DNA damage and repair under different UV fluence location as affecting in different genotypes (see Case Study I) will help to develop a wide knowledge base for developing appropriate cell molecular strategies in plants for acclimation under threatening UV enhancement in the coming years through the process of evolution (see Sect. 4.4.1).

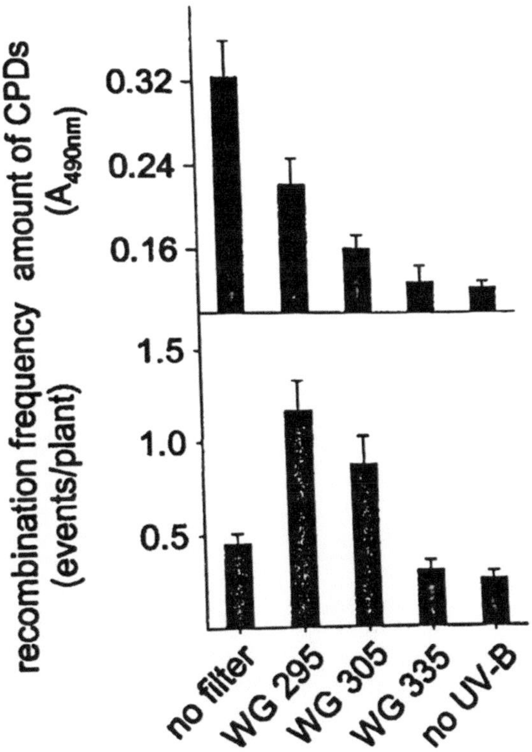

Fig. 4.19 Formation of CPDs and the induction of recombination frequency in dependence of UV-B spectral ranges. Amounts of CPDs (*upper pic*) and recombination frequency (*lower pic*) were determined under these light regimes in an *Arabidopsis* line carrying the GUS substrate as a marker for recombination events. Error bars represent the SE (Taken from Ries et al. 2000b)

4.3.4.3 Nucleotide Excision Repair

This is a nonspecific DNA repair mechanism that is involved in repairing both UV-induced as well as other mutagen-caused DNA damage. CPDs and 6-4 PPs are repaired through this process of nucleotide excision repair (NER). In bacteria (prokaryotes), the process of NER is completed by only three proteins, named UvrA, UvrB, and UvrC, whereas in eukaryotic cells, NER is carried out via four discrete steps: detection of damage, excision of the section of DNA that includes and surrounds the error, filling in of the resulting gap, and sealing of the nicks between the newly synthesized and older DNA by DNA polymerase. The four steps described above are represented in Fig. 4.20.

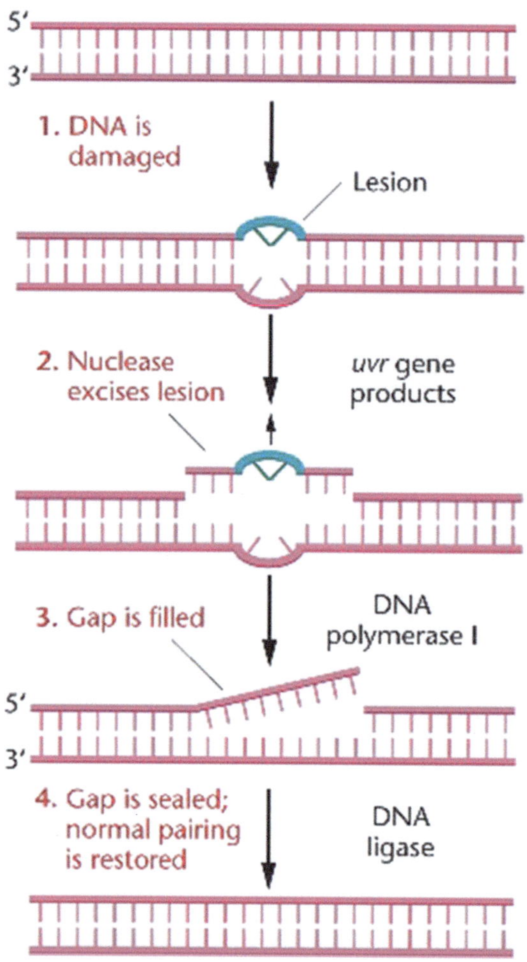

Fig. 4.20 Four steps of nucleotide excision repair in eukaryotes (Taken from web)

In eukaryotes, this complex process relies on the products of approximately 30 genes. Defects in some of these genes have been shown to cause the human disease xeroderma pigmentosum (XP), as well as other conditions that share a risk of skin cancer. The XPA protein is recruited to the damage removal site; therefore, cells deficient in this protein are unable to repair the photoproducts.

4.4 Acclimatory Strategies for Photosynthesis

A number of studies on photosynthesis under natural environmental conditions including PAR and realistically increased levels of UV-B have demonstrated inhibition of photosynthesis (Bornman 1989; Jordan 1993; Tevini et al. 1991). Such studies have identified chloroplast as the major site of damage by UV-B (Bornman 1989). Studies focusing on the molecular mechanisms underlying UV-B sensitivity of photosynthesis (Strid et al. 1994; Baker et al. 1997; Mackerness et al. 1999) have pointed to changes in gene expression. In this context Mpoloka (2010) demonstrated that reduction in expression and synthesis of key photosynthetic genes including RUBISCO (rbc S and rbc L), D1 polypeptide of photosystem II (psbA), chlorophyll a/b binding protein (Lhcb), and the ATPase complex and also decline in total RNA activity constitute major events associated with response to supplemental UV-B. UV-induced damaging effects including inhibition of photosynthesis are found to be aggravated under artificial/unnatural light conditions, viz., relatively high UV-B levels under conditions of low photosynthetically active radiation (PAR) such as conditions prevailing in greenhouse/closed chamber.

Under natural UV radiation, flavonoids accumulating in the epidermal layer, by restricting UV entry provide considerable UV acclimation that confers protection to the photosynthetic apparatus. This has been demonstrated in rye seedlings using UV-B pre-irradiated seedlings that accumulate flavonoids during preexposures (Tevini et al. 1991). These pre-irradiated rye plants were found to be less susceptible (than the control plants) to photosynthetic damage when exposed to short wavelength UV-B radiation as monitored by chlorophyll fluorescence. In *Brassica napus* degradation of the photosystem II, D1 protein was found to be reduced in UV-B adapted plants – the extent of reduction in the rate of degradation appeared less in such plants as would be expected of plants with high flavonoid synthesized and accumulated during preexposures related adaptation to high UV radiation (Wilson and Greenberg 1993).

4.5 UV-Regulated Gene Expression for UV Acclimation

While high-intensity UV radiation causes cell-damaging effects, at low dose UV-B induces expression of genes involved in developing UV amelioration in plants. UV-B-specific photomorphogenic responses associated with upregulation of gene expression related to macromolecular (viz., DNA) repair and upregulation by low-dose UV radiation of phenolic compound (viz., flavonoids) synthesis provide internal UV screen as well as antioxidative protection from UV-induced damage in plants; upregulation of these genes conferring acclimation under high intensity is mediated generally by the UV (both UV-B and UV-A) photoreceptor, viz., UVR8 (Rizzini et al. 2011) that perceives low-dose UV radiation.

From extensive studies relating to development of internal UV screens in leaves of grapevine (*Vitis vinifera* cv. malbec) using transcriptome analysis, Pontin et al. (2010) concluded that manifested UV effect varies with the number of genes modulated (i.e., up- and downregulated) by UV light exposure; UV acclimation related (differential) gene expression induced by UV-B has also been reported by Brown et al. (2005), Kilian et al. (2007), and Ulm et al. (2004). Transcriptome analysis demonstrates that UV-B enhances the expression of various genes that are responsible for counteracting oxidative stress (Ulm et al. 2004; Brown et al. 2005; Kilian et al. 2007). Different types of UV-B exposure regulate different sets of genes (Brosche and Strid 2003; Brown and Jenkins 2008; Casati and Walbot 2004): high fluence and short wavelengths of UV-B induce those genes which are usually responsible for defense, wound, or general stress responses in plants (Brosche and Strid 2003; Kilian et al. 2007; Ulm and Nagy 2005), whereas low fluence UV in brief exposures induces a variety of genes, many of which are involved in UV protection or the amelioration of UV damage (Brown and Jenkins 2008; Frohnmeyer et al. 1999; Ulm et al. 2004). After activation by UV-B, UVR8 photoreceptor interacts through a C-terminal 27 amino acid region with the WD40 domain of COP1 in the nucleus (Cloix et al. 2012) for triggering the induction of a key transcription factor, viz., elongated hypocotyl5 (HY5), that controls several UV-B-responsive genes associated with overall UV-B acclimation (Heijde and Ulm 2012). The role of PAL as a key enzyme-activating phenylpropanoid biosynthesis in response to UV-B radiation has been studied in several plants/systems: increased PAL expression after UV-B exposure has been reported in mesophyll cells of *Pisum sativum* (Kalbin et al. 2001), leaves of Petunia (Ryan et al. 2002), *Cucumis sativus* (Cantarello et al. 2005), and in rosette leaves of *Arabidopsis thaliana* (Ulm et al. 2004); stable PAL expression is needed for the accumulation of myricetin-3-galactoside; the transcription factor HYH has been reported to regulate the expression of genes associated with the synthesis of quercetin and kaempferol rhamnosides (Ulm et al. 2004). UV-B exclusion has been reported to lower PAL activity in *L. sativa* (Krizek et al. 1998).

Solar UV-A exhibits a major role in regulating the expression of most UV-B-regulated genes (Morales et al. 2010, 2011). The authors have also demonstrated that different doses of solar UV-A and UV-B radiation differentially regulate gene expression and the accumulation of flavonoids. These studies were conducted on birch leaves that normally grow as a pioneer species where high expression of PAL and HYH associated with the accumulation of flavonoids in young unfolded leaves would be useful for plant establishment under adverse conditions. Under high PAR irradiance and prolonged UV exposure outdoors, the UV-B irradiances regulate the expression of the genes PAL and HYH (Morales et al. 2010). The authors also demonstrated that the concentrations of six flavonoids, viz., myricetin-3-galactoside, quercetin-3-galactoside, quercetin-3-glucuronide, quercetin-3-arabinopyranoside, quercetin-3-rhamnoside, and kaempferol-3-rhamnoside, decreased in a linear relationship with the exclusion of UV-B. The concentrations of quercetin-3-galactoside and quercetin-3-arabinopyranoside also decreased linearly with attenuation of UV-A.

In addition to providing protection from deleterious effect of UV radiation, this radiation through formation of hydrogen peroxide, a species of ROS that, though generally known to cause oxidative damage, also induces pathogenesis-related protein (PR1) that protect plants from pathogen attack. These effects as well as effects of upregulation of PAL and HYH in birch seedlings contribute to our understanding of the ecological roles of UV radiation and help us to predict how plant responds to UV in its natural habitat, and the resultant systemic changes in plants is important to elucidate changes at the ecosystem level (Caldwell et al. 2007).

4.5.1 Transcriptional Control

Most genes regulated by UV-B are controlled at the level of transcription, and there is direct evidence for this in several studies (Chappell and Hahlbrock 1984; Hartmann et al. 1998; Safrany et al. 2008; Ulm et al. 2004). It is therefore important to identify DNA sequence elements and transcription factors involved in these responses and to understand how transcriptional regulation is coupled to UV-B signaling pathways.

A significant proportion of the genes that are most rapidly induced by UV-B include genes that encode transcription factors (Kilian et al. 2007; Ulm et al. 2004) that play key roles in UV-B responses. Some UV-B-induced transcription factors are involved in controlling the biosynthesis of UV-protective phenolic compounds (Cloix and Jenkins 2008; Cominelli et al. 2008; Piazza et al. 2002). Morales et al. (2010) demonstrated that transcripts of PAL and HYH linearly decreased with exclusion of UV-B, indicating that UV-B regulates the expression of these genes at transcript level. In *A. thaliana*, HYH expression was stimulated after short-term UV-B exposure and, together with its homologous bZIP transcription factor HY5, HYH plays a key role in regulating low fluence UV-B responses. Supporting evidence is obtained from studies on *Arabidopsis* (Ulm et al. 2004; Brown and Jenkins 2008). The observation that low transcript accumulation of HYH with the exclusion of UV-B indicates that HYH expression might be important during prolonged exposure of birch leaves (the "pioneer" species) to solar UV-B during plant accession to unfavorable conditions. UV-A is also reported to regulate the expression of HYH in a nonlinear response, without affecting PAL activity (Morales et al. 2011).

4.5.2 Posttranscriptional (Epigenetic) Regulation

MicroRNA (miRNA)-mediated posttranscriptional regulation under stress conditions is considered an important epigenetic regulation for fine-tuning regulatory pathways. Many miRNAs target mRNAs that encode transcription factors thereby regulating the expression of many downstream genes. Zhao et al. (2007) have reported that in *Arabidopsis thaliana*, the 21 miRNA genes in 11 miRNA families are upregulated under UV-B stress condition.

Case Study I

4.6 UV-Induced Cellular Damage and Protection in Dehydrated Plant System (viz., Rice Seed)

Survival of orthodox (dry) seed under UV radiation in ambient farmer's storage presents a system where countering of UV-induced cell molecular damage is essential for maintaining seed viability – the extent of damage control being a function of UV-induced cell molecular damage that affects UV acclimation processes in postharvest "dry" plant cells. Since maturing seeds in field-grown plants synthesize and accumulate UV (under genetic control) screening compounds (Mazza et al. 2000), the only enzymatic activity required in stored seed relates to macromolecular repair for perpetuation of embryonic axis. In enzymatic processes remaining precluded in embryonic axis during dry storage of postharvest mature seed (mc < 8%), vigor-viability trait (status) would be a reflection of damage, particularly of DNA, incurred under UV fluence during dry storage.

Since the ozone layer is naturally thinner (allowing higher UV-B fluence on earth's surface) at the equatorial tropics (US Environmental Protection Agency 1995), the main location of rice production, a study on UV-induced damage vs UV screening potential in seed, would provide useful information on integrity of DNA and other macromolecules as well as membrane integrity in mature orthodox (dry) seed. A case study in this regard is presented here. Studies in this regard is particularly relevant for exploring inherent survival strategy in plant systems for tiding over low moisture condition as in dry postharvest seed storage during periods of storage of genotype during intermittent periods between cultivation seasons. Under the assumption that solar UV-B radiation will reach peak levels on Earth's surface in the next few years (Kakani et al. 2003) and considering the reports of higher UV-B fluxes in the area around the equator, such study would not only provide basic understanding into the cell molecular processes but would also be relevant for attempts in translational biology for determining appropriate sites for cultivation for different crops and should be carried out in the tropical region.

A study in this context was carried out (Shyam Chowdhury and Sen Mandi 2012) using several varieties. Protective compounds, viz., flavonoids, synthesized in the embryonic axis during seed maturation on mother plant in the open field, constitute nonenzymatic antioxidants that remain stored in dehydrated seed, countering UV-induced cellular degradation during postharvest storage in farmers' field; a major role of flavonoids in this context is to provide internal shield to restrain (albeit partially) UV penetration into the seed. Cultivars that are capable of synthesizing and accumulating high amount of flavonoid (that serve as UV screen as well as nonenzymatic antioxidants) exhibit better protection of macromolecular entities through dry seed aging under spontaneous UV radiation during postharvest storage. These varieties exhibit high vigor-viability trait. Varieties with lower capability of synthesizing (in embryonic axis during seed maturation) the protective compounds exhibit lower vigor-viability traits – the damage being proportional to the genetic difference. Duration of exposure to UV radiation in farmers' dry seed storage determines the extent of direct damage by UV – the extent being proportional to the protective potential of accumulated flavonoid-induced acclimation through countering of effects due to fluencing UV radiation. Depending on the genetic control for determining the balance between damage and repair/protection of the cellular entities, a seed through postharvest aging would exhibit loss of vigor proportional to the extent of macromolecular damage.

At seed germination emergence of coleoptile occurs only after repair processes have been completed as has been demonstrated in *Arabidopsis* (Biever et al. 2014). Alternatively however if the seeds remain in dry storage for too long, cell molecular damage will continue under fluencing UV in open field; this is manifested as proportional loss of seed vigor, subject to varietal difference in the repair/acclimation processes. The experiments described below relate to cell molecular damages vis-à-vis status of acclimation (antioxidative potential and UV screening by flavonoids) as a function of genetic difference between seed stocks stored under UV radiation in farmers' seed storage in open field. Out of 22 varieties grown on pots (Fig. 6.3), six varieties were selected. Among them Patnai-23, Pankaj, and Matla (redesignated as A, B, and C, respectively, in the experiment) showed high growth rate, and Jaya, Jogen, and Lalat (redesignated as D, E, and F, respectively, in the experiment) showed low growth rate. A portion of the freshly harvested seeds were stored at 4 °C in small containers with parafilm (needle-pinched) covers, placed over $CaCl_2$. This storage condition simulated a condition of no cell molecular change thereby simulating seeds at "0" age condition. These were thawed before opening to use for experiments as "unaged"/"0 year aged" seeds. The rest of the seeds was left under ambient UV radiation (and temperature); portions of the seeds were withdrawn after 1 year and again after 2 years. These were stored at 4 °C over $CaCl_2$ till used for experiment – being marked as "1 year old" and "2 years old" seeds, respectively.

Being stored in the dry state, neither of these sets of seeds could undertake metabolic (synthetic/repair) activity. Any cell molecular change in these seeds with progression of time in storage would be due to UV radiation that would enter through the UV transparent husk. Variation in macromolecular damage through the storage period would be a reflection of genetically determined accumulation of protective compounds (viz., flavonoids, serving as internal UV screen and cellular antioxidants) synthesized as a diversion of primary metabolic pathway to the secondary metabolic pathway for acclimation processes (as an attribute of cellular plasticity) developed in maturing seed on mother plant in an open field. The experimental material used is shown in Figs. 6.3 and 6.4.

UV radiation-induced cell damage may be either by direct impinging (see Chap. 3) or through signal transduction (see Chap. 5). In dry embryonic cells (of mature orthodox seed) with precluded metabolic activity, cellular damage is also caused by UV-induced ROS (in absence of metabolic ROS) mediating damage of macromolecular entities such as DNA and antioxidative enzymes.

4.6.1 Lipid Peroxidation

Lipid peroxidation during post-harvest storage gives an indication of the status of membrane integrity under conditions of metabolic halt due to cell dehydration in mature post-harvest seeds aged under UV radiation in ambient storage. Oxidative damage in membrane lipids with progression of storage life (i.e., aging) brings about disruption of membrane integrity – measured by malondialdehyde (a product of lipid peroxidation reaction) concentration.

Correlative study on reduction of seedling growth rate indicating progressive loss of seed vigour–viability (Fig. 6.4) and lipid peroxidation with age in different varieties that adversely affect membrane integrity and thus cell functioning and growth has been observed in different varieties with time during dry storage under UV radiation. Extensive destruction of membrane integrity with increasing time of dry seed storage would result in embryonic cell death and resultant loss of seed viability; the rate of cellular degradation being, in turn, a function of varietal (genetic) difference that regulates acclimation strategy in plants.

The data (Fig. 4.21) shows age-associated enhancement in lipid peroxidation in dry-stored seeds – the data showing variation with difference in genotype. The high vigour varieties incur less oxidative damage due to higher content of effective antioxidants (Fig. 4.22) in the dry embryonic cells. The higher lipid peroxidation observed in low vigour varieties contain low amount of such protective compounds resulting in more lipid peroxidation. Lipid peroxidation in dry embryonic axis of rice, where enzymatic degradation is restrained due to lack of water of hydration, must be a manifestation of UV-induced oxidative damage. With progressive aging under UV environment, lipid peroxidation increases in all varieties – peroxidation through the aging period being higher in low vigour seeds – the phenomenon being a function of varietal difference. The data suggests that the high vigour varieties possess better inherent cell protection/acclimation mechanism provided by flavonoids through UV screening as well as antioxidative role of flavonoids. This matches the lipid peroxidation model of seed aging. Kramer et al. (1991) have demonstrated similar result of lipid peroxidation due to UV-B radiation-related oxidative stress in cucumber. Wilson and McDonald (1986a, b) have also suggested that the lipid peroxidation is the primary cause of seed deterioration under ambient storage. The deteriorative events accumulate with time (under dry storage where repair/turnover events are precluded) leading to a loss of vigour and eventually viability. Flood and Sinclair (1981) have reported decrease in unsaturated fatty acid during aging in 20 years aged clover seed; the

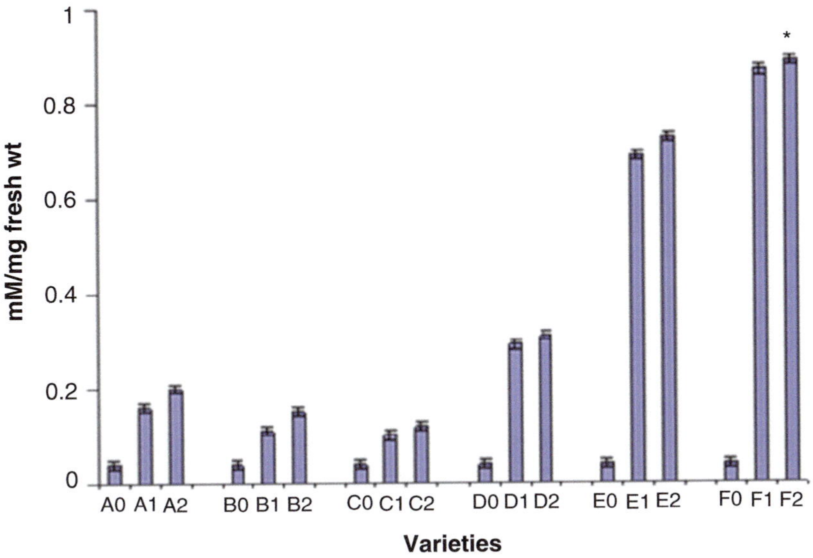

Fig. 4.21 Histogram of lipid peroxidation [(measured by mM (malondialdehyde concentration)/mg fresh wt.] Unaged seeds designated with 0- and 1-year aged seeds designated with 1- and 2-year aged seeds designated with 2. A = Patnai-23, B = Pankaj, C = Matla, D = Jaya, E = Jogen, F = Lalat (S.D calculated for $n = 6$, $P < 0.001$)

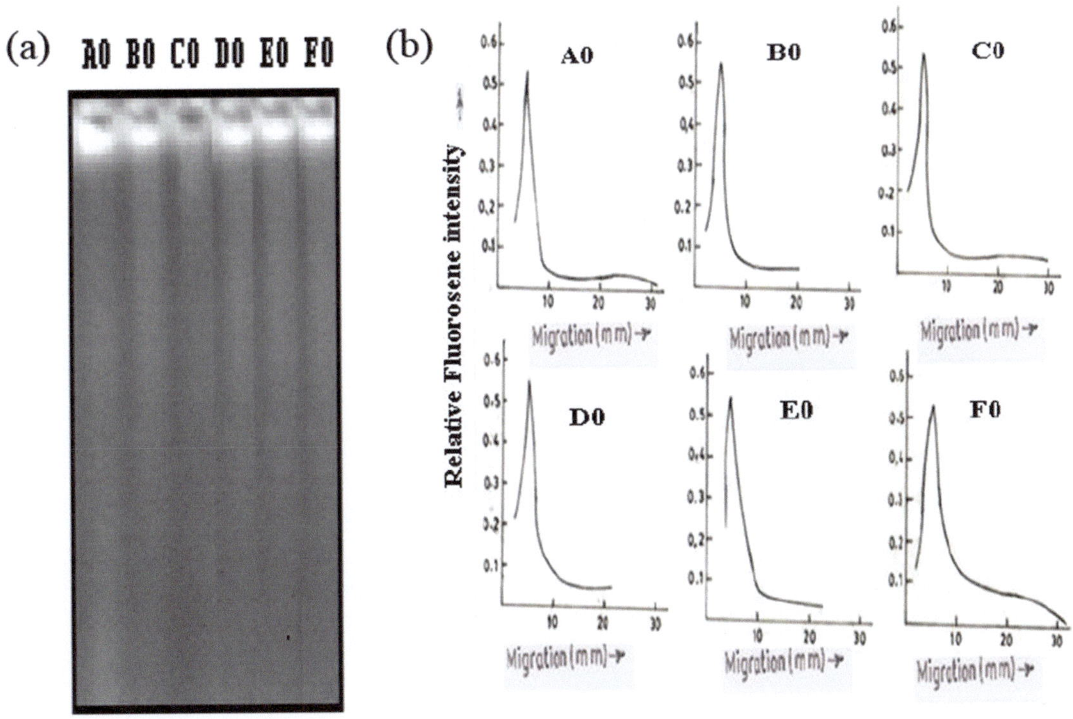

Fig. 4.22 (**a**) Alkaline agarose gel (1 %) of denatured DNA samples obtained from freshly harvested six rice varieties. A = Patnai-23, B = Pankaj, C = Matla, D = Jaya, E = Jogen, F = Lalat. (**b**) Image analysis scans of ethidium bromide fluorescence of DNA extracted from fresh stocks of rice varieties. A = Patnai-23, B = Pankaj, C = Matla, D = Jaya, E = Jogen, F = Lalat (0 is designated for freshly harvest unaged seeds)

authors conclude that free radical and auto-oxidation (oxidative stress) are involved in age-associated membrane damage. Chauhan et al. (2011) working on six different varieties of wheat have suggested that the membrane damage/lipid peroxidation in aged seed is a varietal phenomenon.

4.6.2 Assay of DNA Damage and Loss of Integrity

Another cellular damage that is central to loss of cellular activity and particularly causes stallation of transcription, cell division, and associated inhibition of cell growth, manifested as loss of seed/embryonic growth in seed vigour–viability studies, is DNA damage.

In the present study, a collection of rice seed varieties differing (on the basis of germination potential) in their genetic/varietal attribute of seed vigour–viability were used. Freshly harvested, i.e., unaged (0-year aged) (A0, B0, C0, D0, E0, and F0) and 2 years aged (A2, B2, C2, D2, E2, F2), seeds were used to study extent of DNA damage (Figs. 4.22 and 4.23).

Studies on structural integrity of DNA in agarose gel pictures (Figs. 4.22a and 4.23a) and image analysis scans of the density of ethidium bromide fluorescence of DNA extracted from fresh harvest stocks and 2-year aged seeds (Figs. 4.22b and 4.23b) reveal that in studies on

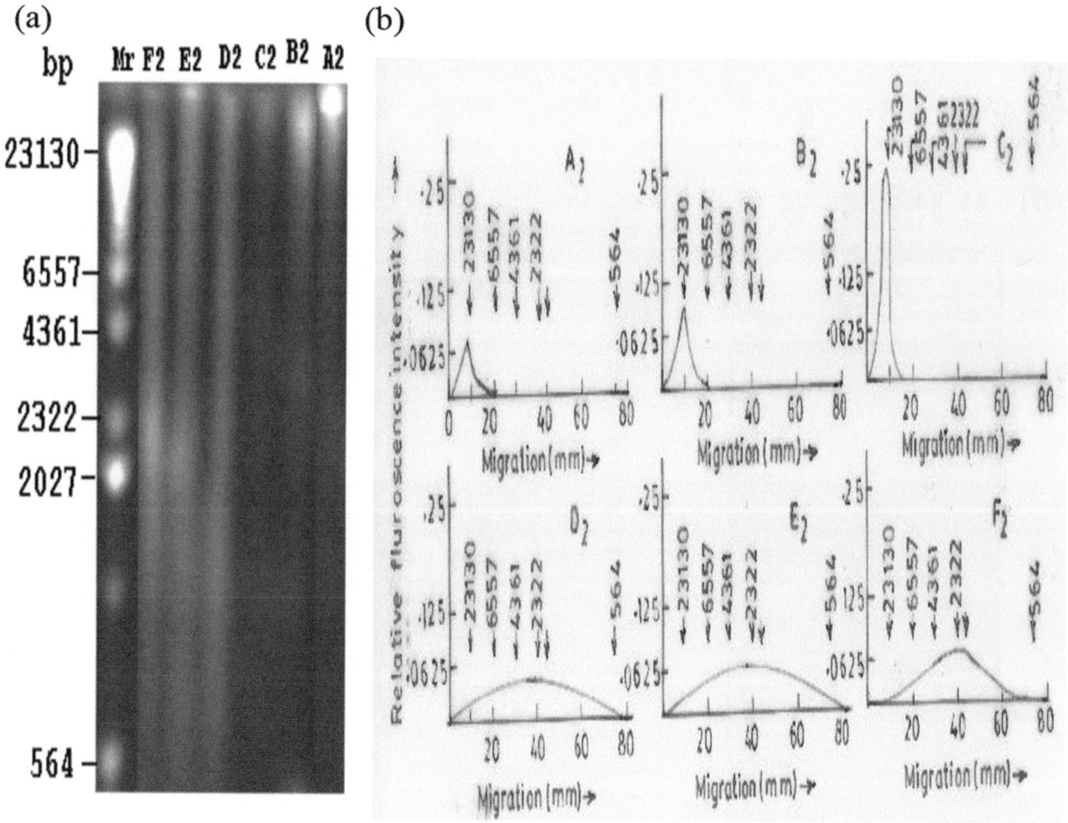

Fig. 4.23 (**a**) Alkaline agarose gel (1 %) of denatured 2-year aged DNA samples obtained from six rice varieties. A = Patnai-23, B = Pankaj, C = Matla, D = Jaya, E = Jogen, F = Lalat. (Two is designated for 2-year aged seed lots) (Mr = λ DNA/Hind III Digest). (**b**) Image analysis scans of the density of ethidium bromide fluorescence of DNA extracted from 2-year aged stocks of rice varieties. DNA fractionated by alkaline agarose gel (1 %) electrophoresis. A = Patnai-23, B = Pankaj, C = Matla, D = Jaya, E = Jogen, F = Lalat (Two is designated for 2 years aged seed lots)

unaged seeds of genetically determined high as well as low vigour varieties, DNA remains at the origin of the gel indicating absence of DNA breaks in DNA of freshly harvested seeds regardless of their genetically conferred vigour status (Fig. 4.22a, b). The same study in aged seeds shows that DNA remain almost near the origin in high vigour-viable varieties, indicating higher DNA integrity even in 2-year aged seeds, while in the low vigour-viable seeds, the DNA appears as a smear on the gel (Fig. 4.23a, b) due to incomplete separation of overlapping size DNA fragments generated due to single-stranded breaks. Image analysis scans of the density of ethidium bromide fluorescence of the same gel (Figs. 4.22b and 4.23b) corroborate the above findings, i.e., 2-year aged stocks of low vigour-viable varieties show higher degree of DNA deterioration than that of the high vigour-viable varieties. The similar data was demonstrated by Elder et al. (1987) and Prakash et al. (2007) working with different varieties of soybean (*Glycine max*). They have demonstrated that the DNA damage (generated from oxidative stress) was varietal. Such studies demonstrate that biomolecular deterioration is the cause for deteriorated seed vigour–viability trait.

A comparison between Figs. 4.22 and 4.23 shows that DNA damage during post-harvest aging is a manifestation of varietally controlled molecular mechanism for cellular protection/acclimation strategy for post-harvest seed survival under UV radiation through the years. These damages need to be repaired at seed imbibition before rapid seedling growth may be initiated. This phenomenon is comparable with delayed hypocotyl elongation allowing time for DNA repair at resumption of growth of seedling/hypocotyl (Biever et al. 2014). The low vigour varieties apparently correlated with low DNA integrity in aged seeds (adversely affected by UV during dry storage under conditions of precluded (repair) enzyme activity of low vigour seed. This is in agreement with the report of Hidema and Kumagai (2006) that CPD photolyase activity is a crucial factor determining the difference in UV-B sensitivity between rice cultivars.

4.6.3 Assessment of Total Antioxidant Potential

Cell-protective mechanism consists mainly of antioxidative potential in cells. Enzymatic as well as nonenzymatic antioxidants account for total antioxidant in the cell. This assay will evaluate both of these kinds of antioxidant content present in the cell.

This data (Fig. 4.24) shows that A, B, and C varieties contain high amount of antioxidant activity in comparison to D, E, and F varieties. With aging amount of antioxidant potential gets reduced proportionally (as a function of varietal difference). This reduction reflects the age-related reduction in enzymatic antioxidant as different from the nonchanging content of nonenzymatic antioxidant. A measure of total antioxidant activity (Fig. 4.24) reflecting activity of both enzymatic (i.e., superoxide dismutase, catalase, ascorbate peroxidase, glutathione reductase) and nonenzymatic antioxidants (including flavonoids, glutathione, ascorbate) reveals a clear difference in antioxidant activity between high vigour and low vigour varieties at freshly harvested, i.e., unaged state indicating differential genetic makeup. In aged seeds, a reduction in antioxidant activity was observed in all the varieties – the reduction being higher in the low vigour varieties.

4.6.4 Enzymatic Antioxidants in Post-harvest Orthodox (Dry) Seed

Enzymatic antioxidants as well as the tripeptide glutathione are subject to loss of structural integrity and thus activity (Fig. 4.25) under UV radiation during post-harvest storage in the open.

Studies on SOD activity in non-denaturing native 10.5 % polyacrylamide gels (representative gel picture and densitometric scan at 638.2 nm of specific isozyme) show (Fig. 4.25) that the extent of loss of enzyme activity in aged (1 year) seeds is lower in the high vigour varieties (A:33 ± 0.01 %, B:25 ± 0.02 %, C:15.2 ± 0.01 %) compared to that in low vigour varieties (D:41.66 ± 0.01 %,

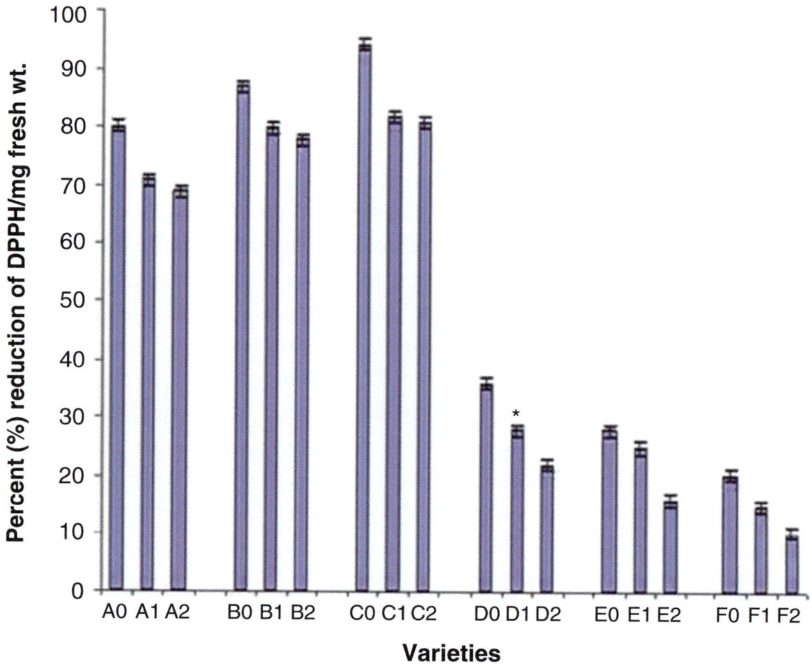

Fig. 4.24 Histogram for assay of total antioxidant activity [recorded as percent (%) reduction of DPPH/mg fresh wt.] in six rice varieties. Unaged seeds designated with 0- and 1-year aged seeds designated with 1- and 2-year aged seeds designated with 2. A = Patnai-23, B = Pankaj, C = Matla, D = Jaya, E = Jogen, F = Lalat (S.D calculated for $n = 6$, $P < 0.001$)

E:59.1 ± 0.01 %, F:85 ± 0.01 %). Standard deviation was calculated for six replicates, significance $P < 0.001$.

The data shows a progressive reduction in SOD activity (assessed at early germination) in rice embryonic axis during aging of seed under ambient UV in post-harvest seed storage. The reduction in activity was found to be a function of the varietal difference as modified by period of aging. Similar observation has also been reported by Fontaine et al. (1967) in barley.

Under the circumstance of functional reduction of enzymatic antioxidants, importance of non-enzymatic antioxidants in (dry) post maturation-stored seeds becomes particularly important.

4.6.5 Nonenzymatic Antioxidant, viz., Flavonoid

Under conditions where enzymatic antioxidants are functionally ineffective, for survival of embryonic axis during seed storage, recourse is taken to nonenzymatic antioxidants in embryonic axis of stored seed. Flavonoids represent the stable antioxidants in dehydrated embryonic axis of orthodox seed for protection during ambient seed storage under UV fluence. These groups of antioxidants are the largest and most studied classes of compounds that function as nonenzymatic antioxidants (in addition to serving as internal UV screens) in plant cells. This is particularly important for dehydrated plant systems where enzyme (including enzymatic antioxidants) activity is precluded due to structural degradation in absence of enzymatic repair/metabolism under low moisture content as in mature (orthodox) seeds, viz., rice. Flavonoids serving as UV screens and also as nonenzymatic antioxidants protect seeds from degeneration (i.e., loss of vigour–viability) during post-harvest storage of orthodox seeds.

Among all the antioxidants (both enzymatic and nonenzymatic both), only flavonoids, i.e., the nonenzymatic antioxidants, were found not to be reduced during the course of aging (Fig. 4.26). Being varietally controlled for synthesis, these antioxidants remain the same in content in each variety through seed storage – higher content is observed in high vigour-viable varieties and lower content in low vigour-viable varieties.

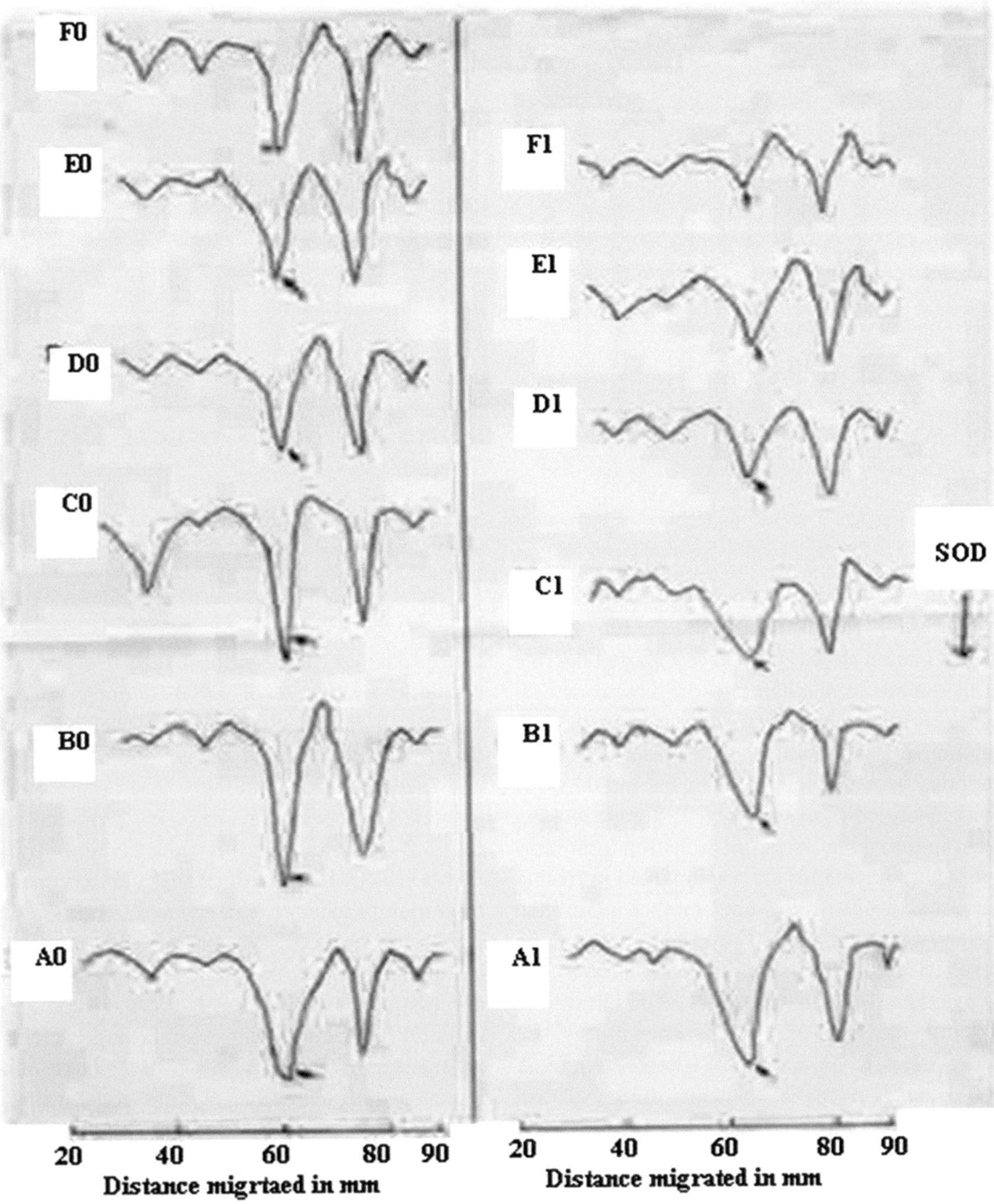

Fig. 4.25 Representative densitometric scan of superoxide dismutase isozyme reflecting enzyme damage during aging evidently due to UV-caused damage of peptide bonds, showing marked reduction in activity in different varieties/genotypes through 1 year periods in farmers' seed storage under ambient UV radiation; UV radiation on 10.5 % native polyacrylamide gel of six different rice varieties. A = Patnai-23, B = Pankaj, C = Matla, D = Jaya, E = Jogen, F = Lalat. *Left Panel*: unaged seed. *Right Panel*: 1-year aged seeds

Varieties A, B, and C exhibiting higher growth rate (Fig. 6.3) thus contain higher amount of flavonoids. It is also important to note that the amount of total flavonoid content was not decreased during the periods of 1 and 2 years of aging (Fig. 4.26). It is pertinent to mention here

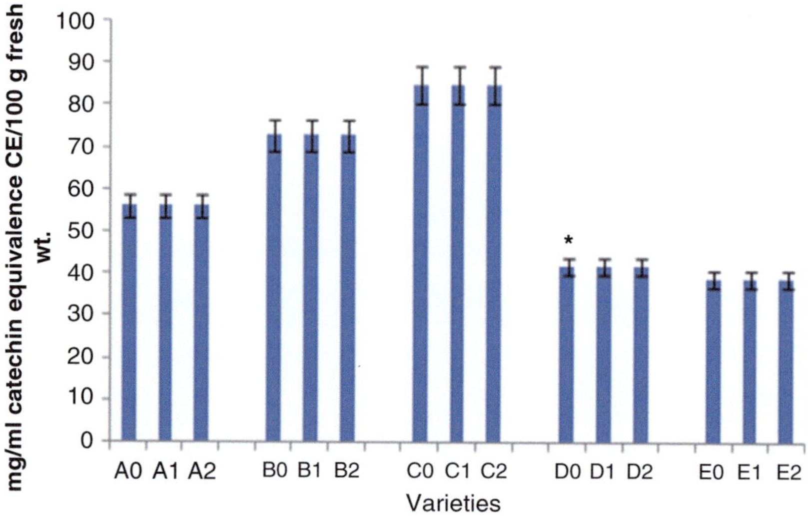

Fig. 4.26 Histogram for spectrophotometric analysis of total flavonoids studied against [(mg/ml catechin equivalence (CE)/100 g fresh wt.)] in six rice varieties. Seed stocks representing aged seeds designated with 0- and 1-year aged seeds designated with -1 and 2-year aged seeds designated with 2. A = Patnai-23, B = Pankaj, C = Matla, D = Jaya, E = Jogen (S.D calculated for $n = 6$, $P < 0.001$)

that Debaujon et al. (2000) have reported the role of flavonoids in maintaining seed vigour as well as longevity of *Arabidopsis* during storage. High amount of flavonoids protect these high vigour varieties from UV-induced damage in DNA and other macromolecules in ambient storage making them more viable during germination after storage. These flavonoids are accumulated in freshly harvested seeds from their mother plant and thereby accumulation of flavonoids depends on genetic/varietal difference. In this context it may be mentioned that enzymes involved in flavonoid biosynthetic pathway are known to be upregulated by UV radiation (Kliebenstein et al. 2002) that is available to plants during growth and seed production in the field.

The data presented here demonstrate the importance of nonenzymatic antioxidants (viz., flavonoids) in countering UV-induced oxidative damage – the protective function being a feature of dehydrated cells of mature embryonic axis in orthodox (dry) seed where enzymatic antioxidant activity remains precluded. Flavonoids, which in addition to serving as nonenzymatic antioxidants also functions as internal UV screen, restrict cellular (viz., DNA) damage (Fig. 4.22a, b). In this context, variation in extent of DNA damage among varieties indicate occurrence of varietal/genetic difference in potential of flavonoid synthesis and accumulation among varieties. Experimental evidence for this is provided in Fig. 4.26. Variation in flavonoid content in different varieties occurs in embryonic axis during seed maturation in mother plants (Mazza et al. 2000). Presence of such cell-protective compounds serves to retain viability in orthodox (dry) seed during post-harvest storage – the extent of protection being varietally/genetically controlled. Ballare et al. (2011) have also reported the occurrence of both constitutively highly tolerant and moderately tolerant populations of *Arabidopsis thaliana* from the same altitude – a phenomenon indicating varietal difference regarding acclimation potential to UV radiation.

This study on rice was conducted to evaluate acclimation processes that are (a) developed during full cellular functioning during seed maturation processes (subject to varietal difference) and (b) effective during dry seed storage when repair (enzyme) activity is precluded. During post-harvest storage "dry" rice seeds (with UV transparent husk) are subjected to ravages of UV radiation. The seeds would remain viable as long as macromolecular damages are not extensive and the seeds are still capable of germination as a resultant of UV-damaging effect being countered by acclimation effect rendered by nonenzymatic antioxidant (Fig. 4.24) and flavonoid content (Fig. 4.26) and DNA repair activities in the different varieties. It may be noted that through seed storage, the nonenzymatic antioxidants are particularly relevant for restricting oxidative cell damage in seeds through post-harvest dry seed storage. This is because enzymatic antioxidants are themselves subject to UV-induced damage without associated repair ability in dry axis of stored mature seed; this parameter exhibits appreciable damage in aged seeds (Fig. 4.25). As long as DNA integrity is not lost extensively (Fig. 4.23), embryonic axis of inhibited seed takes up DNA repair process during early imbibition. For giving time for this damage to be repaired, seeds at this status of cellular damage will initiate germination later (cf. Biever et al. 2014) thereby exhibiting low vigour performance. The extent of germination ability, i.e., vigour and viability status in stored seed, is thus proportional to the extent of damage incurred in cells during seed storage in open field. A comparison of Figs. 4.21 and 4.22 shows oxidative damage in two experiments, viz., lipid peroxidation and DNA damage undergo damage under UV radiation (through dry seed storage). The damage is greater in varieties D, E, and F exhibiting low vigour–viability trait (Fig. 6.3). For all the data presented, a varietal control is clearly evident. Depending on the extent of damage incurred during the storage period, the seed will take proportionate time for emerging from the seed cover (after having completed seed repair process). Protection of embryonic cells in stored seeds is rendered by flavonoid synthesized during seed development by diversion of the primary metabolic pathway to secondary metabolic pathway.

Case Study II

4.7 UV-Induced Cellular Damage and Acclimation in Hydrated Plant System (Viz. Tea leaf)

A commonly occurring UV acclimation strategy for countering UV-induced oxidative damage, viz., lipid peroxidation in hydrated plant system, i.e., plants growing in the open field, is upregulation of flavonoid biosynthesis that restrict UV-induced oxidative stress in plants – the extent of synthesis/accumulation of the compound being proportional to the environment (altitude-/latitude-related variation of UV fluence) X gene (reflected as genetic variation) between plants under study. An interesting point worth exploring in this context would be assessing flavonoid content and associated variation in flavonoid-related UV protection/quality features, viz., aroma conferred by flavonoid glycosides (Fig. 6.17, Fig. 4.8) in clones of plant species growing under natural variation of UV fluence (as in regions of different altitude). Toward this end experiments presented in this section were conducted on (tea) plants growing in open field at different altitudes within the same latitude that provided a situation of natural variation in UV fluence. The experimental material consisted of leaf of tea clones of different varieties growing in Darjeeling (1371.6 m a.s.l) and Assam (107 m a.s.l), in eastern India, both places at 27^0N. A comparative analysis of plants growing in such contrasting environments was undertaken (Shyam Chowdhury and Sen Mandi 2011) in an attempt to understand molecular rearrangement-associated cellular events for production of cell-protective compounds, viz., flavonoids, that are generally known to be associated with acclimation strategy in plants under high UV environment. In view of the fact that flavonoid derivatives (flavonoid glycosides, Fig. 6.16) confer quality traits in tea (Carrao-Panizi and Bordington 2000), the experiments were conducted on tea clones of three different

Table 4.1 Environmental factors in Darjeeling and Assam at the tea cultivation sites. S.D calculated for the table is for three replicates. $P < 0.01$

Location latitude and altitude	Darjeeling (27°03′N and 1371.6 m a.s.l)	Assam (26.75°N and 107 m a.s.l)
Rain fall (average)	1700 ± 0.5–2500 ± 0.32 mm	1867 ± 0.1 mm
Temperature (average)	22 ± 0.6 °C	25 ± 0.5 °C
Humidity	Very high with ~80 ± 0.1 % (mean over the year) (Dutta 2009)	76.33 ± 0.1 % mean humidity over the year
Sunshine (average)	2–4 h/day	2–4 h/day
Cloud cover (Octa unit)	During June–September period (rainy season) the sky is normally overcast – 7 ± 1 Octa, dry periods are observed from November–May – 4 ± 2 Octa	During monsoon (June–September) cloud cover-6 ± 2 Octa and during dry periods (November–May) 5 ± 1 Octa

varieties (viz., genotypes). The experimental fields used for the study at both the altitudes were tea industry-owned cultivation sites where soil and cultivation practices are those normally used by the tea industry. The plant parts used were apical leaves (two leaves and a bud) plucked from the selected tea clones during April–June coinciding with the Northern Hemisphere summer (associated with solstice) representing the months of highest UV radiation over this area (Fig. 1.5) and also the time when leaves of first flush and second flush emerge and are collected for commercial tea manufacture. The experiments were conducted for three successive years to assess the effect of UV radiation at different altitudes (along the same latitude, viz., 27^0N), viz., at Darjeeling and Assam. Both the locations are traditionally and commercially successful tea plantation sites with long history of cultivation. Environmental factors varying at these regions are given in Table 4.1.

These factors/atmospheric conditions may affect UV fluence thereby affecting oscillations between high and low dose of the radiation – flavonoid biosynthesis being upregulated by low dose to counter cell-damaging effects induced by high-dose UV radiation.

4.7.1 Lipid Peroxidation

UV-B radiation frequently triggers oxidative stress through the formation of reactive oxygen species (ROS), which in turn cause enhanced lipid and protein oxidation (Yanarelli et al. 2006a). Oxidative stress assessment vis-à-vis cell-protective arrangement to counter the damaging effect was undertaken in the experimental tea plants at different altitude, i.e., under different extent of UV fluence was done by assaying extent of lipid peroxidation in the same genotypes growing in experimental sites at different altitude.

The data shows lower values of lipid peroxidation recorded in leaf cells of high-altitude (viz., Darjeeling) tea clones compared to the values observed in low altitude (Fig. 4.27). A possible explanation/hypothesis could be higher cell protection efficiency in high-altitude growing plants. The data also exhibit varietal difference, reflecting gene X environment manifestation of the damaging effect varying between altitudes. It is evident thus that the cell damage observed, viz., lipid peroxidation, is a genetically controlled/varietal phenomenon.

Lipid peroxidation in mountain environmental (oxidative stress) condition was also reported by Wang et al. 2009 in leaves of *Polygonum viviparum*. Altitude-related (Chile, 605 m a.s.l; Bolivia, 219 m a.s.l.) varietal variation in lipid peroxidation was also recorded by Lizana et al. (2006) when working with different varieties of common bean. These reports however have not provided a comparative study on lipid peroxidation at different altitudes.

4.7.2 Assessment of Total Antioxidant Potential

Using leaves from the genetically identical tea varieties (clones) from the contrasting locations (high- and low-altitude), total antioxidant

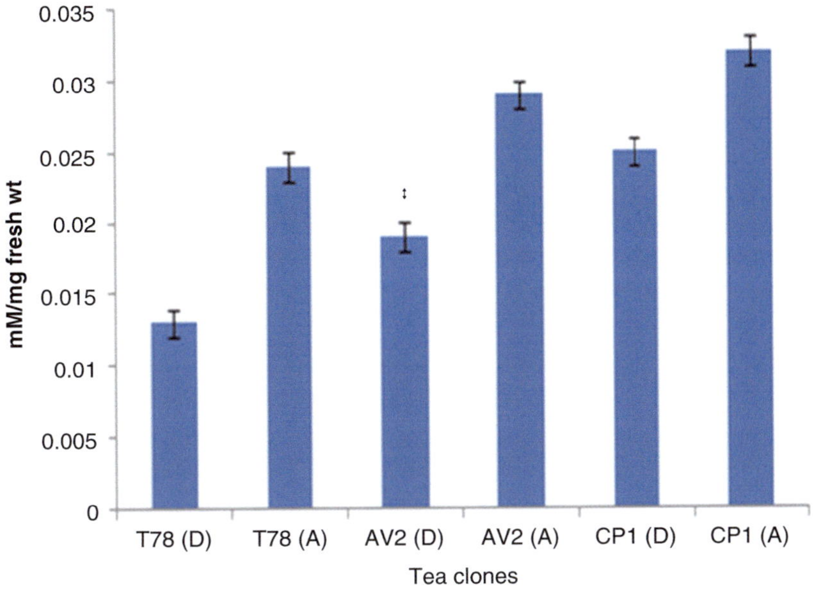

Fig. 4.27 Histogram of lipid peroxidation (given as mM malondialdehyde concentration/mg fresh wt.) in different tea clones, viz., T78, AV2, and CP1. *D* Darjeeling, *A* Assam (S.D calculated for $n = 6$, $P < 0.001$)

Table 4.2 Total antioxidant potential (representing enzymatic as well as nonenzymatic antioxidants) assay (S.D calculated for six replicates, $P < 0.001$) of tea clones showing variation in antioxidant content correlated with altitude

Tea clones	Antioxidant potential (% of DPPH reduction)
T78 (D)	98 ± 0.001
T78 (A)	24.24 ± 0.001
AV2 (D)	87 ± 0.001
AV2 (A)	22.1 ± 0.001
CP1 (D)	85.1 ± 0.001
CP1 (A)	19.8 ± 0.001

potential (both enzymatic and nonenzymatic) was studied to check the hypothesis (proposed above) that the low value of lipid peroxidation at high altitude is due to high peroxidation countering mechanism, viz., high antioxidant potential conferred by high flavonoid content in high-altitude plants. The data (Table 4.2) shows variation between genotypes, each genotype being proportionally affected by altitude-related UV radiation.

In the present study, plants at high altitude indeed showed higher antioxidant potential (Table 4.2) evidently due to variation in altitude-related UV fluence. The high antioxidant value in high-altitude tea clones presents a reflection of upregulated biosynthesis of phenolic compounds that serve to cope with enhanced oxidative stress of UV at high altitude. The high antioxidant potential being higher at high altitude also matches with the low lipid peroxidation value observed at high altitude shown in Fig. 4.27. Polle and Rennenberg (1992) have similarly reported proportionately high content of antioxidant in Norway spruce trees growing at different altitudes. Studies were undertaken in a forest on the Wank Mountain in the Bavarian Calcareous Alps at three altitudes (180 m, 1270 m, 1700 m). The data shows variation in antioxidant potential as a function of varietal difference in plants at high and low altitude. Similar to this study, Fujita et al. (2003) working on tartary buckwheat flour from high altitudes of Nepal, Bhutan, and China demonstrated an altitude-related (the variation of altitude was related with 1620–3880 m a.s.l.) varietal difference in antioxidant activity that again exhibits a gene X environment relationship in manifestation of traits.

Antioxidant potential in the Darjeeling-grown clones is higher than the Assam-grown clones

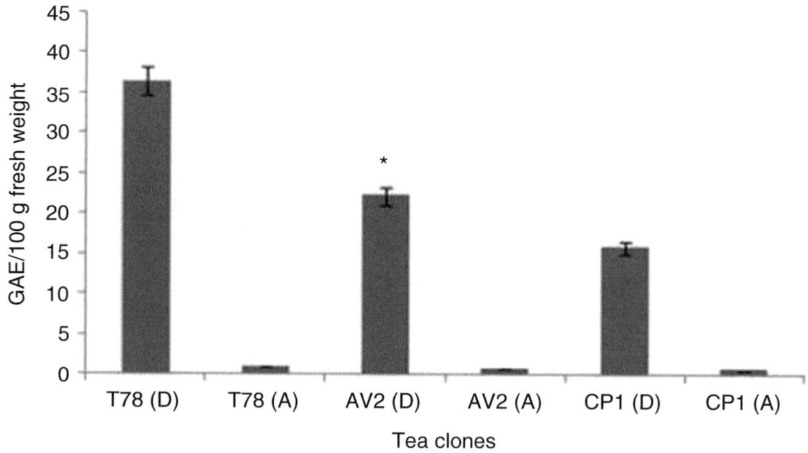

Fig. 4.28 Histogram of total phenolics (GAE/100 g fresh wt.) in different tea clones, viz., T78, AV2, and CP1. *D* Darjeeling, *A* Assam ($P < 0.001$, for S. D $n = 6$)

matching with the universally accepted statement the "Darjeeling tea has high antioxidant potential and is thus considered to be medicinally elite tea leaf." Total phenolics were assayed in clones of three tea varieties grown in Darjeeling and in Assam and presented in Fig. 4.28.

The data showing varietal difference provide information on content of phenolic compounds that constitute the most important group of UV-screening compounds for acclimation in the high- and low-altitude growing plants. The variation in data also shows variation in phenolic content between high- and low-altitude growing plants. Variation in antioxidant potential matching with variation in content of phenolics in the different clones at the different altitudes suggests altitude-related cell protection that is conferred by phenolic components at high altitude. In both sets of plants of high and low altitude, a genotypic difference was evident, thereby indicating a gene X (UV) environment-related manifestation of these compounds.

4.7.3 Assessment of Flavonoid Content

Flavonoids in plants are usually present as flavonoid glycosides (Zhang and Björn 2009).

Flavonoid glycoside contents in tea leaf from both high altitude (viz., Darjeeling) and low altitude (viz., Assam) are shown in Figs. 4.29, 4.30, and 4.31 with rutin (a flavonoid glycoside) used as standard (Fig. 4.32).

The data shows higher flavonoid content in high-altitude plants as a function of varietal difference. In each altitude the data reflects a gene X (UV) environment (altitude-related) relationship showing higher biosynthesis of flavonoid (UV-screening compounds) in tea clones at high altitude compared to that in the same clones growing at low altitude. This provides an example of acclimation strategy in plants under high UV environment. Better acclimation strategy is evident from high-altitude growing plants that exhibit high UV-screening potential in the high-altitude clones.

Higher content of phenolics as well as flavonoid (which acts as UV screen as well as nonenzymatic antioxidant) in high-altitude (viz., Darjeeling)-grown tea plants (Figs. 4.28 and 4.29) is found to be correlated with lower UV-induced damage (evidenced as higher lipid peroxidation in Fig. 4.27) in comparison to tea plants of low altitude (viz., Assam). The high phenolics/flavonoid content (Figs. 4.28 and 4.29) at high altitude matches with the low lipid peroxidation value observed at high altitude (Fig. 4.27).

Oxidative damage of macromolecules by UV at high altitude is reported by Wang et al. (2009) on their studies on DNA damage at high altitude. Lower lipid peroxidation value observed in our

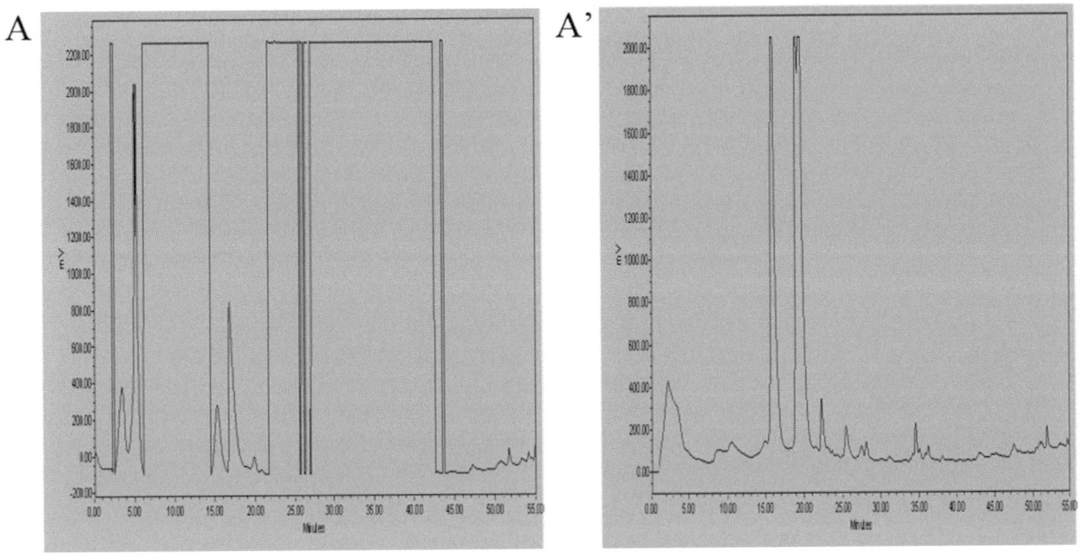

Fig. 4.29 Representative HPLC scan of flavonoid glycosides analysis recorded at 285 nm at 0–55 min retention time in hypersil ODS2 column. A = T78(D), A′ = T78(A). D = Darjeeling, A = Assam compared with rutin as standard

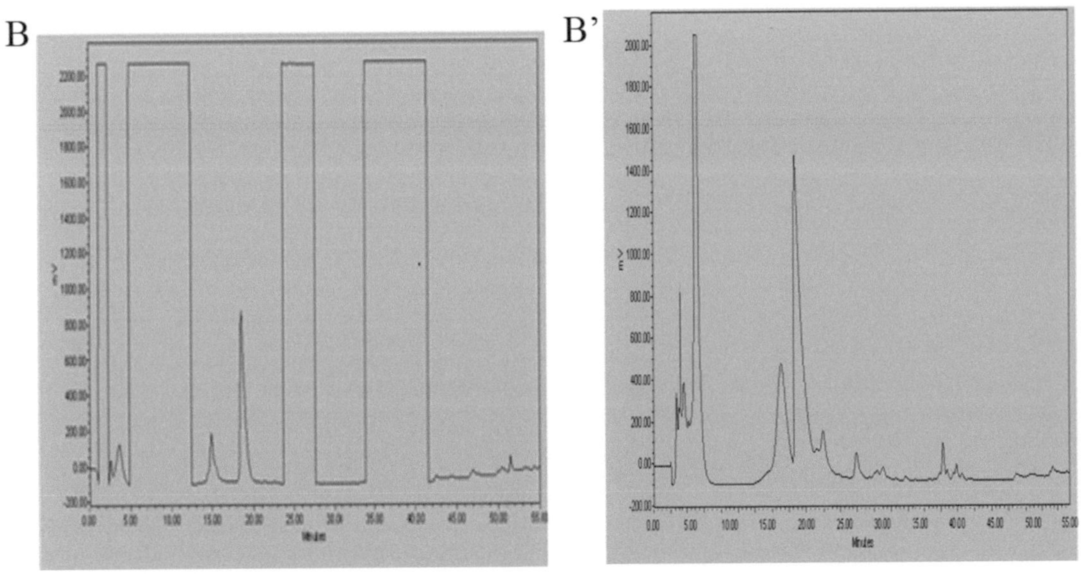

Fig. 4.30 Representative HPLC scan of flavonoid analysis recorded at 285 nm at 0–55 min retention time in hypersil ODS2 column. B = AV2(D), B′ = AV2(A). D = Darjeeling, A = Assam

C C’

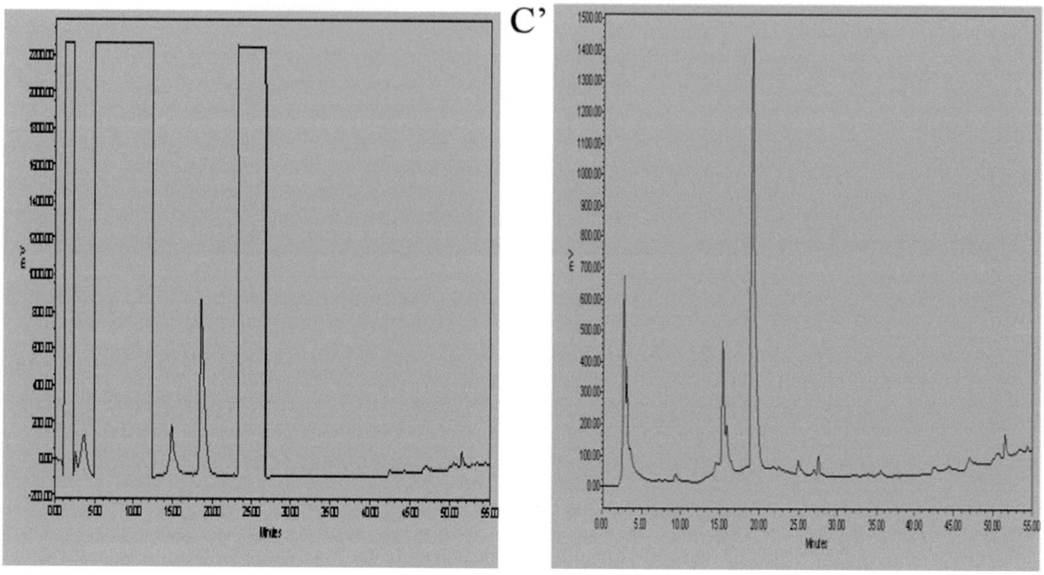

Fig. 4.31 Representative HPLC scan of flavonoid analysis recorded at 285 nm at 0–55 min retention time in hypersil ODS2 column. C = CP1 (D), C′ = CP1(A). D = Darjeeling and A = Assam

D

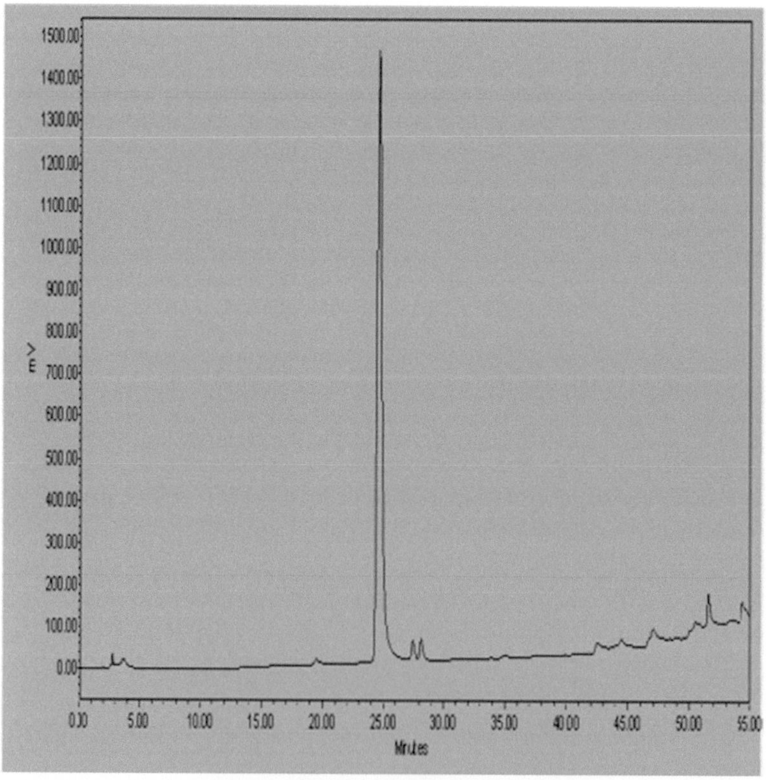

Fig. 4.32 Representative HPLC scan for flavonoid analysis recorded at 285 nm at 0–55 min retention time in hypersil ODS2 column. D = Standard rutin

study (Fig. 4.27) also establishes this fact. In spite of this enhanced damage at high altitude, UV is not considered as a stress response to plants due to the fact that plants at high UV fluence synthesize more UV-protective secondary metabolites (viz., flavonoids) which are a part of their acclimation strategies. Similar studies exhibiting UV-protective/acclimation effects by other types of protective compounds have been conducted by Filella and Peñuelas (1999) on *Quercus ilex* trees of the higher location. The authors have reported that as these plants have higher NDPI (normalized difference pigment reflectance index) values, i.e., higher carotenoids–chlorophyll-a ratio, these plants are protected by carotenoids from detrimental activity of UV-B. Other studies also established that enhanced UV-B radiation increases carotenoids–chlorophyll-a ratio (Nikolopoulos et al. 1995; Stephanou and Manetas 1997).

Bibliography

Adamse P, Britz SJ (1992) Amelioration of UV-B damage under high irradiance. I. Role of photosynthesis. Photochem Photobiol 56:645–650

Agati G, Tattini M (2010) Multiple functional roles of flavonoids in photoprotection. New Phytol 186: 786–793

Agati G, Galardi C, Gravano E, Romani A, Tattini M (2002) Flavonoid distribution in tissues of *Phillyrea latifolia* as estimated by microspectrofluorometry and multispectral fluorescence microimaging. Photochem Photobiol 76:350–360

Agati G, Stefano G, Biricolti S, Tattini M (2009) Mesophyll distribution of 'antioxidant' flavonoid glycosides in Ligustrum vulgare leaves under contrasting sunlight irradiance. Ann Bot 104:853–861

Agati G, Biricolti S, Guidi L, Ferrini F, Fini A, Tattini M (2011) The biosynthesis of flavonoids is enhanced similarly by UV radiation and root zone salinity in *L. vulgare* leaves. J Plant Physiol 168:204–212

Agati G, Azzarello E, Pollastri S, Tattini M (2012) Flavonoids as antioxidants in plants: location and functional significance. Plant Sci 196:67–76

Agati G, Brunetti C, Di Ferdinando M, Ferrini F, Pollastri S, Tattini M (2013) Functional roles of flavonoids in photoprotection: new evidence, lessons from the past. Plant Physiol Biochem 72:35–45

Agrawal SB, Rathore D (2007) Changes in oxidative stress defense in wheat (*Triticum aestivum* L.) and mung bean (*Vigna radiate* L.) cultivars grown with or without mineral nutrients and irradiated by supplemental ultraviolet-B. Environ Exp Bot 59:21–27

Ahmad M, Jarillo JA, Klimczak LJ, Landry LG, Peng T, Last RL, Cashmore AR (1997) An enzyme similar to animal type II photolyases mediates photoreactivation in Arabidopsis. Plant Cell 9:199–207

Ajavon AN, Albritton DL, Watson RT (2007) Scientific assessment of ozone depletion: 2006. Global ozone research and monitoring project, Report No. 50. World Meteorological Organization (WMO), Geneva, p 572

Albert A, Sareedenchai V, Heller W, Seidlitz HK, Zidorn C (2009) Temperature is the key to altitudinal variation of phenolics in Arnica montana L. cv. ARBO. Oecologia 160:1–8

Ålenius CM, Vogelmann TC, Bornman JF (1995) A three dimensional representation of the relationship between penetration of UV-B radiation and UV-screening pigments in leaves of *Brassica napus*. New Phytol 131:297–302

Allen DJ, Nogue's S, Morison JIL, Greenslade PD, McLeod AR, Baker NR (1999) A 30% increase in UV-B has no impact on photosynthesis in well-watered and droughted pea plants in the field. Glob Chang Biol 5(2):235–244

Alscher RG, Donahue JL, Cramer CL (1997) Reactive oxygen species and antioxidants: relationships in green cells. Physiol Plant 100(2):224–233

Alvero-Bascos EM, Ungson LB (2012) Ultraviolet-B (UV-B) radiation as an elicitor of flavonoid production in callus cultures of Jatropha (*Jatropha curcas* L.). Philipp Agric Sci 95(4):335–343

Apel K, Hirt H (2004) Reactive oxygen species: metabolism, oxidative stress, and signal transduction. Annu Rev Plant Biol 55:373–399

Babu S, Akhtar TA, Lampi MA, Tripuranthakam S, Dixon GR, Greenberg BM (2003) Similar stress responses are elicited by copper and ultraviolet radiation in the aquatic plant Lemma gibba: implication of reactive oxygen species as common signals. Plant Cell Physiol 44:1320–1329

Baker NP, Nogues S, Allen DJ (1997) Photosynthesis and photoinhibition. In: Lumsden P (ed) Plants and UV-B: responses to environmental change. Cambridge University Press, Cambridge, pp 95–111

Ballaré CL, Caldwell MM, Flint SD, Robinson SA, Bornman JF (2011) Effects of solar ultraviolet radiation on terrestrial ecosystems. Patterns, mechanisms, and interactions with climate change. Photochem Photobiol Sci 10:226–241

Barnes PW, Flint SD, Caldwell MM (1990) Morphological responses of crop and weed species of different growth forms to ultraviolet-B radiation. Am J Bot 77:1354–1360

Barnes PW, Ballare' CL, Caldwell MM (1996) Photomorphogenic effects of UV-B radiation on plants: consequences for light competition. Plant Physiol 148:15–20

Baroniya SS, Kataria S, Pandey GP, Guruprasad KN (2013) Intraspecific variations in antioxidant defense responses and sensitivity of soybean varieties to ambient UV radiation. Acta Physiol Plant 35(5):1521–1530

Bar-Peled M, Lewinsohn E, Fluhr R, Gressel J (1991) UDP-rhamnose: flavanone-7-O-glucoside-2″-O-rhamnosyltransferase: purification and characterization of an enzyme catalyzing the production of bitter compounds in citrus. J Biol Chem 266:20953–20959

Bassman JH (2004) Ecosystem consequences of enhanced solar ultraviolet radiation: secondary plant metabolites as mediators of multiple trophic interactions in terrestrial plant communities. Photochem Photobiol 79: 382–398

Batschauer A (1999) Light perception in higher plants. Cell Mol Life Sci 55:153–165

Baudry A, Heim MA, Dubreucq B, Caboche M, Weisshaar B, Lepiniec L (2004) TT2, TT8, and TTG1 synergistically specify the expression of BANYULS and proanthocyanidin biosynthesis in *Arabidopsis thaliana*. Plant J 39:366–380

Beggs CJ, Wellmann E (1994) Photocontrol of flavonoid biosynthesis. In: Kendrick RE, Kronenberg GHM (eds) Photomorphogenesis in plants. Kluwer Academic Publishers, Dordrecht, pp 733–751

Beggs CJ, Wellmann E, Griesbach H (1986) Photocontrol of flavonoid biosynthesis. In: Kendrick RE, Kronenberg GMH (eds) Photomorphogenesis in plants. Martinus, Nijhoff, Dordrecht, pp 467–491

Behn H, Albert A, Marx F, Noga G, Ulbrich A (2010) Ultraviolet-B and photosynthetically active radiation interactively affect yield and pattern of monoterpenes in leaves of peppermint (Mentha x piperita L). J Agric Food Chem 58:7361–7367

Berkelsar EJ, Ormrod DP, Hale BA (1996) The influence of photosynthetically active radiation on the effects of ultraviolet-B radiation on arabidopsis thaliana. Photochem Photobiol 64(1):110–116

Berli FJ, Moreno D, Piccoli P, Hespanhol-Viana L, Silva MF, Bressan-Smith R et al (2010) Abscisic acid is involved in the response of grape (*Vitis vinifera* L.) cv. Malbec leaf tissues to ultraviolet-B radiation by enhancing ultraviolet-absorbing compounds, antioxidant enzymes and membrane sterols. Plant Cell Environ 33:1–10

Beveridge CA, Mathesius U, Rose RJ, Gresshoff PM (2007) Common regulatory themes in meristem development and whole-plant homeostasis. Curr Opin Plant Biol 10:44–51

Biever JJ, Brinkman D, Gardener G (2014) UV-B inhibition of hypocotyl growth in etiolated *Arabidopsis thaliana* seedlings is a consequence of cell cycle arrest initiated by photodimer accumulation. J Exp Bot 65(11):2949–2961

Bieza K, Lois R (2001) An Arabidopsis mutant tolerant to lethal ultraviolet-B levels shows constitutively elevated accumulation of flavonoids and other phenolics. Plant Physiol 126:1105–1115

Biggs RH, Kossuth SV, Teramura AH (1981) Response of 19 cultivars of soybeans to ultraviolet-B irradiance. Physiol Plant 53:19–26

Bilger W, Johnsen T, Schreiber U (2001) UV-excited chlorophyll fluorescence as a tool for the assessment of UV-protection by the epidermis of plants. J Exp Bot 52(363):2007–2014

Björn LO, Callaghan TV, Johnsen I et al (1997) The effects of UV-B radiation on European heathland species. Plant Ecol 128:252–264

Blumthaler M, Amback W (1990) Indication of increasing solar UV-B radiation flux in alpine regions. Science 248:206–208

Boccalandro HE, Mazza CA, Mazzella MA, Casal JJ, Ballare' CL (2001) Ultraviolet B radiation enhances a phytochrome-B-mediated photomorphogenic response in *Arabidopsis*. Plant Physiol 126(2):780–788

Bornman JF (1989) Target sites of UV-B radiation in photosynthesis of higher plants. Photochem Photobiol 4:145–158

Bornman JF, Vogelmann TC (1991) Effect of UV-B radiation on leaf optical properties measured with fibre optics. J Exp Bot 42:547–554

Bornman JF, Barnes PW, Robinson SA, Ballaré CL, Flint SD, Caldwell MM (2015) Solar ultraviolet radiation and ozone depletion-driven climate change: effects on terrestrial ecosystems. Photochem Photobiol Sci 14: 88–107

Britt AB (1999) Molecular genetics of DNA repair in higher plants. Trends Plant Sci 4:20–24

Britt AB (2004) Repair of DNA damage induced by solar UV. Photosynth Res 81:105–112

Britt AB, May GD (2003) Re-engineering plant gene targeting. Trends Plant Sci 8:90–95

Brosche M, Strid A (2003) Molecular events following perception of ultraviolet-B radiation by plants. Physiol Plant 117:1–10

Brown BA, Jenkins GI (2008) UV-B signaling pathways with different fluence-rate response profiles are distinguished in mature *Arabidopsis* leaf tissue by requirement for UVR8, HY5, and HYH. Plant Physiol 146: 576–588

Brown DE, Rashotte AM, Murphy AS, Normanly J, Tague BW, Peer WA, Taiz L, Muday GK (2001) Flavonoids act as negative regulators of auxin transport in vivo in *Arabidopsis*. Plant Physiol 126: 524–535

Brown BA, Cloix C, Jiang GH, Kaiserli E, Herzyk P et al (2005) A UV-B-specific signaling component orchestrates plant UV protection. Proc Natl Acad Sci U S A 102:18225–18230

Buer CS, Djordjevic MA (2009) Architectural phenotypes in the transparent testa mutants of *Arabidopsis thaliana*. J Exp Bot 60:751–763

Burchard P, Bilger W, Weissenböck G (2000) Contribution of hydroxycinnamates and flavonoids to epidermal shielding of UV-A and UV-B radiation in developing rye primary leaves as assessed by ultraviolet-induced chlorophyll fluorescence measurements. Plant Cell Environ 23:1373–1380

Burger J, Edwards GE (1996) Photosynthetic efficiency, and photodamage by UV and visible radiation, in red versus green leaf coleus varieties. Plant Cell Physiol 37:395–399

Caldwell MM, Robberecht R, Flint SD (1983a) Internal filters: prospects of UV-acclimation in higher plants. Physiol Plant 58:445–450

Caldwell MM, Gold WG, Harris G, Ashurst CW (1983b) A modulated lamp system for solar UV-B (280–320 nm) supplementation studies in the field. Photochem Photobiol 37:479–485

Caldwell MM, Teramura AH, Tevini M (1989) The changing solar UV climate an the ecological consequences for higher plants. Trends Ecol Evol 4: 363–367

Caldwell MM, Flint SD, Searles PS (1994) Spectral balance and UV-B sensitivity of soybean: a field experiment. Plant Cell Environ 17:267–276

Caldwell MM, Teramura AH, Tevini M, Bornman JF, Bjorn LO, Kulandaivellu G (1995) Effects of increased solar ultraviolet radiation on terrestrial plants. Ambio 24:166–173

Caldwell MM, Bornman JF, Ballare' CL, Flint SD, Kulandaivelu G (2007) Terrestrial ecosystems, increased solar ultraviolet radiation, and interactions with other climate change factors. Photochem Photobiol Sci 6:252–266

Cantarello C, Volpe V, Azzolin C, Bertea C (2005) Modulation of enzyme activities and expression of genes related to primary and secondary metabolism in response to UV-B stress in cucumber (Cucumis sativus L.). J Plant Interact 1(3):151–161

Cao G, Sofic E, Prior RL (1997) Antioxidant and prooxidant behavior of flavonoids: structure-activity relationships. Free Radic Biol Med 22:749–760

Carletti P, Masi A, Wonisch A, Grill D, Tausz M, Ferretti M (2003) Changes in antioxidant and pigment pool dimensions in UV-B irradiated maize seedlings. Environ Exp Bot 50:149–157

Carrao-Panizzi MC, Bordingnon JR (2000) Activity of beta-glucosidase and levels of isoflavone glucosides in soybean cultivars affected by the environment. Pesq Agrop Bras 35:873–878

Casati P, Walbot V (2003) Gene expression profiling in response to UV radiation in maize genotypes with varying flavonoid content. Plant Physiol 132: 1739–1754

Casati P, Walbot V (2004) Crosslinking of ribosomal proteins to RNA in maize ribosomes by UV-B and its effects on translation. Plant Physiol 136: 3319–3332

Casati P, Morrow DJ, Fernandes JF, Walbot V (2011) UV-B signaling in maize: transcriptomic and metabolomic studies at different radiation times. Plant Signal Behav 6:1926–1931

Cen Y-P, Bornman JF (1993) The effect of exposure to enhanced UV-B radiation on the penetration of monochromatic and polychromatic UV-B radiation in leaves of *Brassica napus*. Physiol Plant 87: 249–255

Cerovic ZG, Ounis A, Cartelat A, Latouche G, Goulas Y, Meyer S, Moya I (2002) The use of chlorophyll fluorescence excitation spectra for the non-destructive in situ assessment of UV-absorbing compounds in leaves. Plant Cell Environ 25:1663–1676

Chappell J, Hahlbrock K (1984) Transcription of plant defence genes in response to UV light or fungal elicitor. Nature 311:76–78

Chauhan DS, Deswal DP, Dahiya OS, Punia RC (2011) Change in storage enzymes activities in natural and accelerated aged seed of wheat (*Triticum aestivum*). Ind J Agric Sci 81(11):1037–1040

Chen YP (2009) Response of antioxidant defense system to laser radiation apical meristem of *Isatis indigotica* seedlings exposed to UV-B. Plant Signal Behav 4: 571–573

Chen JJ, Mitchell DL, Britt AB (1994) A light-dependent pathway for the elimination of UV-induced pyrimidine (6–4) pyrimidinone photoproducts in *Arabidopsis*. Plant Cell 6(9):1311–1317

Cloix C, Jenkins GI (2008) Interaction of the *Arabidopsis* UV-B-specific signaling component UVR8 with chromatin. Mol Plant 1:118–128

Cloix C, Kaiserli E, Heilmann M, Baxter KJ, Brown BA, O'Hara A, Smith BO, Christie JM, Jenkins GI (2012) C-terminal region of the UV-B photoreceptor UVR8 initiates signaling through interaction with the COP1 protein. Proc Natl Acad Sci U S A 109: 16366–16370

Close DC, McArthur C (2002) Rethinking the role of many plant phenolics – protection from photodamage not herbivores? Oikos 99:166–172

Cockell CS, Knowland J (1999) Ultraviolet radiation screening compounds. Biol Rev 74:311–345

Cominelli E, Gussmaroli G, Allegra D, Galbiati M, Wade HK, Jenkins GI et al (2008) Expression analysis of anthocyanin regulation genes in response to different light qualities in *Arabidopsis thaliana*. J Plant Physiol 165:886–894

Conklin PL, Saracco SA, Norris SR, Last RL (1996) Identification of ascorbic acid-deficient Arabidopsis thaliana mutants. Genetics 154:847–856

Corlett JE, Stephen J, Jones HG, Woodfin R, Mepsted R, Paul ND (1997) Assessing the impact of UV-B radiation on the growth and yield of field crops. In: Lumsden PJ (ed) Plants and UV-B: responses to environmental change Society for Experimental Biology, vol 64, Seminar Series. Cambridge University Press, Cambridge

Day TA (1993) Relating UV-B radiation screening effectiveness of foliage to absorbing compound concentration and anatomical characteristics in a diverse group of plants. Oecologia 95:542–550

Day TA, Vogelmann TC, De Lucia EH (1992) Are some plant life forms more effective than others in screening out ultraviolet -B radiation? Oecologia 92:513–519

Debaujon I, Leon-Kllosterziel MK, Koornneef M (2000) Influence of the testa on seed dormancy, germination, and longevity in Arabidopsis. Plant Physiol 122(2): 403–414

Debeaujon I, Peeters AJM, Léon-Kloosterziel KM, Koornneef M (2001) The transparent testa12 gene of Arabidopsis encodes a multidrug secondary transporter–like protein required for flavonoid sequestration in vacuoles of the seed coat endothelium. Plant Cell 13:853–872

Deckmyn G, Impens I (1997) The ratio UV-B photosynthetically active radiation (PAR) determines the sensitivity of rye to increased UV-B radiation. Environ Exp Bot 37:3–12

Dixon P, Weinig C, Schmitt J (2001) Susceptibility to UV damage in Impatiens capensis (Balsaminaceae): testing for opportunity costs to shade-avoidance and population differentiation. Am J Bot 88:1401–1408

Du H, Liang Y, Pei K, Ma K (2011) UV radiation-responsive proteins in rice leaves: a proteomic analysis. Plant Cell Physiol 52(2):306–316

Dubest S, Gallego ME, White CI (2002) Role of the AtRad1p endonuclease in homologous recombination in plants. EMBO Rep 3:1049–1054

Dutta TK (2009) Advantage, constrains and key success factors in establishing origin and tradition linked quality signs: the case of Darjeeling Tea, India: case study on quality products linked to geographical origin in Asia carried out for FAO. Food and Agricultural Organization, FAO

Ebel J, Hahlbrock K (1982) Biosynthesis. In: Harbome J, Mabry TJ (eds) The flavonoids, advances in research. Chapman and Hall, London, pp 641–679

Ehmann B, Ocker B, Schidfer E (1991) Development and light-dependent regulation of the expression of two different Chalcone synthase transcripts in mustard cotyledons. Planta 183:416–422

Elder RH, Dell'Aquila A, Mezzina M, Sarasin A, Osborne DJ (1987) DNA ligase in repair and replication in the embryos of rye, Secale cereale. Mutat Res 181:61–71

Essen LO, Klar T (2006) Light driven DNA repair by photolyase. Cell Mol Life Sci 63:1266–1277

Falcone Ferreyra ML, Rius S, Emiliani J, Pourcel L, Feller A, Morohashi K et al (2010) Cloning and characterization of a UV-B-inducible maize flavonol synthase. Plant J 62:77–91

Fedina I, Hidema J, Velitchkova M, Georgieva K, Nedeva D (2010) UV-B induced stress responses in three rice cultivars. Biol Plant 54(3):571–574

Filella I, Peñuelas J (1999) Altitudinal differences in UV absorbance, UV reflectance and related morphological traits of Quercus ilex and Rhododendron ferrugineum in the Mediterranean region. Plant Ecol 145:157–165

Flood RG, Sinclair A (1981) Fatty acid analysis of aged permeable and impermeable seeds of *Trifolium subterraneum* (subterraneum clover). Seed Sci Technol 9:475–477

Fontaine O, Billard JP, Huault C (1967) Effect of glutathione on dormancy breakage in barley seeds. Plant Growth Regul 67(1):55–58

Frederick JE, Snell HE (1990) Tropospheric influence on solar ultraviolet radiation: the role of clouds. J Clim 3:373–381

Frohnmeyer H, Staiger D (2003) Ultraviolet-B radiation-mediated responses in plants. Balancing damage and protection. Plant Physiol 133:1420–1428

Frohnmeyer H, Loyall L, Blatt MR, Grabov A (1999) Millisecond UV-B radiation evokes prolonged elevation of cytosolic-free Ca^{2+} and stimulates gene expression in transgenic parsley cell cultures. Plant J 20: 109–117

Fuglevand G, Jackson JA, Jenkins GI (1996) UV-B, UV-A and blue light signal transduction pathways interact synergistically to regulate chalcone synthase gene expression in *Arabidopsis*. Plant Cell 8: 2347–2357

Fujita K, Inoue N, Yang Z, Hagiwara S, Hagiwara M (2003) Varietal differences of antioxidant activity in Tartary buckwheat flour as evaluated by chemiluminescence. Fagopyrum 20:47–52

Ganguli S, Das G, Sen-Mandi S (1992) Plant emergence and productivity of different vigour of wheat (*Triticum aestivum*) seeds. Ind J Agr Sci 62(3):224–227

Gao Q, Zhang L (2008) Ultraviolet-B-induced oxidative stress and antioxidant defense system responses in ascorbate-deficient vtc1 mutants of Arabidopsis thaliana. J Plant Physiol 165:138–148

Gardner G, Lin C, Tobin EM, Loehrer H, Brinkman D (2009) Photobiological properties of the inhibition of etiolated Arabidopsis seedling growth by ultraviolet-B radiation. Plant Cell Environ 32:1573–1583

Ghosh S, Sen Mandi S (2015) SNP in Chalcone synthase gene is associated with variation of 6-gingerol content in contrasting landraces of *Zingiber officinale*. Roscoe Gene 566(2):184–188

Gil M, Bottini R, Piccoli P, Pontin M (2010) Crosstalk between MAV and MEP pathways in vitro grape plants exposed to UV-B radiation. Biocell 35(1):A40

Giordanoi CV, Mori T, Sala OE, Scopel AL, Caldwell MM, Ballaré CL (2003) Functional acclimation to solar UV-B radiation in *Gunnera magellanica*, a native plant species of southernmost Patagonia. Plant Cell Environ 26:2027–2036

Grammatikopoulos G, Petropoulou Y, Manetas Y (1999) Site dependent differences in transmittance and UV-B-absorbing capacity of isolated leaf epidermis and mesophyll in Urginea maritima (L.) Baker. J Exp Bot 50:517–521

Grotewold E (2004) The challenges of moving chemicals within and out of cells: insights into the transport of plant natural products. Planta 219:906–909

Hahlbrock K, Scheel D (1989) Physiology and molecular biology of phenylpropanoid metabolism. Annu Rev Plant Mol Biol 40:347–369

Harborne JB, Williams CA (2000) Advances in flavonoid research since 1992. Phytochemistry 55:481–504

Hartmann U, Valentine WJ, Christie JM, Hays J, Jenkins GI, Weisshaar B (1998) Identification of UV/blue light-response elements in the *Arabidopsis thaliana* chalcone synthase promoter using a homologous protoplast transient expression system. Plant Mol Biol 36: 741–754

Hectors K, Prinsen E, De Coen W, Jansen MA, Guisez Y (2007) *Arabidopsis thaliana* plants acclimated to low dose rates of ultraviolet B radiation show specific changes in morphology and gene expression in the absence of stress symptoms. New Phytol 175(2): 255–270

Heijde M, Ulm R (2012) UV-B photoreceptor-mediated signaling in plants. Trends Plant Sci 17:230–237

Heijde M, Ulm R (2013) Reversion of the *Arabidopsis* UV-B photoreceptor UVR8 to the homodimeric ground state. PNAS 110(3):1113–1118

Herna'ndez I, Alegre L, van Breusegem F, Munne'-Bosch S (2009) How relevant are flavonoids as antioxidants in plants? Trends Plant Sci 14:125–132

Herrmann KM, Weaver LM (1999) The shikimate pathway. Annu Rev Plant Physiol Plant Mol Biol 50:473–503

Hideg E, Jansen MA, Strid A (2013) UV-B exposure, ROS, and stress: inseparable companions or loosely linked associates? Trends Plant Sci 18:107–115

Hidema J, Kumagai T (1998) UVB-induced cyclobutyl pyrimidine dimer and photorepair with progress of growth and leaf age in rice. J Photochem Photobiol B Biol 43:121–127

Hidema J, Kumagai T (2006) Sensitivity of rice to ultraviolet-B radiation. Ann Bot 97(6):933–942

Hidema J, Kumagai T, Sutherland JC, Sutherland BM (1997) Ultraviolet B–sensitive rice cultivar deficient in cyclobutyl pyrimidine dimer repair. Plant Physiol 113:39–44

Hidema J, Taguchi T, Ono T, Teranishi M, Yamamoto K, Kumagai T (2007) UV-B resistance and CPD photolyase activity in rice. Plant J 50:70–79

Hoffman PD, Batschauer A, Hays JB (1996) PHH1, a novel gene from *Arabidopsis thaliana* that encodes a protein similar to plant blue-light photoreceptors and microbial photolyases. Mol Gen Genet 253: 259–265

Hoffmann RW, Campbell BD, Bloor SJ, Swinny EE, Markham KR, Ryan KG, Fountain DF (2003) Responses to UV-B radiation in *Trifolium repens* L. – physiological links to plant productivity and water availability. Plant Cell Environ 26:603–612

Horowitz RM (1986) Taste effects of flavonoids. In: Cody V, Middleton E, Harbome JB (eds) Plant flavonoids in biology and medicine. Alan R Liss, New York, pp 163–175

Hutchinson F (1987) A review of some topics concerning mutagenesis by ultraviolet light. Photochem Photobiol 45:897–903

Hutzler P, Fischbach R, Heller W, Jungblut TP, Reuber S, Schmitz R, Veit M, Weissenbo¨ ck G, Schnitzler JP (1998) Tissue localization of phenolic compounds in plants by confocal laser scanning microscopy. J Exp Bot 49:953–965

Ibdah M, Krins A, Seidlitz HK, Heller W, Strack D, Vogt T (2002) Spectral dependence of flavonol and betacyanin accumulation in Mesembryanthemum crystallinum under enhanced ultraviolet radiation. Plant Cell Environ 25:1145–1154

Inoue K (2011) Emerging roles of the chloroplast outer envelope membrane. Trends Plant Sci 16:550–557

Jacobs M, Rubery PH (1988) Naturally occurring auxin transport regulators. Science 241:346–349

Janknegt JP, de Graaff CM, van de Poll WH, Visser RJW, Helbling EW, Buma AGJ (2009) Antioxidative responses of two marine microalgae during acclimation to static and fluctuating natural UV radiation. Photochem Photobiol 85:1336–1345

Jansen MAK (2002) Ultraviolet-B radiation effects on plants: induction of morphogenic responses. Physiol Plant 116:423–439

Jansen MAK, Gaba V, Greenberg BM (1998) Higher plants and UV-B radiation: balancing damage, repair and acclimation. Trends Plant Sci 3:131–135

Jansen MAK, Gaba V, Greenberg BM, Mattoo AK, Edelman M (1996) Low threshold levels of ultraviolet-B, in a background of photosynthetically active radiation trigger rapid degradation of the D2 protein of photosystem II. Plant J 9:693–699

Jansen MAK, Hectors K, O'Brien NM, Guisez Y, Potters G (2008) Plant stress and human health: do human consumers benefit from UV-B acclimated crops? Plant Sci 175:449–458

Jansen MAK, Hideg E, Lidon FJC (2012) UV-B radiation: "When does the stressor cause stress?". Emir J Food Agric 24(6):21–33

Jende-Strid B (1993) Genetic control of flavonoid biosynthesis in barley. Hereditas 119:187–204

Jenkins GI (2009) Signal transduction in responses to UV-B radiation. Annu Rev Plant Biol 60:407–431

Jetter R, Schäffer S (2001) Chemical composition of the *Prunus laurocerasus* leaf surface: dynamic changes of the epicuticular wax film during leaf development. Plant Physiol 126:1725–1737

Jia X, Ren L, Chen QJ, Li R, Tang G (2009) UV-B-responsive microRNAs in *Populus tremula.* J Plant Physiol 166:2046–2057

Johnson CB, Kirby J, Naxakis G, Pearson S (1999) Substantial UV-B-mediated induction of essential oils in sweet basil (*Ocimum basilicum* L). Phytochemistry 51(4):507–510

Jordan BR (1993) The molecular biology of plants exposed to ultraviolet B radiation and the interaction with other stresses. In: Jackson MB, Black CR (eds) Interacting stresses on plants in a changing climate, vol 16, NATO ASI Series. Springer, Berlin, pp 153–170

Jordan BR (1996) The effects of ultraviolet-B radiation on plants: a molecular perspective. Adv Bot Res 22:97–162

Jordan BR (2002) Molecular response of plant cells to UV-B stress. Funct Plant Biol 29:909–916

Kakani VG, Reddy KR, Zhao D, Mohammed AR (2003) Effects of ultraviolet-B radiation on cotton (*Gossypium hirsutum* L.) morphology and anatomy. Ann Bot 91:817–826

Kaku S, Iwaya-Inoue M, Toki K (1992) Anthocyanin influence on water proton NMR relaxation times and water contents in leaves of evergreen woody plants during the winter. Plant Cell Physiol 33:131–137

Kalbin G, Hidema J, Brosché M, Kumagai T, Bornman JF, Strid Å (2001) UV-B-induced DNA damage and expression of defence genes under UV-B stress: tissue-specific molecular marker analysis in leaves. Plant Cell Environ 24:983–999

Kang H-S, Hidema J, Kumagai T (1998) Effects of light environment during culture on UV-induced cyclobutyl pyrimidine dimers and their photorepair in rice (*Oryza sativa* L.). Photochem Photobiol 68(1):71–77

Karousou R, Grammatikopoulos G, Lanaras T, Manetas Y, Kokkini S (1998) Effect of enhanced UV-B radiation on Mentha spicata essential oil. Phytochemistry 49:2273–2277

Khoroshilova EV, Repeyev YA, Nikogosyan DN (1990) UV photolysis of aromatic amino acids and related dipeptides and tripeptides. J Photochem Photobiol 7: 159–172

Kilian J, Whitehead D, Horak J, Wanke D, Weinl S, Batistic O, D'Angelo C, Bornberg-Bauer E, Kudla J, Harter K (2007) The AtGenExpress global stress expression data set: protocols, evaluation and model data analysis of UV-B light, drought and cold stress responses. Plant J 50:347–363

Kitamura S (2006) Transport of flavonoids: from cytosolic synthesis to vacuolar accumulation. In: Grotewold E (ed) The science of flavonoids. Springer Science, LLC, New York, pp 123–146

Klein RM (1978) Plants and near-ultraviolet radiation. Bot Rev 44:1–127

Kliebenstein DJ, Lim JE, Landry LG, Robert L (2002) Arabidopsis UVR8 regulates ultraviolet-B signal transduction and tolerance and contains sequence similarity to human regulator of chromatin condensation I. Plant Physiol 130(1):234–243

Koes R, Verweij W, Quattrocchio F (2005) Flavonoids: a colorful model for the regulation and evolution of biochemical pathways. Trends Plant Sci 10(5): 236–242

Kolb CA, Laser MA, Kopecky J, Zotz G, Riederer M, Pfundel EE (2001) Effects of natural intensities of visible and UV radiation on epidermal UV-screening and photosynthesis in grape leaves (vitis vinifera cv. Silvaner). Plant Physiol 127:863–875

Kolb CA, Kpecký J, Riederer M, Pfündel EE (2003) UV screening by phenolics in berries of grapevine (Vitis vinifera). Funct Plant Biol 30:1177–1186

Korkina LG (2007) Phenylpropanoids as naturally occurring antioxidants: from plant defence to human health. Mol Cel Biol 53:15–25

Kotilainen T, Tegelberg R, Julkunen-Tiitto R, Lindfors A, Aphalo PJ (2008) Metabolic specific effects of solar UV-A and UV-B on alder and birch leaf phenolics. Glob Chang Biol 14:1–11

Kotilainen T, Venäläinen T, Tegelberg R, Lindfors A, Julkunen-Tiitto R, Sutinen S, O'Hara RB, Aphalo PJ (2009) Assessment of biological spectral weighting functions for phenolic metabolites and growth responses in silver birch seedlings. Photochem Photobiol 85:1346–1355

Kramer GF, Norman HL, Krizek DT, Mirecki RM (1991) Influence of UV-B radiation on polyamines, lipid peroxidation and membrane lipids in cucumber. Phytochemistry 30:2101–2108

Kramer GF, Krizek DT, Mirecki RM (1992) Influence of photosynthetically active radiation and spectral quality on UV-B induced polyamine accumulation in soybean. Phytochemistry 31:1119–1125

Krause GH, Galle' A, Gademann R, Winter K (2003) Capacity of protection against ultraviolet radiation in sun and shade leaves of tropical forest plants. Funct Plant Biol 30:533–542

Kravets EA, Zelena LB, Zabara EP, Blume YB (2012) Adaptation strategy of barley plants to UV-B radiation. Emir J Food Agric 24:632–645

Krizek DT, Bridtz SJ, Mirecki RM (1998) Inhibitory effects of ambient levels of solar UV-A and UV-B radiation on growth of cv. New Red Fire Lettuce Physiol Plant 103:1–7

Krol M, Gray GR, Hurry VM, Öquist G, Malek L, Huner NPA (1995) Low-temperature stress and photoperiod affect an increased 46 tolerance to photo inhibition in *Pinus banksiana* seedlings. Can J Bot 73:1119–1127

Kubasek WL, Shirley BW, Mckillop A, Goodman HM, Briggs W, Ausubel FM (1992) Regulation of flavonoid biosynthetic genes in germinating *Arabidopsis* seedlings. Plant Cell 4(10):1229–1236

Kunz BA, Anderson HJ, Osmond MJ, Vonarx E (2005) Components of nucleotide excision repair and DNA damage tolerance in *Arabidopsis thaliana*. Environ Mol Mutagen 45:115–127

Lam RYY, Woo AYH, Leung PS, Cheng CHK (2007) Antioxidant actions of phenolic compounds found in dietary plants on low-density lipoprotein and erythrocytes in vitro. J Am Coll Nutr 26(3):233–242

Landry LG, Chapple CCS, Last RL (1995) *Arabidopsis* mutants lacking phenolic sunscreens exhibit enhanced ultraviolet-B injury and oxidative damage. Plant Physiol 109(4):1159–1166

Landry LG, Stapleton AE, Lim J, Hoffman P, Hays JB, Walbot V, Last RL (1997) An *Arabidopsis* photolyase mutant is hypersensitive to ultraviolet-B radiation, of special interest. Proc Natl Acad Sci U S A 94:328–332

Larson RA (1988) The antioxidants of higher plants. Phytochemistry 27:969–978

Lazar G, Goodman HM (2006) MAX1, a regulator of flavonoid pathway, controls vegetative bud outgrowth in *Arabidopsis*. Proc Natl Acad Sci U S A 103: 472–476

Lee DW, Lowry JB, Stone BC (1978) Abaxial anthocyanin layer in leaves of tropical rainforest plants: enhancer of light capture in deep shade. Biotropica 11:70–77

Lee DW, Brammeier S, Smith AP (1987) The selective advantages of anthocyanins in developing leaves of mango and cacao. Biotropica 19:40–49

Lercari B, Sodi F, Lipucci di Paola M (1990) Photomorphogenic responses to UV radiation: involvement of phytochrome and UV photoreceptors in the control of hypocotyl elongation in *Lycopersicon esculentum*. Physiol Plant 79:668–672

Lewinsohn E, Britsch L, Mazur Y, Gressel J (1989) Flavanone glycoside biosynthesis in Citrus: chalcone synthase, UDP-glucose:flavanone-7-O-glucosyl-transferase and rhamnosyltransferase activities in cell-free extracts. Plant Physiol 91:1323–1328

Li J, Ou-Lee T, Raba R, Amundson RG, Last RL (1993) *Arabidopsis* flavonoid mutants are hypersensitive to UV-B radiation. Plant Cell 5:171–179

Liu L, Gitz DC, McClure JW (1995) Effects of UV-B on flavonoids, ferulic acid, growth and photosynthesis in barley primary leaves. Physiol Plant 93:725–733

Liu Z, Hossain GS, Islas-Osuna MA, Mitchell DL, Mount DW (2000) Repair of UV damage in plants by nucleotide excision repair: Arabidopsis UVH1 DNA repair gene is a homolog of Saccharomyces cerevisiae Rad1. Plant J 21:519–552

Lizana C, Wentworth M, Martinez JP et al (2006) Differential adaptation of two varieties of common bean to abiotic stress. I. Effects of drought on yield and photosynthesis. J Exp Bot 57:685–697

Lyddon J, Teramura AH, Coffman CB (1987) UV-B radiation effects on photosynthesis, growth and cannabinoid production of two *Cannabis sativa* chemotypes. Photochem Photobiol 46:201–206

Mackerness SAH (2000) Plant response to ultraviolet-B (UV-B: 280–320) stress: what are the key regulators? Plant Growth Regul 32:27–39

Mackerness AHS, Surplus SL, Blake P, John CF, Buchana-Wollaston V, Jordan BR, Thomas B (1999) Ultraviolet-B-induced stress and changes in gene expression in *Arabidopsis thaliana*: role of signaling pathways controlled by jasmonic acid, ethylene and reactive oxygen species. Plant Cell Environ 22:1413–1423

Madronich S, de Gruijl FR (1994) Stratospheric ozone depletion between 1979 and 1992: implications for biologically active ultraviolet-B radiation and non-melanoma skin cancer incidence. Photochem Photobiol 59:541–546

Madronich S, McKenzie RL, Caldwell MM, Bjorn LO (1995) Changes in ultraviolet radiation reaching the Earth's surface. Ambio 24:143–152

Maffei M, Scannerini S (2000) UV-B effect on photomorphogenesis and essential oil composition in peppermint (Mentha piperita L). J Essent Oil Res 12(5):523–529

Markham KR, Tanner GJ, Caasi-Lit M, Whitecross MI, Nayudu M, Mitchell KA (1998) Possible protective role for 3′,4′-dihydroxyflavones induced by enhanced UV-B in a UV-tolerant rice cultivar. Phytochemistry 49:1913–1919

Marksta¨dter C, Queck I, Baumeister J, Riederer M, Schreiber U, Bilger W (2001) Epidermal transmittance of leaves of Vicia faba for UV radiation as determined by two different methods. Photosynth Res 67:17–25

Martinez-Luscher J, Morales F, Delrot S, Sanchez-Diaz M, Gomes E, Aguirreolea J et al (2013) Short- and long-term physiological responses of grapevine leaves to UV-B radiation. Plant Sci 213:114–122

Mazza CA, Boccalandro HE, Giordano CV, Battista D, Scopel AL, Ballare' CL (2000) Functional significance and induction by solar radiation of ultraviolet-absorbing sunscreens in field-grown soybean crops. Plant Physiol 122:117–125

McIntosh CA, Latchinian L, Mansell RL (1990) Flavanone-specific 7-O-glucosyltransferase activity in Citrus paradisi seedlings: purification and characterization. Arch Biochem Biophys 282:50–57

Mewis I, Schreiner M, Nguyen CN, Krumbein A, Ulrichs C, Lohse M, Zrenner R (2012) UV-B radiation changes specifically the secondary metabolite profile in broccoli sprouts: induced signaling overlaps with defense response to biotic stressors. Plant Cell Physiol 53(9):1546–1560

Middleton E (1984) The flavonoids. Trends Pharmacol Sci 5:335–338

Middleton Jr (1998) Effect of plant flavonoids on immune and inflammatory cell function. Adv Exp Med Biol 439:175–182

Middleton EM, Teramura AH (1993) The role of flavonol glycosides and carotenoids in protecting soybean from ultraviolet-B damage. Plant Physiol 103: 741–752

Mirecki RM, Teramura AH (1984) Effects of ultraviolet-B irradiance on soybean. The dependence of plant sensitivity on the photosynthetic photon flux density during and after leaf expansion. Plant Physiol 74: 475–480

Moellering ER, Muthan B, Benning C (2010) Freezing tolerance in plants requires lipid remodeling at the outer chloroplast membrane. Science 330:226–228

Moldau H (1999) Ozone detoxification in the mesophyll cell wall during a stimulated oxidative burst. Free Radic Res 31:S19–S44

Molinier J, Lechner E, Dumbliauskas E, Genschik P (2008) Regulation and role of Arabidopsis CUL4-DDB1A-DDB2 in maintaining genome integrity upon UV stress. PLoS Genet 4:23–29

Morales LO, Tegelberg R, Brosché M, Keinänen M, Lindfors A, Aphalo PJ (2010) Effects of solar UV-A and UV-B radiation on gene expression and phenolic accumulation in Betula pendula leaves. Tree Physiol 30:923–934

Morales LO, Tegelberg R, Brosché M, Lindfors A, Siipola S, Aphalo PJ (2011) Temporal variation in epidermal flavonoids due to altered solar UV radiation is moderated by the leaf position in Betula pendula. Physiol Plant 143:261–270

Mpoloka SW (2008) Effects of prolonged UV-B exposure in plants. Afr J Biotechnol 7(25):4874–4883

Mpoloka SW (2010) Residual effects of prolonged UV-B exposure on photosynthetic gene expression. Sci Res Essay 5(16):2119–2127

Müller-Xing R, Xing Q, Goodrich J (2014) Footprints of the sun: memory of UV and light stress in plants. Front Plant Sci 5:474

Mullineaux PM, Karpinski S (2002) Signal transduction in response to excess light: getting out of the chloroplast. Curr Opin Plant Biol 5:43–48

Nakajima S, Sugiyama M, Iwai S, Hitomi K, Otoshi E, Kim S-T, Jiang C-Z, Todo T, Britt AB, Yamamoto K (1998) Cloning and characterization of a gene (UVR3) required for photorepair of 6–4 photoproducts in *Arabidopsis thaliana*. Nucl Acids Res 26(2):638–644

Nawkar GM, Maibam P, Park JH, Sahi VP, Lee SY, Kang CH (2013) UV-induced cell death in plants. Int J Mol Sci 14(1):1608–1628

Nikolopoulos D, Petropoulou Y, Kyparissis A, Manetas Y (1995) Effects of enhanced UV-B radiation on the drought semideciduous Mediterranean shrub *Phlomis fruticosa* under field conditions are season-specific. Aust J Plant Physiol 22:737–745

Olsson LC, Veit M, Weissenböck G, Bornman JF (1998) Differential flavonoid response to enhanced UV-B-radiation in *Brassica napus*. Phytochemistry 49: 1021–1028

Olsson LC, Veit M, Bornman JF (1999) Epidermal transmittance and phenolic composition of atrazine-tolerant and atrazine-sensitive cultivars of *Brassica napus* grown under enhanced UV-B radiation. Physiol Plant 107:259–266

Panagopoulos I, Bornman F, Björn LO (1990) Effect of ultraviolet radiation and visible light on growth, fluorescence induction, ultra weak luminescence and peroxidase activity in sugar beet plants. J Photochem Photobiol B Biol 8:73–87

Pattison DI, Davies MJ (2006) Action of ultraviolet light on cellular structures. In: Bignold LP (ed) Cancer: cell structures, carcinogens and genomic instability. Birkhäuser, Virlag, Basel

Peer WA, Murphy AS (2007) Flavonoids and auxin transport: modulators or regulators? Trends Plant Sci 12: 556–563

Peer WA, Bandyopadhyay A, Blakeslee JJ, Makam SN, Chen R, Mason P, Murphy A (2004) Variation in expression and protein localization of the PIN family of auxin efflux facilitator proteins in flavonoid mutants with altered auxin transport in *Arabidopsis thaliana*. Plant Cell 16:898–911

Perkins MC, Roberts CJ, Briggs D, Davies MC, Friedmann A, Hart CA, Bell GA (2005) Surface morphology and chemistry of *Prunus laurocerasus* L. leaves: a study using X-ray photoelectron spectroscopy, time-of-flight secondary-ion mass spectrometry, atomic-force microscopy and scanning-electron icroscopy. Planta 221:123–134

Piao X-M, Chung J-W, Lee G-A, Jung-Ro Lee J-R et al (2014) Variation in antioxidant activity and flavonoid aglycones in eggplant (*Solanum melongena* L.) germplasm. Plant Breed Biotechnol 2:396–403

Piazza P, Procissi A, Jenkins IG, Tonelli C (2002) Members of the C1/pl1 regulatory gene family mediate the response of maize aleurone and mesocotyl to different light qualities and cytokinins. Plant Physiol 128:1077–1086

Pollastri S, Tattini M (2011) Flavonols: old compound for old roles. Ann Bot 108:1225–1233

Polle A, Rennenberg H (1992) Field studies on Norway spruce trees at high altitudes: II. Defence system against oxidative stress in needles. New Phytol 121: 635–642

Polster J, Dithmar H, Burgemeister R, Friedemann G, Feucht W (2006) Flavonoids in plant nuclei: detection by laser microdissection and pressure catapulting (LMPC), in vivo staining, and UV-visible spectroscopic titration. Physiol Plant 126:163–174

Pontin MA, Piccoli PN, Francisco R, Bottini R, Martinez-Zapater JM, Lijavetzky D (2010) Transcriptome changes in grapevine (*Vitis vinifera* L.) cv. Malbec leaves induced by ultraviolet-B radiation. BMC Plant Biol 10:224

Prakash D, Upadhyay G, Singh BN, Singh HB (2007) Antioxidant and free radical-scavenging activities of seeds and agri-wastes of some varieties of soybean (Glycine max). Food Chem 102(4):783–790

Quail PH (2002) Phytochrome photosensory signalling networks. Nat Rev Mol Cell Biol 3:85–93

Rastogi RP, Richa KA, Tyagi MB, Sinha RP (2010) Molecular mechanisms of ultraviolet radiation-induced DNA damage and repair. J Nucleic Acids 56:1–32

Rausher MD (2006) The evolution of flavonoids and their genes. In: Grotewold E (ed) The science of flavonoids. Springer, New York, pp 175–211

Rautenberger R, Wiencke C, Bischof K (2013) Acclimation to UV radiation and antioxidative defence in the endemic Antarctic brown macroalga Desmarestia anceps along a depth gradient. Polar Biol 36:1779–1789

Renger G, Vo¨lker M, Eckert HJ, Fromme R, Hohm-Veit S, Gra¨ber P (1989) On the mechanism of photosystem II deterioration by UV-B radiation. Photochem Photobiol 49:97–105

Reuber S, Bornman JF, Weissenbo¨ck G (1996) A flavonoid mutant of barley (*Hordeum vulgare* L.) exhibits increased sensitivity to UV-B radiation in the primary leaf. Plant Cell Environ 19:593–601

Rice-Evans CA, Miller NJ, Paganga G (1996) Structure-antioxidant activity relationships of flavonoids and phenolic acids. Free Radic Biol Med 20(7):933–956

Rice-Evans CA, Miller NJ, Papanga G (1997) Antioxidant properties of phenolic compounds. Trends Plant Sci 2:152–159

Ries G, Buchholz G, Frohnmeyer H, Hohn B (2000a) UV-damage-mediated induction of homologous recombination in *Arabidopsis* is dependent on photosynthetically active radiation. Proc Natl Acad Sci U S A 97: 13425–13429

Ries G, Heller W, Puchta H, Sandermann H, Seidlitz HK, Hohn B (2000b) Elevated UV-B radiation reduces genome stability in plants. Nature 406:98–101

Rizzini L, Favory J-J, Cloix C, Faggionato D, O'Hara A, Kaiserli E, Baumeister R, Schäfer E, Nagy F, Jenkins GI, Ulm R (2011) Perception of UV-B by the Arabidopsis UVR8 protein. Science 332:103–106

Robberecht R, Caldwell MM (1982b) Leaf epidermal transmittance of ultraviolet radiation and its implication for plant sensitivity to ultraviolet-radiation induced injury. Oecologia 32:277–287

Robberecht R, Caldwell MM, Billings WD (1980) Leaf ultraviolet optical properties along a latitudinal gradient in the arctic-alpine life zone. Ecology 61:612–619

Roleda MY, Wiencke C, Hanelt D (2006a) Thallus morphology and optical characteristics affect growth and DNA damage by UV radiation in juvenile Arctic *Laminaria* sporophytes. Planta 223:407–417

Roleda MY, Hanelt D, Wiencke C (2006b) Growth and DNA damage in young Laminaria sporophytes exposed to ultraviolet radiation implication for depth zonation of kelps on Helgoland (North Sea). Mar Biol 148:1201–1211

Rousseaux MC, Ballare CL, Scopel AL, Searles PS, Caldwell MM (1998) Solar ultraviolet-B radiation affects plant–insect interactions in a natural ecosystem of Tierra del Fuego. Oecologia 116:528–535

Rozema J, Chardonnens A, Tosserams M, Hafkenscheid R, Bruijnzeel S (1997) Leaf thickness, and UV-B absorbing pigments of plants in relation to an elevational gradient along the Blue Mountains. Jamaica Plant Ecol 128:150–159

Rozema J, Bjorn OL, Bornman JF et al (2002) The role of UV-B radiation in aquatic and terrestrial ecosystems – an experimental and functional analysis of the evolution of UV-B absorbing compounds. J Photochem Photobiol B Biol 66:2–12

Ruhland CT, Xiong FS, Clark WD, Day TA (2005) The influence of ultraviolet-B radiation on hydroxycinnamic acids, flavonoids and growth of *Deschampsia antarctica* during the springtime ozone depletion season in Antarctica. Photochem Photobiol 81:1086–1093

Ryan KG, Markham KR, Bloor SJ, Bradley JM, Mitchell KA, Jordan BR (1998) UV-B radiation induces increase in quercetin: kaempferol ratio in wild-type and transgenic lines of *Petunia*. Photochem Photobiol 68:323–330

Ryan KG, Swinny EE, Winefield C, Markham KR (2001) Flavonoids and UV photoprotection in *Arabidopsis* mutants. Z Naturforsch C 56:745–754

Ryan KG, Swinny EE, Markham KR, Winefield C (2002) Flavonoid gene expression and UV photoprotection in transgenic and mutant Petunia leaves. Phytochemistry 59:23–32

Sáfrány J, Haasz V, Máté Z, Ciolfi A, Fehér B, Oravecz A, Stec A, Dallmann G, Morelli G, Ulm R, Nagy F (2008) Identification of a novel cis-regulatory element for UV-B induced transcription in *Arabidopsis*. Plant J 54:402–414

Sancar A (1994) Structure and function of DNA photolyase. Biochemistry 33(1):2–9

Saslowsky DE, Warek U, Winjel BSJ (2005) Nuclear localization of flavonoid enzymes in *Arabidopsis*. J Biol Chem 280:23735–23740

Schmelzer E, Jahnen W, Hahlbrock K (1988) In situ localisation of light induced chalcone synthase mRNA, chalcone synthase, and flavonoid end products in epidermal cells of parsley leaves. Proc Natl Acad Sci U S A 85:2989–2993

Schmitz-Hoerner R, Weissenböck G (2003) Contribution of phenolic compounds to the UV-B screening capacity of developing barley primary leaves in relation to DNA damage and repair under elevated UV-B levels. Phytochemistry 64:243–255

Schnitzler HP, Junglubt TP, Heller W, Kofferlein K, Hutzler P, Heinzmann U, Schmelzer E, Ernst D, Langebartels C, Sandermann H Jr (1996) Tissue localization of UV-B screening pigments and chalcone synthase mRNA in needles of Scots pine seedlings. New Phytol 132:247–258

Schweikert K, Sutherland JES, Hurd CL, Burritt DJ (2011) UV-B radiation induces changes in polyamine metabolism in the red seaweed Porphyra cinnamomea. Plant Growth Regul 65(2):389–399

Sharma R (2001) Impact of solar UV-B on tropical ecosystems and agriculture. Case study: effect of UV-B on rice. Proc SEAWPIT98 SEAWPIT2000 1:92–101

Sheahan JJ (1996) Sinapate esters provide greater UV-B attenuation than flavonoids in *Arabidopsis thaliana* (Brassicaceae). Am J Bot 83:679–686

Shinkle JR, Edwards MC, Koenig A, Shaltz A, Barnes PW (2010) Photomorphogenic regulation of increases in UV-absorbing pigments in cucumber (*Cucumis sativus*) and *Arabidopsis thaliana* seedlings induced by different UV-B and UV-C wavebands. Physiol Plant 138:113–121

Shirley BW (1996) Flavonoid biosynthesis: 'new' functions for an old pathway. Trends Plant Sci 1:377–382

Shyam Choudhury S, Sen Mandi S (2011) Natural ultra violet irradiance related variation in antioxidant and aroma compounds in tea (*Camelia sinensis* L. Kuntze) plants grown in two different altitudes. Int J Environ Biol 2(1):1–6

Shyam Choudhury S, Sen Mandi S (2012) Natural ultra violet radiation on field grown rice (*Oryza sativa* L.) plants confer protection against oxidative stress in seed during storage under subtropical ambience. Environ Pollut 1:2

Sichel G, Corsaro C, Scalia M, di Bilio AJ, Bonomo RP (1991) In vitro scavenger activity of some flavonoids and melanins against O^{2-}. Free Radic Biol Med 11:1–8

Singh R, Singh S, Tripathi R, Agrawal SB (2011) Supplemental UV-B radiation induced changes in growth, pigments and antioxidant pool of bean (*Dolichos lablab*) under field conditions. J Environ Biol 32: 139–145

Solovchenko A (2010) Photoprotection in plants, vol 14, Springer Series in Biophysics. Springer, Berlin, pp 1–168

Stafford HA (1990) Flavonoid metabolism. CRC Press, Boca Raton

Stafford HA (1991) Flavonoid evolution: an enzymic approach. Plant Phys 96:680–685

Stapleton AE, Walbot V (1994) Flavonoids can protect maize DNA from the induction of ultraviolet-radiation damage. Plant Physiol 105:881–889

Stephanou M, Manetas Y (1997) The effects of seasons, exposure, enhanced UV-B radiation and water stress on leaf epicuticular and internal UV-B absorbing

capacity of *Cistus creticus*: a Mediterra-nean field study. J Exp Bot 48:1977–1985
Strack D, Meurer B, Weissenböck G (1982) Tissue-specific kinetics of flavonoid accumulation in primary leaves of rye (*Secale cereale* L.). Z Pflanzenphysiol 108:131–141
Strack D, Heilemann J, Mo¨mken M, Wray V (1988) Cell wall-conjugated phenolics from Coniferae leaves. Phytochemistry 27:3517–3521
Stracke R, Favory JJ, Gruber H, Bartelniewoehner L, Bartels S, Binkert M, Funk M, Weisshaar B, Ulm R (2010) The *Arabidopsis* bZIP transcription factor HY5 regulates expression of the PFG1/MYB12 gene in response to light and ultraviolet-B radiation. Plant Cell Environ 33:88–103
Strid A, Chow WS, Anderson JM (1994) UV-B damage and protection at the molecular level in plants. Photosynth Res 39:475–489
Surplus SL, Jordan BR, Murphy AM, Carr JP, Thomas B, Mackerness SA-H (1998) UV-B induced response in Arabidopsis thaliana: role of salicylic acid and ROS in the regulation of transcripts and acidic PR proteins. Plant Cell Environ 21:685–694
Takahashi A, Takeda K, Ohnishi T (1991) Light-induced anthocyanin reduces the extent of damage to DNA in UV-irradiated *Centaurea cyanus* cells in culture. Plant Cell Physiol 32:541
Tarrant AM, Reitzel AM, Kwok CK, Jenny MJ (2014) Activation of the cnidarian oxidative stress response by ultraviolet radiation, polycyclic aromatic 3 hydrocarbons and crude oil. J Exp Biol 217:1444–1453
Tattini M, Gravano E, Pinelli P, Mulinacci N, Romani A (2000) Flavonoids accumulate in leaves and glandular trichomes of *Phillyrea latifolia* exposed to excess solar radiation. New Phytol 148:69–77
Tattini M, Galardi C, Pinelli P, Massai R, Remorini D, Agati G (2004) Differential accumulation of flavonoids and hydroxycinnamates in leaves of *Ligustrum vulgare* under excess light and drought stress. New Phytol 163:547–561
Taylor LP, Grotewold E (2005) Flavonoids as developmental regulators. Curr Opin Plant Biol 8:317–323
Teramura AH (1983) Effects of ultraviolet-B radiation on the growth and yield of crop plants. Physiol Plant 58:415–427
Tevini M (1994) UV-B effects on terrestrial plants and aquatic organisms. Prog Bot 55:174–190
Tevini M, Teramura AH (1989) UV-B effects on terrestrial plants. Photochem Photobiol 50:479–487
Tevini M, Braun J, Fieser G (1991) The protective function of the epidermal layer of rye seedlings against ultraviolet- B radiation. Photochem Photobiol 533: 29–333
Torabinejad J, Caldwell MM (2000) Inheritance of UV-B tolerance in seven ecotypes of *Arabidopsis thaliana* L. Heynh. and their F_1 hybrids. J Hered 91:228–233
Tournaire C, Croux S, Maurette M-T, Beck I, Hocquaux M, Oliveros E, Braun AM (1993) Antioxidant activity of flavonoids: efficiency of singlet oxygen ($^1\Delta_g$) quenching. J Photochem Photobiol B 19: 205–215
Tsormpatsidis E, Henbest RGC, Davis FJ, Battey NH, Hadley P, Wagstaffe A (2007) UV irradiance as a major inûuence on growth, development and secondary products of commercial importance in LolloRosso lettuce 'Revolution' grown under polyethylene ûlms. Environ Exp Bot 63:232–239
Tsormpatsidis E, Henbest RGC, Battey NH, Hadley P (2010) The influence of ultraviolet radiation on growth, photosynthesis and phenolic levels of green and red lettuce: potential for exploiting effects of ultraviolet radiation in a production system. Ann Appl Biol 156(3):357–366
Tzin V, Galili G (2010) The biosynthetic pathways for shikimate and aromatic amino acids in *Arabidopsis thaliana*. Arabidopsis Book 8:e0132. doi:10.1199/tab.0132
Ulm R, Nagy F (2005) Signalling and gene regulation in response to UV light. Curr Opin Plant Biol 8:477–482
Ulm R, Baumann A, Oravecz A, Mate Z, Adam E, Oakeley J, Schafer E, Nagy F (2004) Genome-wide analysis of gene expression reveals HY5 function in the UV-B response of *Arabidopsis*. Proc Natl Acad Sci U S A 101:1397–1402
United States Environmental Protection Agency (USEPA) (1995) Environmental effects of ozone depletion: 1995 assessment
Vierstra RD, John TR, Poff KI (1982) Kaempferol 3-O-galactoside 7-O-rhamnoside is the major green fluorescing compound in the epidermis of *Vicia faba*. Plant Physiol 69:522–525
Vispe S, Cazaux C, Lesca C, Defais M (1998) Overexpression of Rad51 protein stimulates homologous recombination and increases resistance of mammalian cells to ionizing radiation. Nucl Acids Res 26:2859–2864
Vogelmann TC (1993) Plant tissue optics. Annu Rev Plant Physiol Plant Mol Biol 44:231–251
Vogt T (2010) Phenylpropanoid biosynthesis. Mol Plant 3:2–20
Wang Y, He W, Huang H, An L, Wang D, Zhang F (2009) Antioxidative responses to different altitudes in leaves of alpine plant Polygonum viviparum in summer. Acta Physiol Plant 31:839–848
War AR, Paulraj MG, Ahmad T, Buhroo AA, Hussain B, Ignacimuthu S, Sharma HC (2012) Mechanisms of plant defense against insect herbivores. Plant Signal Behav 7(10):1306–1320
Warner CW, Caldwell MM (1983) Influence of photon flux density in the 400–700 nm waveband of inhibition of photosynthesis by UV-B (280–320 nm) radiation in soybean leaves: separation of indirect and immediate effects. Photochem Photobiol 38:341–346
Waterworth WM, Jiang Q, West CE, Nikaido M, Bray CM (2002) Characterization of *Arabidopsis* photolyase enzymes and analysis of their role in protection

from ultraviolet-B radiation. J Exp Bot 53:1005–1015

Weinig C, Gravuer KA, Kane NC, Schmitt J (2004) Testing adaptive plasticity to UV: costs and benefits of stem elongation and light-induced phenolics. Evolution 58:2645–2656

White AL, Jhanke LS (2002) Contrasting effects of UV-A and UV-B on photosynthesis and photoprotection of β-carotene in two *Dunaliella* spp. Plant Cell Physiol 43(8):877–884

Wilson MI, Greenberg BM (1993) Specificity and photomorphogenic nature of ultraviolet-B-induced cotyledon curling in *Brassica napus* L. Plant Physiol 102:671–677

Wilson DO, McDonald MB Jr (1986a) A convenient volatile aldehyde assay for measuring seed vigour. Seed Sci Technol 14:259–268

Wilson DO Jr, McDonald MB Jr (1986b) The lipid peroxidation model of seed aging. Seed Sci Technol 18: 269–300

Wilson KE, Wilson MI, Greenberg BM (1998) Identification of the flavonoid glycosides that accumulate in *Brassica napus* L. cv. Topas specifically in response to ultraviolet-B radiation. Photochem Photobiol 67: 547–553

Wilson KE, Thompson JE, Huner NPA, Greenberg BM (2001) Effects of ultraviolet-A exposure on ultraviolet-B-induced accumulation of specific flavonoids in Brassica napus. Photochem Photobiol 73:678–684

Wolf L, Rizzini L, Stracke R, Ulm R, Rensing SA (2010) The molecular and physiological responses of *Physcomitrella patens* to ultraviolet-B radiation. Plant Physiol 153:1123–1134

Wondrak GT, Jacobson MK, Jacobson EL (2006) Endogenous UV-A-photosensitizers: mediators of skin photodamage and novel targets for skin photoprotection. Photochem Photobiol Sci 5:215–237

Yamasaki H, Sakihama Y, Ikehara N (1997) Flavonoid–peroxidase reaction as a detoxification mechanism of plant cells against H_2O_2. Plant Physiol 115: 1405–1417

Yao LH, Jiang YM, Shi J et al (2004) Flavonoids in food and their health benefits. Plant Food Hum Nutr 59(3): 113–122

Zhang WJ, Bjo¨rn LO (2009) The effect of ultraviolet radiation on the accumulation of medicinal compounds in plants. Fitoterapia 80(4):207–218

Zhao J, Dixon R (2009) The 'ins' and 'outs' of flavonoid transport. Trends Plant Sci 15:72–80

Zhao B, Liang R, Ge L, Li W, Xiao H, Lin H, Ruan K, Jin Y (2007) Identification of drought-induced microRNAs in rice. Biochem Biophys Res Commun 354: 585–590

5 UV Signal Transduction for Countering Deleterious Effects of UV Radiation in Plant

5.1 Introduction

UV radiation, in addition to causing high intensity related direct cell molecular damage, functions at low intensity in a positive role manifested as conferring UV acclimation; the cell molecular processes associated with such process involves signal transduction from site of perception to the site of action in the genome. A number of specialized photoreceptors (Jenkins 2009; Mckenzie et al. 2003) epigenetically signals up regulation of appropriate UV protective genes. UV at low dose by signal transduction via specific photoreceptors:

(a) upregulate genes for synthesis of: (i) cell protective compounds viz. flavonoids that due to structural specialization function in dual role for providing internal UV screen (Wargent et al. 2009) as well as non enzymatic antioxidants (Rice-Evans et al. 1997), (ii) enzymes for DNA repair (viz. Photolyase – Hutchinson 1987) (iii) proteins, such as Pathogen Related-1(PR1) protein (Neill et al. 2002) that protect plant cells from pathogen attack and (iv) antioxidative enzymes (Nawkar et al. 2013).
(b) down regulate LhcB1 gene that is responsible for reduced production of the major chlorophyll binding protein that regulates redox potential in chloroplasts, peroxisomes and mitochondria (Foyera and Noctor 2003). In addition to specific photoreceptors, non specific cellular ROS also function in UV signal transduction (Apel and Hirt 2004). In a different mode, UV induced damaged DNA may also initiate a signaling cascade for DNA repair (this process involves sensor proteins for damage detection – Culligan et al. 2004, 2006).

Cell molecular processes conferring UV acclimation in plants are mediated by a number of specific photoreceptors; of these UVR-8, generally known to be associated with UV-B induces upregulation of specific genes viz. (i) DNA photolyase for repair of UV induced DNA damage (Hutchinson 1987) and (ii) genes for synthesis of internal UV screening compounds viz. flavonoids (Close and McArthur 2002; Brown et al. 2005), that due to their structural specialization, also function as antioxidants in plant cells (as well as in animal cells through the dietary route in animal cells). In humans the role of UV screening compound is served by melanin that is reported to be signaled by UV induced damaged DNA (Gilchrest and Eller 1999; Eller and Gilchrest 2000).

The UV-activated photoreceptor UVR8 plays a pivotal role in developing UV acclimation in plants (Brosche and Strid 2003; Ulm and Nagy 2005). Recurrent intermittent environmental variation related low-dose UV-signaled upregulation of cell's internal UV-screening compounds results

S. Sen Mandi, *Natural UV Radiation in Enhancing Survival Value and Quality of Plants*,
DOI 10.1007/978-81-322-2767-0_5

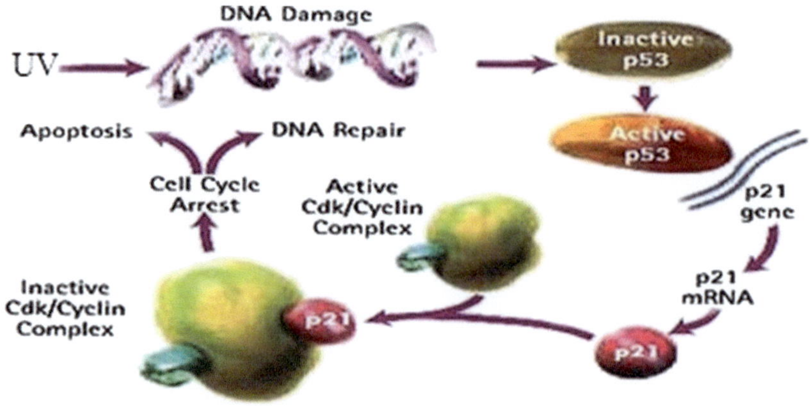

Fig. 5.1 UV-induced p53 activation and subsequent cell cycle arrest leading to apoptosis or cell repair (Taken from Sancar et al. 2004)

in accumulation of an armory of UV protective compounds. This feature, manifested as UV-adapted trait in plants, ensures plant survival under prolonged span of high UV environment.

5.2 UV-Damaged DNA-Mediated Signaling

Studies on *Arabidopsis* established that repair of UV induced DNA damage occurs through a signal transduction process, involving damaged DNA being recognized by specific cellular proteins; this initiates the signaling cascade. UV-induced SSB/DSB in DNA is sensed by orthologs of ATM (ataxia telangiectasia-mutated) and ATR (ATM-Rad3-related) protein kinases, respectively (Culligan et al. 2004, 2006). *Arabidopsis* mutants that lack a functional ATR appear hypersensitive to UV-B (Culligan et al. 2004), thereby establishing the fact that ATR plays a significant role in resumption of cell cycle progression subsequent to UV-B-induced DNA damage related halting of cell cycle.

In animals, photodimers, viz., CPD, a product of DNA damage under UV radiation act as signals for activation of the tumor suppressor gene p53 that has significant role in generation of apoptotic keratinocytes (studied in vivo as sunburn cells in the epidermis – Boccalandro et al. 2001); p53 in turn induces transcription of p21 gene; the newly formed p21 protein interacts with active cdk/cyclin complex and renders the complex inactive. As a result, cell cycle undergoes a temporary halt until the DNA is repaired (Sancar et al. 2004). If however the amount of damage is severe, then apoptosis (programmed cell death) occurs. This phenomenon is schematically represented in Fig. 5.1 and is described in the following flowchart: UV → pyrimidine dimers → p53 activation → cell cycle arrest at G1 → excision repair of DNA → normal cell cycle. This understanding is based on the observation of Boccalandro et al. (2004) in experiments where mice lacking functional p53 developed significantly fewer sunburn cells after UV exposure in comparison to UV-irradiated wild-type mice. This study suggests that sunburn cells, i.e., keratinocytes which fail to efficiently repair damaged DNA, undergo apoptosis; this helps to escape the risk of conversion of normal cells into malignant cell. Thus, formation of sunburn cells appears to be a p53-controlled scavenging phenomenon that protects individuals from developing UV-induced skin cancer (Bornman et al. 1997).

5.3 Photoreceptor-Mediated UV Signaling

Photosensory system that plants use to monitor their surrounding light including UV environment includes known classes of informational photoreceptors. For photoreceptors to be effective in signal transduction process in cells, the radiation must first be absorbed by a cellular photoreceptor that consists of a chromophore attached to a protein moiety. Such molecules transduce the energy of fluencing low-dose UV radiation through a biochemical signaling pathway to induce changes in gene expression (Jordan 1996).

Various lines of evidence indicate the presence of several photoreceptors that absorb in UV-B (280–315 nm) and in UV-A (315–390 nm) (Briggs and Lino 1983; Short and Briggs 1994; Jenkins et al. 1995). Interesting information on UV photoreceptor comes from studies on *Cyanobacteria* that respond to a broad spectrum of light, ranging from near UV to far-red light (Mullineaux 2001; Montgomery 2007). For sensing visible light of different intensity as well as radiation in the UV spectral range (100–390 nm), *Cyanobacteria* use highly specialized photoreceptors, viz., cyanobacteriochromes (CBCRs), that constitute well characterized phytochrome-related photoreceptors. Here, though the phytochromes (phy) are active in the red (600–700 nm) and far-red (700–750 nm) regions (Kendrick and Kronenberg 1994; Cashmore et al. 1999; Somers and Fujiwara 2009) as conventionally known, the phytochrome-related photoreceptors exhibit unique and diverse photochemical properties that are distinct from the functioning of the generally known red ↔ far-red reversible photoconversion in plants (Ikeuchi and Ishizuka 2008).

UV-A, in addition to having specific photoreceptors, viz., cryptochrome (CRY) (Christie 2007), phototropins (PHOT) (Demarsy and Fankhauser 2009), and members of the Zeitlupe family (ztl, fkf, and lkp2) that are active in the violet/blue region (390–500 nm), also recognizes, to some extent, UVR8 as its photoreceptor (Christie et al. 2012a) which was for long known to be the only photoreceptor for UV-B. This UVR8 is a unique photoreceptor lacking a prosthetic chromophore and possessing light sensing ability intrinsic to the molecule. This photoreceptor was originally identified from genetic screening experiments using *Arabidopsis* mutants that were hypersensitive to UV-B (Klibenstein et al. 2002).

Different types of photoreceptors and their region of absorbance is shown in Fig. 5.2. All these photoreceptors perform distinctive photosensory cell physiological functions including gene expression in plants (Cashmore et al. 1999; Smith 2000). The combined activities of these photoreceptors enable plants to detect and respond to the presence, absence, variation in color, intensity, as well as direction and diurnal duration of impinging light signals including UV.

5.3.1 UV-B-Related Photoreceptor

For UV-B, till date only one specific photoreceptor (UVR8) has been identified to absorb UV-B and function in transduction of the UV-B signal

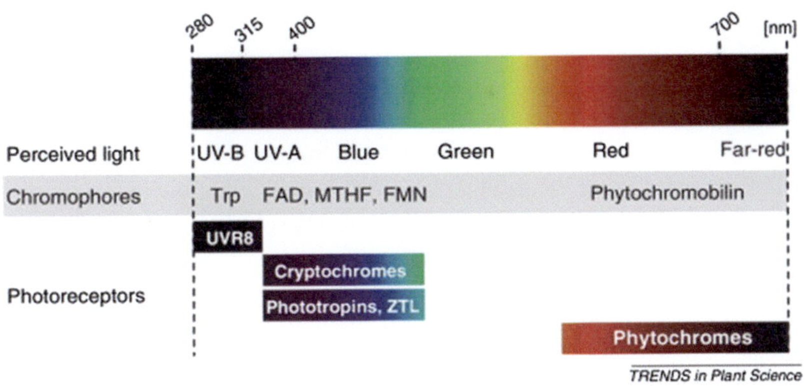

Fig. 5.2 Different types of photoreceptors and their region of absorbance (Taken from Heijde and Ulm 2012)

for upregulation of important cell molecular events in plant cells. Recent studies report that UV-A is also absorbed by UVR8 (Christie et al. 2012b). Distinct responses to UV-B radiation have been reported in different plant species (Wellmann 1976; Ballare et al. 1991; Beggs and Wellmann 1994), with particular reference to *Arabidopsis* (Christie and Jenkins 1996; Brown et al. 2005; Favory et al. 2009). Signaling response of UV-B modifies the biochemical composition of plant tissues thereby affecting UV acclimation as a genetic trait. Favory et al. (2009) demonstrated UV-B tolerance related upregulation of appropriate enzymes relating to UV acclimation, as illustrated by the sensitivity of uvr8 mutants compared to wild type under simulated sunlight. Flavonoids, product of UVR8-mediated gene upregulation by UV-B, in addition to conferring UV acclimation (through development of internal UV screen and macromolecular repair processes – see Chap. 4), also helps to deter pests and pathogens (Tevini and Teramura 1989; Frohnmeyer and Staiger 2003; Stratmann 2003).

At the molecular level, low fluence UV (perceived by UVR8) regulates induction of genes related to UV protection involving photorepair of photodimer formation-related DNA damage and phenolic compound accumulation that functions as internal UV screen that confer UV acclimation of plants. Photomorphogenic effects, viz., inhibition of hypocotyl elongation, phototropic curvature, regulation of leaf growth, and stomatal differentiation in plants, represent UV protective (photomorphogenic) strategies for allowing time toward developing UV induced flavonoids that constitute internal UV screen for conferring UV acclimation specifically in plants (Jenkins 2009; Wargent et al. 2009) and are mediated by the UV-B photoreceptor UVR8 (Rizzini et al. 2011).

The recent characterizations of UVR8 as a UV-B photoreceptor (Rizzini et al. 2011) and elucidation of the mechanisms for UV-B absorption by UVR8 (Wu et al. 2011, 2012; Christie et al. 2012a) have enhanced the understanding relating to the mechanism of UV-B perception in plants. UVR8 is formed of seven-bladed β-propeller proteins which have sequence similarity with human regulator of chromatin condensation1 (RCC1 – Kliebenstein et al. 2002), although UVR8 and RCC1 differ in their activity, function (Jenkins 2009; Rizzini et al. 2011), and monomeric topology (Wu et al. 2011; Christie et al. 2012b). UVR8 exists in plants as a dimer, but after UV-B perception, the dimer is split into monomers due to disruption of the salt bridges in the molecules (Christie et al. 2012a; Wu et al. 2012); this leads to the molecule's monomerization (Christie et al. 2012b; Rizzini et al. 2011). Following monomerization, UVR8 accumulates in the nucleus where it interacts with a protein called constitutively photomorphogenic 1 (COP1), an E3 ubiquitin ligase (generally known to cause degradation of key transcription factors – Oravecz et al. 2006). The interaction between UVR8-COP1 occurs through the C terminal region of UVR8 and WD40 domain of COP1 (Cloix et al. 2012). COP1 after its association with UVR8 acts as a positive regulator of UVR8-mediated UV-B signaling through activation of transcription factors relating to genes associated with UV protection in plants (Jenkins 2009). Transcriptome analyses of uvr8 mutants after radiation with low-fluence UV-B have demonstrated that UVR8 is required for the induction of genes involved in UV protection (flavonoid and alkaloid pathways) and photorepair (Brown et al. 2005; Favory et al. 2009).

This transcription factor is stabilized by HY5 protein. After stabilization UVR8 associates with the chromatin (through histones – Brown et al. 2005; Cloix and Jenkins 2008) containing the HY5 gene and also several other genes under its regulation. UVR8 in turn promotes the recruitment or activation of transcription factors and also other proteins that facilitate transcriptional regulation. UVR8-COP1 complex activates many UV responsive genes whose products are essential in UV protection (e.g., phenylpropanoid biosynthesis pathway, including CHS and FLS) and damage repair (e.g., photolyases PHR1 and UVR3) (Fig. 5.4). This complex also induces transcription and subsequent translation of RUP1 (repressor of photomorphogenesis 1) and RUP2 (repressor of photomorphogenesis 2) proteins that associate with UVR8 monomer

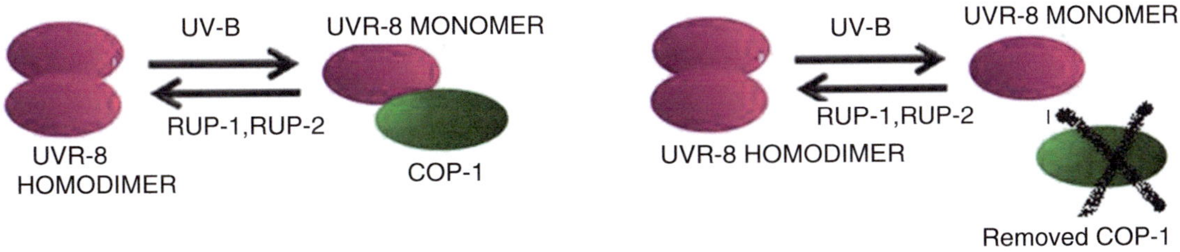

Fig. 5.3 Schematic diagram showing RUP1- and RUP2-mediated redimerization of UVR8 is independent of COP1 (Taken from Heijde and Ulm 2012)

(Grubera et al. 2010; Cloix et al. 2012) and redimerize UVR8 monomers into dimers (without involvement of COP1) by negative feedback mechanism (Heijde and Ulm 2013) (Fig. 5.3). UVR8 redimerization prevents the interaction between UVR8 monomer and COP1; this stops signaling through UVR8. The transcription factor salt tolerance (STO/BBX24) and the stress regulator radical-induced cell death1 (RCD1) are also proposed negative regulators of UV-B signaling by impinging on HY5 (Jiang et al. 2009, 2012).

The cascade of events responsible for UVR8-mediated signaling causing gene upregulation for protection from high-dose UV fluence is shown in Fig. 5.4.

The interaction between UVR8 and COP-1 is essential for transmitting the signal causing gene expression in plants (Favory et al. 2009). Three tryptophan amino acids (W233, W285, and W337) which are located in the β-propeller core play significant role in stabilizing the structure and function of UVR8. Of these three amino acids, W285 is the most important for functioning of UVR8; the other two, viz., W233 and W337, though not directly involved are necessary for proper functioning of UVR8. Arginines, located at three positions (A146, A286, and A338) around the four tryptophan moieties (arranged as a pyramid like configuration), participate in salt bridges and form an extensive network of cation-Π interactions with the tryptophans (Christie et al. 2012a; Wu et al. 2012). Of this arrangement, A286 is particularly significant; this is demonstrated by the fact that mutation of A286 to alanine ($UVR8^{R286A}$) creates a constitutive UVR8 monomer (Christie et al. 2012b; Wu et al. 2012).

5.3.2 UV-A-Related Photoreceptor

UV-A photoreceptor was first identified in *Arabidopsis thaliana* by Ahmad and Cashmore (1993) who named it as cryptochrome 1 (CRY1). This photoreceptor known to absorb blue light mediates expression of blue light-regulated genes (Short and Briggs 1994; Ahmad et al. 1995; Jackson and Jenkins 1995). CRY1 has been shown to mediate inhibition of hypocotyl elongation, anthocyanin accumulation, leaf and cotyledon expansion, elongation of petioles, and responsiveness to night breaks. The CRY 1 photoreceptor protein is coded by hy4 gene and has sequence similarity with microbial DNA photolyases (Ahmad and Cashmore 1993). These cryptochromes are flavoproteins which carry two chromophores, a pterin or a deazaflavin at one site and FAD at another with properties of a UV-A/blue light photoreceptor (Ahmad and Cashmore 1993, 1996; Lin et al. 1995a; Malhotra et al. 1995). Subsequently another member of the cryptochrome family, viz., CRY2, was identified in *A. thaliana* (Ahmad et al. 1998). These two chromophores (proteins) overlap in function but differ in stability: CRY2 alone is rapidly degraded under light fluences (green, blue, and UV) that is known to activate the photoreceptor. CRY1 is relatively stable (Ahmad et al. 1998), and CRY2 in association with CRY1 exhibits better stability and synergistically acts as UV-A

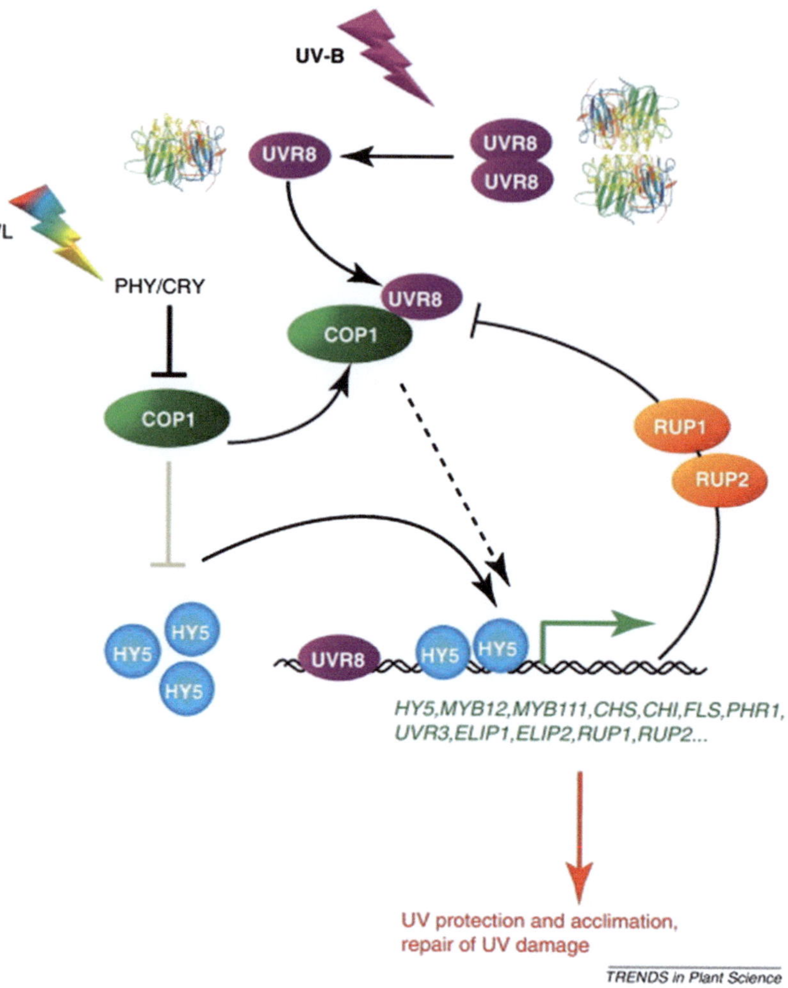

Fig. 5.4 Model of UVR8-mediated signaling. *CHI* chalcone isomerase, *CHS* chalcone synthase, *CRY* cryptochrome, *COP1* constitutively photomorphogenic 1, *ELIP1* and *ELIP2* early light-inducible protein 1 and 2, *FLS* flavonol synthase, *HY5* elongated hypocotyl 5, *MYB12* and *MYB111* MYB domain protein 12 and 111, *PHR1* photolyase 1, *PHY* phytochrome, *RUP1* and *RUP2* repressor of UV-B photomorphogenesis 1 and 2, *UV-B* ultraviolet-B radiation, *UVR3* UV repair defective 3, *UVR8* UV resistance locus 8, *WL white* light (Taken from Heijde and Ulm 2012)

photoreceptor (Ahmad et al. 1998). Cryptochromes are found in various taxa and are thought to have evolved from photolyases though lacking in DNA repair activity (Gyula et al. 2003) that is commonly known for DNA photolyases. Both CRY1 and CRY2 also contain carboxy terminal extensions that exhibit difference in the two photoreceptors. The cryptochromes, like phytochromes (the commonly known photoreceptors in plants absorbing in the visible region, red and far red), also control growth and developmental responses exhibiting photomorphogenesis subject to variations in the wavelength, intensity, and diurnal duration of the radiation (Cashmore et al. 1999; Smith 2000). The hy4 mutant of *Arabidopsis thaliana* (L) (unable to produce CRY1 receptor) is unable to suppress the elongation of hypocotyl and showed lengthy bolted stems, increased petiole extension, and increased leaf width under UV-A/blue light in comparison to the wild type (Jackson and Jenkins 1995). On the other hand, transgenic tobacco plants which are designed to overexpress CRYl photoreceptor show a hyper-responsive phenotype, with shorter hypocotyl under UV-A , blue, and green light (Lin et al. 1995b). In recent times, Christie et al. (2012a) have reported that UVR8, which for long was considered as specific photoreceptor for UV-B, is now known to absorb in the UV-A region.

In addition to cryptochromes, plant cells also contain two other types of UV-A/blue photoreceptors, viz., phototropins (PHOT1, PHOT2) and members of the Zeitlupe family (ZTL, FKF1, and LKP2) (Christie 2007; Demarsy and Fankhauser 2009; Somers and Fujiwara 2009).

Structurally, the phototropins differ from cryptochromes by the fact that they have two flavin mononucleotide (FMN) chromophores in association with two Per–Arnt–Sim (PAS) subdomains in the amino-terminal domain of the molecule (cf. chromophores of cryptochromes). The carboxy terminal domain of phototropins contains a classical serine/threonine protein kinase whose activity is regulated by blue/UV-A absorbed by the amino-terminal domain (Briggs and Huala 1999; Sakai et al. 2001). Phototropins function primarily in controlling directional (phototropic) growth in response to directional light and/or intracellular chloroplast movement in response to light intensity (Briggs and Huala1999; Kagawa et al. 2001; Sakai et al. 2001).

Members of the Zeitlupe (ztl) family play important roles in controlling the degradation and stability of components associated with circadian clock regulation and the photoperiodic control of flowering. Ztl members localize in the cytosol or the nucleus (Takase et al. 2011) and comprise of three members: Zeitlupe (ztl), Flavin-binding kelch repeat F-box 1 (fkf1), and LOV kelch protein 2 (lkp2) (Ito et al. 2012; Suetsugu and Wada 2013). Genetic analysis in *Arabidopsis* indicates that ztl, fkf1, and lkp2 partially overlap in function (Fornara et al. 2009; Baudry et al. 2010; Takase et al. 2011). *Arabidopsis* mutants lacking ztl are primarily impaired in circadian clock function (Somers et al. 2000), whereas fkf1 mutants mostly show alterations in flowering time (Imaizumi et al. 2003).

5.4 Role of Signaling in UV-Related Gene Expression

UVR8, on perception of low-dose UV radiation, regulates many genes (Brown et al. 2005; Favory et al. 2009) related to the secondary metabolite production resulting in enhanced synthesis of phenolics/flavonoids. UV-B can induce rapidly (within minutes) nuclear localization of UVR8 even at low-fluence rates; however, the constitutive nuclear localization of UVR8 is not sufficient to activate UV-B target gene expression (Kaiserli and Jenkins 2007), suggesting more complex regulation, for example, through additional UVR8 subcellular localizations or post-translational modifications. Chromatin immunosuppression assay (ChIP) demonstrated that UVR8 monomer (produced after UV-B perception) binds to chromatin after transportation from cytoplasm to nucleus (Brown et al. 2005; Cloix and Jenkins 2008). Detailed study revealed that UVR8 associates in a UV-independent manner with a 3 kb region around the elongated hypocotyl-5(HY5) locus. This association of UVR8 with chromatin is mediated via histones, particularly H2B (Brown et al. 2005; Cloix and Jenkins 2008). In *Arabidopsis*, it has been demonstrated that UVR8 along with COP-1 and HY5 induces the expression of chalcone synthase (CHS) gene which is the rate limiting enzyme of phenylpropanoid pathway that produces flavonoids as end product (Frohnmeyer et al. 1992; Kubasek et al. 1992; Christie and Jenkins 1996). In *Arabidopsis* Morales et al. (2011) have shown that expression of many flavonoid genes including *CHI* and *DFR* was significantly increased (between five and tenfold) by solar UV exposure for 12 h. Studies on CHS induction by UV demonstrate that after UV radiation HY5 binds to CHS promoter and partially induces CHS production (Ang et al. 1998; Lee et al. 2007; Stracke et al. 2010). Under UV radiation, HY5 is overexpressed which subsequently binds to the activation domain of CHS leading to induction of *CHS* expression (Stracke et al. 2010). UV-A/blue light photoreceptors also induce CHS production by heterotrimeric G-proteins. Though the activation paths for CHS induction are different for different photoreceptors, all pathways involve a light-regulated unit (unit1) of 52 bp size on the CHS promoter which is known to be necessary and sufficient to confer a signal for gene expression by activation of transcription factors binding to these elements (Weißhaar et al. 1991; Kaiser et al. 1995). Calcium ion is also involved in the process of CHS induction by UV-B; it has been demonstrated that after UV radiation for a millisecond, intracellular calcium ion concentration increases, and it lasts for about 20 min; subsequently CHS expression is also increased (Frohnmeyer et al. 1999).

The involvement of calmodulin in UV-B-mediated induction of CHS and the noninvolvement of calmodulin in UV-A-mediated induction of CHS proves that UV-B and UV-A induce CHS gene through different pathways, although both pathways involve reversible protein phosphorylation and require protein synthesis (Schafer et al. 1997). The photoreceptor mediated induction of CHS also depends on plant developmental stage as shown in parsley, mustard, and *Arabidopsis* seedlings (Ehmann et al. 1991; Frohnmeyer et al. 1992); this pattern was also found in transgenic *Arabidopsis* seedlings containing the β-glucuronidase (GUS) reporter gene under control of the CHS promoter (Feinbaum et al. 1991; Kaiser et al. 1995).

The photoreceptor UVR8 plays key roles in UV acclimation of plants by regulating accumulation of gene transcripts involved in the biosynthesis of flavonoids and anthocyanins (DFR, LDOX, AT5MAT, and PAP1) and in defense responses (At4g12490, LURP1, JAZ1, SYR1, WRKY70, JR1, At1g16850, AOC3, ANNAT1, and MKK4). In addition to regulating genes in flavonoid/anthocyanin pathway relating to plant acclimation under UV stress, UVR8 also regulates the expression of genes involved in the biosynthesis and signaling of several other genes (AOS, AOC1, AOC3, WRKY70, JAZ1, SYR1, and GRX480 and OPR3) involved in cell protection mechanism such as jasmonic acid (JA), glucosinolate biosynthesis (ATR4 and SOT17), and auxin (HAT2) and brassinosteroid (AIF1) signaling (Morales et al. 2013).

There is evidence that UV-B signaling may also be perceived in cells independent of UVR8. This is evidenced from the fact that different groups of UV-B-responsive genes are still effective in uvr8, cop1, and hy5 mutants (Brown et al. 2005; Brown and Jenkins 2008; Lang-Mladek et al. 2012). Mitogen-activated protein kinase (MAPK) signaling cascades have also been proposed to be involved in high dose UV-B stress independently of UVR8-mediated UV-B signaling (Ulm et al. 2002; Gonzalez Besteiro et al. 2011).

In addition to the generally studied function of UVR8 as photoreceptors of UV-B, solar UV-A has a major role in regulating the expression of most UV-B-regulated genes (Morales et al. 2010, 2011, 2013) indicating that UV-A radiation could be a modulator of UV-B response. The authors also indicate that UVR8 could mediate solar UV-A responses under natural sunlight exhibiting the impact of UVR8 on solar UV-A-mediated gene expression, acting as a positive and negative regulator of transcript accumulation of several genes (Morales et al. 2011). Christie et al. (2012b) further demonstrated that despite having maximal absorption in the UV-B, UVR8 also absorbs in the UV-A region of the spectrum.

Fuglevand et al. (1996) have investigated which of the inductive responses were mediated by the UV-A-specific CRYl photoreceptor by examining CHS transcript levels in the hy4-2.23N mutant that was generated by fast neutron bombardment (Koornneef et al. 1980). Because of a deletion starting in the third intron (a null mutant), hy4-2.23N produces an aberrant transcript that accumulates in reduced amounts (Fig. 5.5) (Ahmad and Cashmore 1993). Further, immunoblots of hy4-2.23N leaf proteins when probed with an antibody raised against CRY1 expressed in *Escherichia coli* did not reveal CRYI (Lin et al. 1996). Similar results were reported for several other hy4 null mutants (Ahmad et al. 1995). It may be noted that the mutant phenotype of hy4-2.23N is as strong as that of other hy4 null alleles (Ahmad and Cashmore 1993).

From studies on Parsley cell cultures by Harter et al. (1994a), it has been demonstrated that Ser kinases are also involved in UV mediated signal transduction. As ser kinases are known to be regulated by light, they could be associated with transcription factor activation including their transport from cytoplasm to

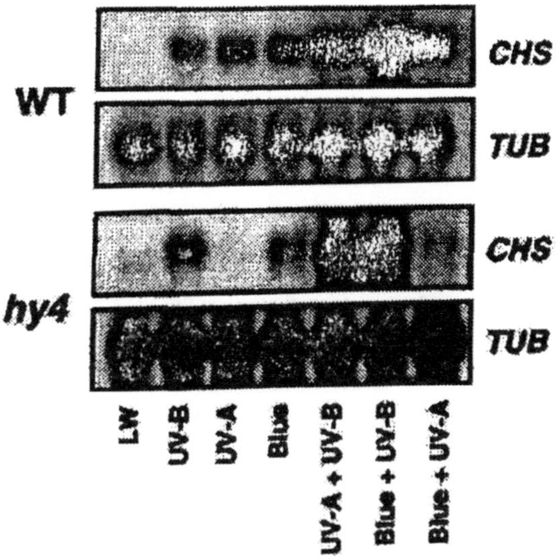

Fig. 5.5 CHS (chalcone synthase) transcript levels in the wild type and hy4-2.23N. Wild type (WT) and hy4-2.23N. TUB (tubulin) was taken as control in this experiment. First lane: CHS transcript from plants grown in 20 μmol m^2 s^{-1}*white* light for 3 weeks (LW). Second lane: after exposure for 16 h in 3 μmol m^2 s^{-1} UV-B. Third lane: after exposure in 75 μmol m^2 s^{-1} UV-A. Fourth lane: after exposure in 75 μmol m^2 s^{-1}*blue* light. Fifth lane: after exposure in 75 μmol m^2 s^{-1} UV-A plus 3 μmol m^2 s^{-1} UV-B. Sixth lane: after exposure in 75 μmol m^2 s^{-1}blue plus 3 μmol m^2 s^{-1} UV-B. Seventh lane: after exposure in 75 μmol m^2 s^{-1} UV-A plus 75 μmol m^2 s^{-1}*blue* light treatments (Taken from Fuglevand et al. 1996)

nucleus and subsequent binding to specific sites on chromosome (Harter et al. 1994a). This signal transduction cascade mediated by ser kinases involves activation by phosphorylation (Harter et al. 1994b). Studies of Holley et al. (2003) on tomato cell cultures in relation to mitogen activated protein kinase activity also strengthen the above hypothesis. It is interesting to note that nitric oxide (NO) which acts as second messenger is also involved in UV-mediated signal transduction and gene expression including CHS in *Arabidopsis* (Mackerness et al. 2001).

5.5 Multiple Pathways Mediate UV-B Responses

Under natural radiation, various photoreceptors are simultaneously activated, and the control of several physiological events can occur through multiple interactions between photoreceptors (Casal 2000; Jenkins et al. 2001; Chen et al. 2004). As different from the commonly known red–far red-mediated phytochrome signal transduction pathway, several distinct pathways are known for UV radiation in plants. Low-fluence UV radiation activates UVR8-dependent pathways relating upregulation of (1) internal UV-screening compounds for acclimation under UV radiation (A in Fig. 5.6), (2) antioxidative proteins for scavenging ROS (B in Fig. 5.6), and (3) DNA repair enzymes for removal of photodimers such as CPDs and 6-4PPs restoring normal configuration of DNA (C in Fig. 5.6). High-fluence UV radiation activates UVR8-independent stress pathway and cell death; the stress pathway involves mitogen-activated protein kinase (MAPK) causing programmed cell death (PCD) or apoptosis. Mitogen-activated protein kinase (MAPK) signaling cascades have been proposed to be involved in high-dose UV-B stress independently of UVR8-mediated UV-B signaling (Ulm et al. 2002; Gonzalez Besteiro et al. 2011). *Arabidopsis* plants with loss of MAP kinase phosphatase1 (mkp1) function are hypersensitive to acute UV-B radiation, but without showing impairment to UV-B acclimation. The MKP1 interacting partners MPK3 and MPK6 are activated by UV-B stress and are hyperactivated in mkp1 mutants suggesting that MKP1/MPK3/MPK6-mediated stress signaling pathway is crucial to UV-B tolerance in plants.

5.6 Nonspecific ROS-Mediated UV-B Signaling

While reactive oxygen species, i.e., ROS, produced in cells under high intensity UV radiation are generally known to cause oxidative damage to macromolecules (see Chap. 3), cellular ROS under low-dose UV fluence serve in the signal transduction pathway for upregulating a number of genes for protective function in cells. ROS influence the expression of a number of genes via the signal transduction pathways for upregulation of cell protective genes, viz., PR1

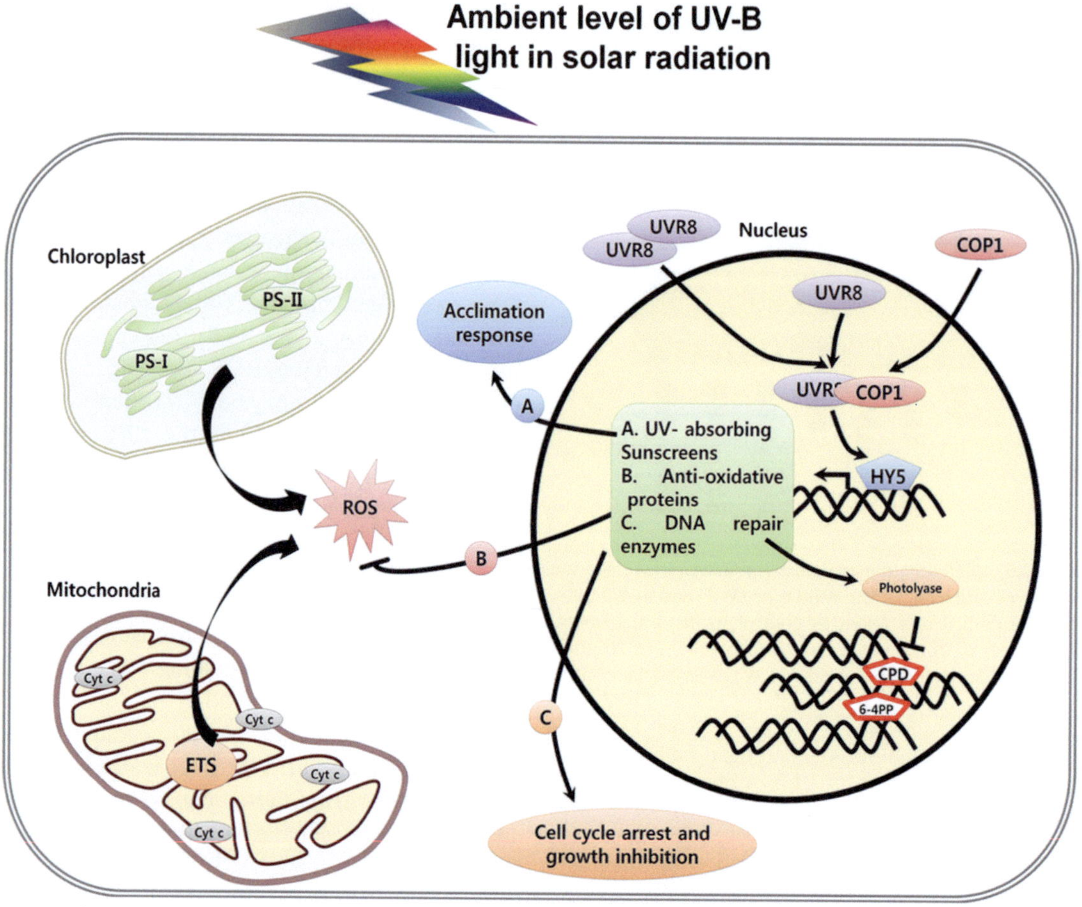

Fig. 5.6 Low-fluence UV radiation activates UVR8-dependent photomorphogenesis (Taken from Nawkar et al. 2013)

(Neill et al. 2002). These observations suggest that cells have evolved strategies to utilize ROS as environmental indicators and biological signals that activate and control various genetic stress response programs (Dalton et al. 1999). Spatial and temporal accumulation of UV induced ROS in cells have been reported by Apel and Hirt (2004) and Mittler et al. (2004) which is subject to dose of fluencing UV radiation (Nawkar et al. 2013). In plants, ROS-induced genes have been identified for receptor kinase (Desikan et al. 2000), annexin (Moseyko et al. 2002), and peroxisome biogenesis (Desikan et al. 2000). Recent approaches using cDNA profiling and DNA microarrays have analyzed large scale gene expression in response to ROS. Following exposure of *Arabidopsis* cells to H_2O_2, a total of 175 genes showed changes in expression levels (Desikan et al. 2001). Of the 113 induced genes, several encoded for proteins with antioxidant functions or were associated with defense responses or other stresses. ROS, especially superoxide which are generated by NADPH oxidase, is an important mediator of defense signaling in plants (Apel and Hirt 2004) through the generation of hydrogen peroxide (A-H-Mackerness 2000; A-H-Mackerness et al. 2001). ROS production, which is increased (as "unaccounted ROS") after UV radiation, causes both upregulation and downregulation of UV responsive genes (Nawkar et al. 2013). For example, in *Arabidopsis*, hydrogen peroxide (functioning as ROS and formed from superoxide and serving as inducer) mediates upregulation of plant pathogenesis-related protein (PR1) as a means of plant protection (Neill et al. 2002). During plant pathogen reactions, the activity and levels of the ROS detoxifying enzymes APX and CAT are suppressed by salicylic acid (SA) and nitrous oxide (NO) (Klessig et al. 2000). Because during the plant pathogen defense response, the plant simultaneously produces more ROS while decreasing its ROS scavenging capacities, accumulation of ROS and activation of PCD occurs.

Hydrogen peroxide is also known to downregulate the transcription of LhcB1 protein (which encodes major chlorophyll-binding protein of chloroplasts) (Foyera and Noctor 2003); this UV-induced ROS mediated adverse effect is in turn countered by ascorbate (Surplus et al. 1998), a scavenger of superoxide radicals (A-H-Mackerness et al. 2001). Ascorbate is thus an important cellular antioxidant, and it has been reported that mutant *Arabidopsis* that lacks ascorbate-vtc1 gene is highly sensitive to UV-B (Conklin et al. 1996; Gao and Zhang 2008).

It is thus established that much of the UV-induced molecular damages are in turn countered by UV-induced signal transduction pathways and associated cell molecular processes that help plants to acclimate under varying intensity and dose of UV radiation. This ensures plant survival in terrestrial habitat through the ages, the extent varying under variation in UV fluence, and genetic variation of plants.

Bibliography

A-H-Mackerness S (2000) Plant responses to UV-B (UV-B: 280–320 nm) stress: what are the key regulators? Plant Growth Regul 32:27–39

A-H-Mackerness S, John CF, Jordan B, Thomas B (2001) Early signaling components in ultraviolet-B responses: distinct roles for different reactive oxygen species and nitric oxide. FEBS Lett 489:237–242

Ahmad M, Cashmore AR (1993) HY4 gene of A. thaliana encodes a protein with characteristics of a blue-light photoreceptor. Nature 366:162–166

Ahmad M, Cashmore AR (1996) Seeing blue: the discovery of cryptochrome. Plant Mol Biol 30:851–861

Ahmad M, Lin C, Cashmore AR (1995) Mutations throughout an Arabidopsis blue-light photoreceptor impair blue-light- responsive anthocyanin accumulation and inhibition of hypocotyl extension. Plant J 8:853–6513

Ahmad M, Jarillo JA, Smirnova O, Cashmore AR (1998) The CRY1 blue light photoreceptor of *Arabidopsis* interacts with phytochrome A in vitro. Mol Cell 1:939–948

Ang LH, Chattopadhyay S, Wei N, Oyama T, Okada K, Batschauer A, Deng XW (1998) Molecular interaction between COP1 and HY5 defines a regulatory switch for light control of Arabidopsis development. Mol Cell 1(2):213–222

Apel K, Hirt H (2004) Reactive oxygen species: metabolism, oxidative stress, and signal transduction. Annu Rev Plant Biol 55:373–399

Ballaré CL, Barnes PW, Kendrick RE (1991) Photomorphogenic effects of UV-B radiation on hypocotyl elongation in wild type and stable-phytochrome-deficient mutant seedlings of cucumber. Physiol Plant 83: 652–658

Baudry A, Ito S, Song YH, Strait AA, Kiba T, Lu S et al (2010) F-box proteins FKF1 and LKP2 act in concert with ZEITLUPE to control Arabidopsis clock progression. Plant Cell 22:606–622

Beggs CJ, Wellmann E (1994) Photocontrol of flavonoid biosynthesis. In: Kendrick RE, Kronenberg GHM (eds) Photomorphogenesis in plants, 2nd edn. Kluwer Academic Publishers, Dordrecht, pp 733–751

Boccalandro H, Mazza C, Mazzella M, Casal J, Ballaré C (2001) Ultraviolet-B radiation enhances a phytochrome-B-mediated photomorphogenic response in Arabidopsis. Plant Physiol 126:780–788

Boccalandro HE, Rossi MC, Saijo Y, Deng XW, Casal JJ (2004) Promotion of photomorphogenesis by COP1. Plant Mol Biol 56:905–915

Bornman JF, Reuber S, Cen YP, Weissenbock G (1997) Ultraviolet radiation as a stress factor and the role of protective pigments. In: Lumsden P (ed) Plants and UV-B: responses to environmental change. Cambridge University Press, Cambridge, pp 156–168

Briggs WR, Huala E (1999) Blue-light photoreceptors in higher plants. Annu Rev Cell Dev Biol 15:33–62

Briggs WR, Lino M (1983) Blue-light-absorbing photoreceptors in plants. Philos Trans R Soc Lond B Biol Sci 303:347–359

Brosche M, Strid A (2003) Molecular events following perception of ultraviolet-B radiation by plants. Physiol Plant 117:1–10

Brown BA, Jenkins GI (2008) UV-B signaling pathways with different fluence-rate response profiles are distinguished in mature Arabidopsis leaf tissue by requirement for UVR8, HY5, and HYH. Plant Physiol 146:576–588

Brown BA, Cloix C, Jiang GH, Kaiserli E, Herzyk P et al (2005) A UV-B-specific signaling component orchestrates plant UV protection. Proc Natl Acad Sci U S A 102:18225–18230

Casal J (2000) Phytochromes, cryptochromes, phototropin: photoreceptor interactions in plants. Photochem Photobiol 71:1–11

Cashmore AR, Jarillo JA, Wu Y-J, Liu D (1999) Cryptochromes: blue light receptors for plants and animals. Science 284:760–765

Chen M, Chory J, Fankhauser C (2004) Light signal transduction in higher plants. Annu Rev Genet 38:87–117

Christie JM (2007) Steric interactions stabilize the signaling state of the LOV2 domain of phototropin 1. Biochemistry 46:9310–9319

Christie JM, Jenkins GI (1996) Distinct UV-B and UV-A/blue light signal transduction pathways induce chalcone synthase gene expression in Arabidopsis cells. Plant Cell 8:1555–1567

Christie JM, Arvai AS, Baxter KJ, Heilmann M, Pratt AJ, O'Hara A, Kelly SM, Hothorn M, Smith BO, Hitomi K et al (2012a) Plant UVR8 photoreceptor senses UV-B by tryptophan-mediated disruption of cross-dimer salt bridges. Science 335:1492–1496

Christie JM, Arvai AS, Baxter KJ, Heilmann M, Pratt AJ, Kelly SM, Hothorn M, Smith BO, Hitomi K, Jenkins GI, Getzoff ED (2012b) Plant UVR8 Photoreceptor senses UV-B by Tryptophan-mediated disruption of cross-dimer salt bridges. Science 335:1492–1496

Cloix C, Jenkins GI (2008) Interaction of the Arabidopsis UV-B-specific signaling component UVR8 with chromatin. Mol Plant 1:118–128

Cloix C, Kaiserli E, Heilmann M, Baxter KJ, Brown BA, O'Hara A, Smith BO, Christie JM, Jenkins GI (2012) C-terminal region of the UV-B photoreceptor UVR8 initiates signaling through interaction with the COP1 protein. Proc Natl Acad Sci U S A 109:16366–16370

Close DC, McArthur C (2002) Rethinking the role of many plant phenolics – protection from photodamage not herbivores? Oikos 99:166–172

Conklin PL, Williams EH, Last RL (1996) Environmental stress sensitivity of an ascorbic acid-deficient Arabidopsis mutant. Proc Natl Acad Sci U S A 93:9970–9974

Culligan K, Tissier A, Britt A (2004) ATR regulates a G2-phase cell-cycle checkpoint in Arabidopsis thaliana. Plant Cell 16(5):1091–1104

Culligan KM, Robertson CE, Foreman J, Doerner P, Britt AB (2006) ATR and ATM play both distinct and additive roles in response to ionizing radiation. Plant J 48(6):947–961

Dalton TP, Shertzer HG, Puga A (1999) Regulation of gene expression by reactive oxygen. Annu Rev Pharmacol Toxicol 39:67–101

Demarsy E, Fankhauser C (2009) Higher plants use LOV to perceive blue light. Curr Opin Plant Biol 12:69–74

Desikan R, Neill SJ, Hancock JT (2000) Hydrogen peroxide-induced gene expression in Arabidopsis thaliana. Free Rad Biol Med 28:773–778

Desikan R, A-H Mackerness S, Hancock JT, Neill SJ (2001) Regulation of the Arabidopsis transcriptome by oxidative stress. Plant Physiol 127:159–172

Ehmann B, Ocker B, Schäfer E (1991) Development- and light-dependent regulation of the expression of two different chalcone synthase transcripts in mustard cotyledons. Planta 183:416–422

Eller MS, Gilchrest BA (2000) Tanning as part of the eukaryotic SOS response. Pigment Cell Res 13:94–97

Favory JJ, Stec A, Gruber H, Rizzini L, Oravecz A, Funk M, Albert A, Cloix C, Jenkins GI, Oakeley EA et al (2009) Interaction of COP1 and UVR8 regulates UV-B-induced photomorphogenesis and stress acclimation in Arabidopsis. EMBO J 28:591–601

Feinbaum R, Storz G, Ausubel F (1991) High intensity and blue-light regulated expression of chimeric chalcone synthase genes in transgenic Arabidopsis thaliana plants. Mol Gen Gent 226:449–456

Fornara F, Panigrahi KC, Gissot L, Sauerbrunn N, Ruhl M, Jarillo JA et al (2009) Arabidopsis DOF transcription factors act redundantly to reduce CONSTANS expression and are essential for a photoperiodic flowering response. Dev C 17:75

Foyera CH, Noctor G (2003) Redox sensing and signalling associated with reactive oxygen in chloroplasts, peroxisomes and mitochondria. Physiol Plant 119:355–364

Frohnmeyer H, Staiger D (2003) Ultraviolet-B radiation-mediated responses in plants. Balancing damage and protection. Plant Physiol 133:1420–1428

Frohnmeyer H, Ehmann B, Kretsch T, Rocholl M, Harter K, Nagatani A, Furuya M, Batschauer A, Hahlbrock K, Schäfer E (1992) Differential usage of photoreceptors for chalcone synthase gene expression during plant development. Plant J 2:899–906

Frohnmeyer H, Loyall L, Blatt MR, Grabov A (1999) Millisecond UV-B radiation evokes prolonged elevation of cytosolic-free Ca2_ and stimulates gene expression in transgenic parsley cell cultures. Plant J 20:109–118

Fuglevand G, Jackson JA, Jenkins GI (1996) UV-B, UV-A, and blue light signal transduction pathways interact synergistically to regulate chalcone synthase gene expression in Arabidopsis. Plant Cell 8 (12):2347–2357

Gao Q, Zhang LX (2008) Ultraviolet-B-induced oxidative stress and antioxidant defense system responses in ascorbate-deficient vtc1 mutants of Arabidopsis thaliana. J Plant Physiol 165:138–148

Gardner G, Lin C, Tobin EM, Loehrer H, Brinkman D (2009) Photobiological properties of the inhibition of etiolated *Arabidopsis* seedling growth by ultraviolet-B radiation. Plant Cell Environ 32:1573–1583

Gilchrest BA, Eller MS (1999) DNA photodamage stimulates melanogenesis and other photoprotective responses. J Investig Dermatol Symp Proc 4:35–40

González Besteiro MA, Bartels S, Albert A, Ulm R (2011) Arabidopsis MAP kinase phosphatase 1 and its target MAP kinases 3 and 6 antagonistically determine UV-B stress tolerance, independent of the UVR8 photoreceptor pathway. Plant J 68:727–737

Grubera H, Heijdea M, Hellerc W, Albertd A, Seidlitzd HK, Ulma R (2010) Negative feedback regulation of UV-B–induced photomorphogenesis and stress acclimation in Arabidopsis. Proc Natl Acad Sci U S A 107 (46):20132–20137

Gyula P, Schafer E, Nagy F (2003) Light perception and signalling in higher plants. Curr Opin Plant Biol 6 (5):446–452

Harter K, Frohnmeyer H, Kircher S, Kunkel T, Muhlbauer S, Schäfer E (1994a) Light induces rapid changes of the phosphorylation pattern in the cytosol of evacuolated parsley protoplasts. Proc Natl Acad Sci U S A 91:5038–5048

Harter K, Kircher S, Frohnmeyer H, Krenz M, Nagy F, Schäfer E (1994b) Light-regulated modification and nuclear translocation of cytosolic G-box binding factors in parsley. Plant Cell 6:545–559

Heijde M, Ulm R (2012) UV-B photoreceptor-mediated signaling in plants. Trends Plant Sci 17:230–237

Heijde M, Ulm R (2013) Reversion of the Arabidopsis UV-B photoreceptor UVR8 to the homodimeric ground state. Proc Natl Acad Sci U S A 110 (3):1113–1118

Holley SR, Yalamanchili RD, Moura DS, Ryan CA, Stratmann JW (2003) Convergence of signaling pathways induced by systemin, oligosaccharide elicitors, and ultraviolet-B radiation at the level of mitogen-activated protein kinases in Lycopersicon peruvianum suspension-cultured cells. Plant Physiol 132:1728–1738

Hutchinson F (1987) A review of some topics concerning mutagenesis by ultraviolet light. Photochem Photobiol 45:897–903

Ikeuchi M, Ishizuka T (2008) Cyanobacteriochromes: a new superfamily of tetrapyrrole-binding photoreceptors in cyanobacteria. Photochem Photobiol Sci 7:1159–1167

Imaizumi T, Tran HG, Swartz TE, Briggs WR, Kay SA (2003) FKF1 is essential for photoperiodic-specific light signalling in Arabidopsis. Nature 426:302–306

Ito S, Song YH, Imaizumi T (2012) LOV domain-containing F-box proteins: light-dependent protein degradation modules in Arabidopsis. Mol Plant 5:573–582

Jackson JA, Jenkins GI (1995) Extension growth responses and expression of flavonoid biosynthesis genes in the Arabidopsis hy4 mutant. Planta 197:233–239

Jenkins GI (2009) Signal transduction in responses to UV-B radiation. Ann Rev Plant Biol 60:407–431

Jenkins GI, Christie JM, Fuglwand G, Long JC, Jackson JA (1995) Plant responses to UV and blue light: biochemical and genetic approaches. Plant Sci 112:117–138

Jenkins GI, Long JC, Wade HK, Shenton MR, Bibikova TN (2001) UV and blue light signalling: pathways regulating chalcone synthase gene expression in Arabidopsis. New Phytol 151:121–131

Jiang L, Wang Y, Björn LO, Li SS (2009) Arabidopsis RADICAL-INDUCED CELL DEATH1 is involved in UV-B signaling. Photochem Photobiol Sci 8:838–846

Jiang L, Wang Y, Li QF, Björn LO, He JX, Li SS (2012) Arabidopsis STO/BBX24 negatively regulates UV-B signaling by interacting with COP1 and repressing HY5 transcriptional activity. Cell Res 22:1046–1057

Jordan BR (1996) The effects of ultraviolet-B radiation on plants: a molecular perspective. Adv Bot Res 22:98–138

Kagawa T, Sakai T, Suetsugu N, Oikawa K, Ishiguro S, Kato T, Tabata S, Okada K, Wada M (2001) Arabidopsis NPL1: a phototropin homolog controlling the chloroplast high-light avoidance response. Science 291:2138–2141

Kaiser T, Emmler K, Kretsch T, Weisshaar B, Schäfer F, Batschauer A (1995) Promoter elements of the mustard CHSI gene are sufficient for light-regulation in transgenic plants. Plant Mol Biol 28:219–229

Kaiserli E, Jenkins GI (2007) UV-B promotes rapid nuclear translocation of the UV-B-specific signaling component UVR8 and activates its function in the nucleus. Plant Cell 19:2662–2673

Kendrick RE, Kronenberg GHM (1994) Photomorphogenesis in plants, 2nd edn. Kluwer, Dordrecht

Klessig DF, Durner J, Noad R, Navarre DA, Wendehenne D et al (2000) Nitric oxide and salicylic acid signaling in plant defense. Proc Natl Acad Sci U S A 97:8849–8855

Kliebenstein DJ, Lim JE, Landry LG, Last RL (2002) Arabidopsis UVR8 regulates ultraviolet-B signal transduction and tolerance and contains sequence similarity to human regulator of chromatin condensation. Plant Physiol 130:234–243

Koornneef M, Rolff E, Spruit CJP (1980) Genetic control of light-inhibited hypocotyl elongation in Arabidopsis thaliana (L.) Heynh. Z Pflanzenphysiol 100:147–160

Kubasek WL, Shirley BW, Mckillop A, Goodman HM, Briggs W, Ausubel FM (1992) Regulation of flavonoid biosynthetic genes in germinating Arabidopsis seedlings. Plant Cell 4(10):1229–1236

Lang-Mladek C, Xie L, Nigam N, Chumak N, Binkert M, Neubert S et al (2012) UV-B signaling pathways and fluence rate dependent transcriptional regulation of ARIADNE12. Physiol Plant 145:527–539

Lee J, He K, Stole V, Lee H, Figueroa P, Gao Y, Tongprasit W, Zhao HY, Lee I, Deng X (2007) Analysis of transcription factor HY5 genomic binding sites revealed its hierarchical role in light regulation of development. Plant Cell 19:731–749

Lin C, Robertson DE, Ahmad M, Raibekas AA, Jorns MS, Dutton PL, Cashmore AR (1995a) Association of flavin adenine dinucleotide with the Arabidopsis blue light receptor CRYI. Science 269:968–970

Lin C, Ahmad M, Gordon D, Cashmore AR (1995b) Expression of an Arabidopsis cryptochrome gene in transgenic tobacco results in hypersensitivity to blue, UV-A and green light. Proc Natl Acad Sci U S A 92:8423–8427

Lin C, Ahmad M, Cashmore AR (1996) Arabidopsis cryptochrome is a soluble protein mediating blue light-dependent regulation of plant growth and development. Plant J 10:893–902

Mackerness S, John CF, Jordan BR, Thomas B (2001) Early signaling components in ultraviolet-B responses: distinct roles for different reactive oxygen species and nitric oxide. FEBS Lett 489:237–242
Malhotra K, Kim ST, Batschauer A, Dawut L, Sancar A (1995) Putative blue-light photoreceptors from Arabidopsis thaliana and Sinapis alba with a high-degree of sequence homology to DNA photolyase contain the 2 photolyase cofactors but lack DNA repair activity. Biochemistry 34:6892–6899
McKenzie RL, Björn LO, Bais A, Ilyas M (2003) Changes in biologically active ultraviolet radiation reaching the Earth's surface. Photochem Photobiol Sci 2:5–15
Mittler R, Vanderauwera S, Gollery M, Van Breusegem F (2004) Reactive oxygen gene network of plants. Trends Plant Sci 9:490–498
Montgomery BL (2007) Sensing the light: photoreceptive systems and signal transduction in cyanobacteria. Mol Microbiol 64:16–27
Morales LO, Tegelberg R, Brosché M, Keinänen M, Lindfors A, Aphalo PJ (2010) Effects of solar UV-A and UV-B radiation on gene expression and phenolic accumulation in Betula pendula leaves. Tree Physiol 30:923–934
Morales LO, Tegelberg R, Brosché M, Lindfors A, Siipola S, Aphalo PJ (2011) Temporal variation in epidermal flavonoids due to altered solar UV radiation is moderated by the leaf position in Betula pendula. Physiol Plant 143:261–270
Morales LO, Brosché M, Vainonen J, Jenkins GI, Wargent JJ, Sipari N, Strid Å, Lindfors AV, Tegelberg R, Aphalo PJ (2013) Multiple roles for UV RESISTANCE LOCUS 8 in regulating gene expression and metabolite accumulation in Arabidopsis under solar UV radiation. Plant Physiol 161(2):744–759
Moseyko N, Zhu T, Chang HS, Wang X, Feldman LJ (2002) Transcription profiling of the early gravitropic response in Arabidopsis using high-density oligonucleotide probe microarrays. Plant Physiol 130:720–728
Mullineaux CW (2001) How do cyanobacteria sense and respond to light? Mol Microbiol 41:965–971
Nawkar GM, Maibam P, Park JH, Sahi VP, Lee SY, Kang CH (2013) UV-induced cell death in plants. Int J Mol Sci 14(1):1608–1628
Neill SJ, Desikan R, Clarke A, Hurst RD, Hancock JT (2002) Hydrogen peroxide and nitrogen oxide as signaling molecules in plants. J Exp Bot 53:1237–1247
Oravecz A, Baumann A, Måte Z, Brzezinska A, Monlinier J et al (2006) Constitutively photomorphogenic1 is required for the UV-B response in Arabidopsis. Plant Cell 18:1975–1990
Rice-Evans CA, Miller NJ, Papanga G (1997) Antioxidant properties of phenolic compounds. Trends Plant Sci 2:152–159
Rizzini L, Favory J-J, Cloix C, Faggionato D, O'Hara A, Kaiserli E, Baumeister R, Schäfer E, Nagy F, Jenkins GI, Ulm R (2011) Perception of UV-B by the Arabidopsis UVR8 protein. Science 332:103–106
Sakai T, Kagawa T, Kasahara M, Swartz TE, Christie JM, Briggs WR, Wada M, Okada K (2001) Arabidopsis nph1 and npl1: blue light receptors that mediate both phototropism and chloroplast relocation. Proc Natl Acad Sci U S A 98:6969–6974
Sancar A, Lindsey-Boltz LA, Unsal-Kacmaz K, Linn S (2004) Molecular mechanisms of mammalian DNA repair and the DNA damage checkpoints. Annu Rev Biochem 73:39–85
Schafer E, Kunkel T, Frohnmeyer H (1997) Signal transduction in the photocontrol of Chalcone synthase gene expression. Plant Cell Environ 20(6):722–727
Short TW, Briggs WR (1994) The transduction of blue light signals in higher plants. Annu Rev Plant Physiol Plant Mol Biol 45:143–171
Smith H (2000) Phytochromes and light signal perception by plants? An emerging synthesis. Nature 407:585–591
Somers DE, Fujiwara S (2009) Thinking outside the F-box: novel ligands for novel receptors. Trend Plant Sci 14:206–213
Somers DE, Schultz TF, Milnamow M, Kay SA (2000) ZEITLUPE encodes a novel clock-associated PAS protein from Arabidopsis. Cell 101:319–329
Stracke R, Favory JJ, Gruber H, Bartelniewoehner L, Bartels S, Binkert M, Funk M, Weisshaar B, Ulm R (2010) The Arabidopsis bZIP transcription factor HY5 regulates expression of the PFG1/MYB12 gene in response to light and ultraviolet-B radiation. Plant Cell Environ 33:88–103
Stratmann J (2003) Ultraviolet-B radiation co-opts defense signaling pathways. Trends Plant Sci 8:526–533
Suetsugu N, Wada M (2013) Evolution of three LOV blue light receptor families in green plants and photosynthetic stramenopiles: phototropin, ZTL/FKF1/LKP2 and ureochrome. Plant Cell Physiol 54:8–23
Surplus SL, Jordan BR, Murphy AM, Carr JP, Thomas B, A-H-Mackerness S (1998) Ultraviolet-B induced responses in Arabidopsis thaliana: role of salicylic acid and ROS in the regulation of transcripts and acidic PR proteins. Plant Cell Environ 21:685–694
Takase T, Nishiyama Y, Tanihigashi H, Ogura Y, Miyazaki Y, Yamada Y et al (2011) LOV KELCH PROTEIN2 and ZEITLUPE repress Arabidopsis photoperiodic flowering under non-inductive conditions, dependent on FLAVIN-BINDING KELCH REPEAT F-BOX1. Plant J 67:608–621
Tevini M, Teramura AH (1989) UV-B effects on terrestrial plants. Photochem Photobiol 50:479–487
Ulm R, Nagy F (2005) Signalling and gene regulation in response to UV light. Curr Opin Plant Biol 8:477–482
Ulm R, Baumann A, Oravecz A, Máté Z, Adám E, Oakeley EJ, Schäfer E, Nagy F (2002) Genome-wide analysis of gene expression reveals function of the bZIP transcription factor HY5 in the UV-B response of *Arabidopsis*. Proc Natl Acad Sci U S A 101:1397–1402
Wargent JJ, Moore JP, Roland Ennos A, Paul ND (2009) Ultraviolet radiation as a limiting factor in leaf expansion and development. Photochem Photobiol 85:279–286

Weißhaar B, Block A, Armstrong G, Herrmann A, Schulze-Lefert P, Hahlbrock K (1991) Regulatory elements required for light-mediated expression of the Petroselinum crispum chalcone synthase gene. EMBO J 10:1777–1786

Wellmann E (1976) Specific ultraviolet effects in plant morphogenesis. Photochem Photobiol 24:659–660

Wu M, Grahn E, Eriksson LA, Strid Å (2011) Computational evidence for the role of Arabidopsis thaliana UVR8 as UV-B photoreceptor and identification of its chromophore amino acids. J Chem Inf Model 51:1287–1295

Wu D, Hu Q, Yan Z, Chen W, Yan C, Huang X, Zhang J, Yang P, Deng H, Wang J et al (2012) Structural basis of ultraviolet-B perception by UVR8. Nature 484:214–219

Zhou B, Li Y, Xu Z, Yan H, Homma S, Kawabata S (2007) Ultraviolet A-specific induction of anthocyanin biosynthesis in the swollen hypocotyls of turnip (Brassica rapa). J Exp Bot 58:1771–1781

6 Genotype X–(UV) Environment Interaction-Based Trait Expression in Plants

6.1 Introduction

Enhanced UV fluence on earth through depleted ozone layer was first reported during the later part of the last century (Blumthaler and Ambach 1990) when an "ozone hole" was detected in the stratosphere over the Antarctic region. This disrupted the equilibrium that persisted between survival of life forms and the earth's atmosphere since the migration of life forms to terrestrial habitat in the Ordovician era. With further ozone thinning detected over the Arctic and also over the mid-latitudes (Atkinson et al. 1989), enhanced fluence of the deleterious (solar) UV on earth became a widespread occurrence that threatened persistence of life on earth. This raised an alarm and stimulated urgent research to explore UV acclimation strategy in plants that due to their unique feature of upregulation of flavonoid (key player in acclimation process – see Chap. 4) synthesis are pivotal in sustaining life on earth.

Cell molecular mechanism that confers UV tolerance in plants, thereby allowing plants to cope with ambient UV, includes upregulation of (a) synthesis of genetically controlled internal UV screening compounds, viz., flavonoid (Filippos et al. 2007), that due to their structural specialization also serve to enhance antioxidative potential in cells (Tarrant et al. 2014; Brosche and Strid 2003) and of (b) repair enzymes, viz., DNA photolyase for DNA repair (Hutchinson 1987), for ameliorating cellular conditions toward supporting cell division and growth in plants under deleterious UV environment. Plants exhibit tolerance to variation in naturally fluencing UV radiation, such as along latitudinal variation (Robberecht and Caldwell 1978) and also along altitudinal gradient (Sullivan et al. 1992) – UV fluence being high at high altitude due to rarefication of air, even within the same latitude.

While plants native to low elevations under low UV fluence have been found to be sensitive when exposed to high UV-B atmosphere, plants with high UV preexposures exhibit tolerance to experimental conditions of high UV-B radiations (Teramura and Sullivan 1991). This is due to the fact that low dose UV, fluencing through aerosol/cloud cover/weather variation in locations of high-intensity UV, induces (through signal transduction) upregulation of genes effective for developing UV acclimation (Filippos et al. 2007; Hutchinson 1987). Accumulation of flavonoids thus synthesized serves to protect plants (acclimated during preexposures to high UV associated temporal low dose UV environment) through location-based variation in ambient UV radiation (Jakkola and Hathhola 2010).

Upregulated synthesis and accumulation of UV protective secondary metabolites such as flavonoids which act as UV screens in the epidermal layer restricting entry of UV to sensitive tissue layers has been shown to vary greatly

S. Sen Mandi, *Natural UV Radiation in Enhancing Survival Value and Quality of Plants*,
DOI 10.1007/978-81-322-2767-0_6

among various plant species reflecting genetic control (Mazza et al. 2000). The extent of variation in accumulated UV protective compounds exhibiting variation with respect to difference in latitude/altitude is finally manifested as a resultant of gene X–UV environment experienced by the genotype – the genetic complement being the factor determining variation of UV related trait development in different genotypes in the same location (Shyam Chowdhury and Sen Mandi 2012). Such acclimation strategies enable plants to establish and survive in the open field under variations of UV atmosphere (Caldwell et al. 1994; Björn et al. 1997; Allen et al. 1999) often at altered (latitudinal/altitudinal) location with associated trait (viz., flavonoid accumulation) variation. A large body of literature is available, from experiments in open field, on plant growth and development including synthesis of internal UV screening compounds, viz., phenolics, such as flavonoids. Due to their structural specialization, flavonoids also function as (nonenzymatic) antioxidants (Chen et al. 2013); another component of UV acclimation process in plants constitute UV induced damage control through repair of macromolecules, viz., DNA, by enzymes such as photolyase. Upregulation of phenolics and related compounds (Hofmann et al. 2000)/ enzymes (Huthinson 1987) for development of tolerance to UV radiation is manifested as trait variation reflecting gene X environment (UV) interaction (Caldwell et al. 1982); this determines plant's ability to survive under locational (altitude/latitude) variation of UV fluence (Tevini and Teramura 1989) – the efficiency for plant survival exhibiting varietal difference (Teramurra et al. 1991). Thus, plants are seen to exhibit trait difference for perpetuating and successfully performing under variations of UV radiation, thereby establishing a trait selective role for UV.

Such studies indicate that manifestation of UV tolerance, exhibited as UV-adapted traits, is developed during preexposure to high UV fluence. Accumulation of flavonoids synthesized as acclimatory strategy during (pre)exposure to high UV radiation serves for cell protection through subsequent exposures to variations in UV fluence (Antonovics 1975). As a corollary, Caldwell et al. (1982) demonstrated that in the Arctic region, with naturally thick stratospheric ozone layer, arctic ecotypes, variants of the same species, viz., *Oxyria digyna*, were consistently more sensitive to high UV radiation due to their preexposures under low UV environment than their counterparts collected from mountains in lower latitudes (with high UV fluence). Such studies demonstrate that the plant's tolerance to UV is determined by the extent of UV previously experienced by the plant.

Early studies on the effect of UV on plants were conducted in closed chambers (Rozema et al. 1997; Corlett et al. 1997); such studies demonstrated damaging effect of UV on plants (Jordan 1996) – the effects often manifested as trait variation (Teramura et al. 1991). Data from experiments in greenhouse/closed chamber, without regard to natural atmospheric balance, however, showed exaggerated detrimental effects on structural and functional aspect of plant cells. In green house grown plants, leaf cuticle that serves to restrict UV penetration into leaf cells appear excessively thick (exaggerated) compared to the cuticle seen in plants in open field (Martin and Juniper 1970; Steinmüller and Tevini 1985). Again, many of the detrimental effects of UV on photosynthesis observed under laboratory conditions are appropriately countered in plants in open field (Fiscus and Booker 1995) where UV and PAR interactions are effective (Potters et al. 2009; Lidon 2012). Evidently, for realistic understanding on UV affected trait development, studies should be conducted in open field (Blumthaler and Amback 1990; Ajavon et al. 2007). In open field, natural UV-B radiation, fluencing as variable dose throughout the day (varying with aerosol/cloud cover as well as other environmental changes), exhibits pleiotropic effects (Frohnmeyer and Staiger 2003) often as developmental (morphology and physiology) photomorphogenic traits reflecting UV acclimation strategies in plants.

Genotypic differences in acclimation to UV-B radiation may contribute to the variation in growth responses has been documented in field grown plants (Ballaré et al. 2011). Such

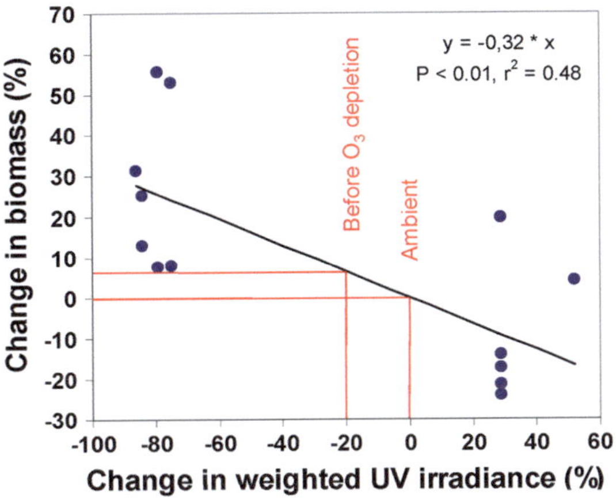

Fig. 6.1 A quantitative estimate of the reduction in the growth of the primary producers that may have occurred as a consequence of increased UV-B radiation between 1978 and 2008. Change in aboveground biomass as a function of experimentally imposed changes in weighted UV irradiance (Taken from Ballaré et al. 2011)

studies provide rough estimates of dose-response functions, where parameters such as leaf area and biomass are plotted against the dose of UV-B radiation received by plants (Fig. 6.1).

6.2 UV-Related Trait Development: Reflection of Gene X–UV Environment

The role of genetic variation in plants, responding to selective agents in the environment (particularly UV radiation), has been studied by many authors (Fischer 1930). For studies relating to survival and trait development under varying UV environment, it is important to focus on the action spectra because of the differential sensitivity of plants as a function of varietal difference vs absorption spectra taking into consideration surface flux of the different wavelength related to the biological impacts of UV-B on plants. Genotypes within a given species (i.e., cultivar) have been reported to differ in their tolerance to UV-B radiation (Correia et al. 1998; Usmanov et al. 1988; Kalbina and Strid 2006). Studying maize in Mediterranean climate, Correia et al. (1998) have reported detrimental effect of UV on dry weight, leaf area, ear length, mean relative growth rate, and mean net assimilation that were significantly reduced by UV-B treatment in some cultivars whereas other cultivars were not much affected by UV. Usmanov et al. (1988) from their studies on 19 lines of *Arabidopsis thaliana* demonstrated intraspecific genetic differences for UV-B sensitivity in survival, developmental rate, plant height, and pod number per plant. The authors found that these 19 genotypes differed significantly in their response to UV-B and that intraspecific differences exceeded average value for both interspecific and intergenic differences. That UV protective traits in plants, developed under UV-B radiation, are genetically controlled have also been demonstrated in ecotypes of *Arabidopsis thaliana* (Kalbina and Strid 2006); *Arabidopsis thaliana* ecotypes C24, Wassilewskija (Ws), and Columbia-0 (Col-0) under supplementary ultraviolet A + B (UV A + B) radiation exhibit ecotype specific differences in expression of the gene for the pathogenesis related protein PR-5. C24 showed an increased expression level of PR-5 (5 and 20-fold higher compared with Col-0 and Ws, respectively).

The unique potential involving plasticity of molecular mechanism operative particularly in plant cells helps in developing UV acclimation in plants for survival under damaging UV radiation. While high dose UV radiation on entry into leaf internal tissue (Sect. 4.2) is scheduled to cause UV induced macromolecular damage, intermittently fluencing low dose UV radiation epigenetically signals synthesis of UV protective compounds viz. flavonoids that constitute internal screen for restraining UV entry into plant cells. This represents a prophylactic measure for protection from damaging effects of high-dose UV radiation. Plants that exhibit environment-related variation in manifestation of UV related effects represent UV adaptive plants.

Such cell molecular processes effective for plant survival involving enhanced synthesis of flavonoids, bypassing stress symptoms (Hectors et al. 2007) affect phenotype/trait development that is distinct from the nonspecific cell damage response observed under stress-inducing high intensity UV-B radiation (Jansen et al. 1998).

Flavonoids representing compounds of dietary and medicinal value in plants often lead to development of elite plants with desirable value added traits.

Chen et al. (2013) reported that leaf phenolics accumulation in the Tibetan Plateau grasslands (altitude 4531 m a.s.l) with higher UV radiation, was significantly higher than that in the inner Mongolian grasslands (altitude 518 m a.s.l) with lower UV radiation. From studies on grasslands of the Tibetan Plateau, Chen et al. (2013) have found that UV radiation was the strongest climatic factor driving the variation of leaf phenolics, a genetic trait specifically operative in plants, across Chinese grasslands, suggesting that higher leaf phenolics could facilitate plants' adaptation under strong light and UV radiation conditions, particularly in the high altitude as seen in the Tibetan Plateau. Their results indicate that leaf phenolics in general may influence vegetation attributes and indirectly influence ecosystem processes by covarying with other leaf functional traits (Chen et al. 2013).

Molecular events for cell protection by rearrangement under variation of UV fluence at different locations, also include enzymatic DNA repair by CPD photolyase (Sect. 4.3). Wang et al. (2014) have reported higher photolyase activity (Sect. 4.3.4.1) in plants at high elevation (with high UV fluence (Sect. 1.5.3)). Plants thus exhibiting malleable traits for adjusting epigenetically with variation in ambient UV environment are significant with respect to evolutionary developments of plants for survival under enhancing variation of UV environment.

Studies of Ren et al. (2007) on two species of *Populus* (with appreciable similarity in genomes) originating from high and low altitude have shown that *Populus canadensis* that originated from high altitude being apparently adapted to high levels of UV-B exhibits greater tolerance to enhanced UV radiation compared to *P. cathayana* originating from lower altitude; these two plants represent ecotypes that on genetic analysis reveal cryptic variation between the genomes (detectable through SNP analysis). These two species apparently represent micro-evolutionary variants. Similar findings have also been recorded by Ghosh and Sen Mandi on ecotypes of *Zingiber officinale* collected from Northeastern India (Ghosh and Sen Mandi 2015). Fiella and Peñuelas (1999) and Baroniya et al. (2011) have demonstrated that Arctic plant species are more sensitive to UV-B, both in terms of growth and reproduction, than alpine species (representing a naturally high UV fluencing region). Baroniya et al. (2011) have reported role of ambient UV radiation in growth and yield characteristics of eight soybean cultivars grown under field conditions. It has also been shown that UV fluence during early plant development markedly affects plant traits in mature plant that may confer plant survival value (Caldwell et al. 1982). In view of the above discussion, a paradigm shift in photobiology connecting UV radiation with molecular plant biology related to plant protection by secondary metabolites and crop productivity achieved by the primary metabolic pathway is now gaining grounds (Barnes et al. 2005).

6.3 Agronomic and Quality Traits in Plants

Biosynthesis of phenolics, viz., flavonoids, by secondary metabolic pathway is central to development of UV-adaptive quality traits that, in addition to providing cellular protection from UV, also affect plant quality traits conferred by secondary metabolites such as color conferred by anthocyanin (Kubasek et al. 1992) and aroma conferred by flavonoid-derived product, viz., flavonoid glycosides (Shyam Choudhury and Sen Mandi 2011). It is important to appreciate that UV related trait development in plants could represent a) temporal epigenetic phenomena reflected as UV adaptive traits or b) UV adapted/constitutive traits when the epigenetic effect is imprinted on the genome (see Chap. 7). Epigenetic effects that relate to temporal trait variation (albeit at the genetic level) as in UV adaptive traits would be subject to variation and may lead to unstable/confusing outcome particularly in breeding experiments that aim to incorporate desirable trait into otherwise desirable varieties. If however the trait represents an UV adapted trait (see Sect. 7.3), reflected as constitutive trait, planned experiment should help to maximally exploit the natural phenomenon of UV adapted traits through selection and utilization of genotypes from various ecological regions for enhancing survival value of plants under different UV fluence. This may allow for directional selection for UV-B tolerance within plant populations in response to long term changes in UV-B levels (Ballaré et al. 2011).

Prior to the organization of the Climatic Impact Assessment Program (CIAP) by the US Department of Transportation (Nachtwey and Rundel 1975), very little analytical studies were conducted on plants growing under different UV fluencing locations. Efforts under this organization subsequently reported large cultivar differences in UV-B response relating to growth and productivity of several species. Such information is now available on a large number of crops such as in soybean (Biggs et al. 1981; Vu et al. 1978), cotton (Ambler et al. 1975), bean (Bennett 1981; Dumpert and Boscher 1982), collard and cabbage (Van and Garrard 1976; Garrard et al. 1976), wheat, barley, corn and rice (Biggs and Kossuth 1978), and spinach (Dumpert and Boscher 1982). Such changes in cell molecular processes are manifested as trait variation as in UV-adapted phenotype, e.g., seed vigor–viability trait in orthodox seed (Talai and Sen Mandi 2010) and also color and aroma status in high-altitude plants/leaf (Shyam Choudhury and Sen Mandi 2011). Planned experiment in this regard should help to maximally exploit the natural phenomenon of UV-adapted traits through selection and study of genotypes from various ecological regions (that differ in UV fluence) for enhancing survival value of plants on earth. This may allow for directional selection for UV-B tolerance within plant populations in response to long-term changes in UV-B levels (Ballaré et al. 2011).

Some instances where gene X (UV) environment interrelationship is manifested as trait development are described below.

6.3.1 Rice

Plants adapted to high radiation at its original site develop higher tolerance to UV compared to plants originating under low UV regions. This is in confirmation with the fact that Asian rice cultivars differ in their responses to elevated UV-B in terms of high/low growth and physiological processes suggesting that the geographical location where the cultivars had been cultivated might influence sensitivity to UV-B radiation (Teramura et al. 1991) – the effect exhibiting intra- and interspecific variability (Bornman and Teramura 1993; Correia et al. 1998; Mazza et al. 2000). Sato and Kumagai (1993) examined the sensitivity to UV-B radiation of 198 rice cultivars, belonging to five Asian rice ecotypes (aus, aman, and boro from the Bengal region and tjereh and bulu from Indonesia) along with Japanese lowland and upland rice groups. The authors found that sensitivity to UV-B radiation varies widely among different cultivars belonging to the same ecotype and the same group. Although genotypes generally originating from high UV regions exhibit higher UV tolerance, the authors however reported that rice cultivars originating from regions with higher ambient UV-B radiation do not necessarily exhibit higher levels of tolerance. Sato and Kumagai (1993) have demonstrated that among Japanese lowland rice cultivars, Sasanishiki (a leading variety) is more resistant to UV-B radiation, while Norin 1, a progenitor of Sasanishiki, is less resistant (Kumagai and Sato 1992) indicating that Norin 1 when first introduced to the Japanese lowland high sunshine region developed UV adaptive trait; this on persistent survival in the same environment developed UV-adapted traits through genetic imprinting of the epigenetic trait, thereby conferring wide ranging UV acclimation, in the genome. This imprinting along with subsequent development of other UV-protective traits may lead to the origin of more resistant variety as Sasanishiki. Many rice cultivars belonging to the Japanese lowland rice group and the Indonesian boro ecotype, i.e., plants of the winter ecotype growing under low-dose UV radiation and thus capable of accumulating more flavonoids were found to be resistant to UV-B, while the more sensitive cultivars were found to belong to the aus and aman ecotypes from the Bengal region.

6.3.2 Grape

UV acclimation related capacity of UV screening simultaneously provide specific traits of commercial value that may provide health benefits in production of wine from grape berries.

Accumulation of flavonoids represents an important (genetically controlled) chemical trait of medicinal importance conferring potent antioxidant potential in grape berries and wines. Researchers from the Universidad de La Rioja, Edificio Científico Tecnológico in Spain identified 121 grape genes that exhibited altered expressions as adapted trait under solar UV radiation (Tello et al. 2015). Analysis of the grape berry skin revealed that genes upregulated by UV radiation were mostly UV-B-responsive secondary metabolism related genes.

Flavonols and hydroxycinnamic acids were identified as the main groups of UV-absorbing phenolics in grape berry (Kolb et al. 2003). While increased flavonol level acts as an efficient UV-A shield, UV-B shielding was found to be incomplete, probably due to decreased level of the UV-B-absorbing hydroxycinnamic acids during exposure (Kolb et al. 2003). Anthocyanin pigment accumulation also increases in the skin of black-skinned grape cultivars as a consequence of UV radiation; although high UV irradiances such as those received at high altitudes seem to be required for triggering high acclimation response of compounds (Berli et al. 2008, 2011; Zhang et al. 2012; Kataoka et al. 2003) that perform mainly photoprotective and antioxidative functions to the extent as appropriate according to their UV radiation absorption capacity. In grapes, UV radiation-responsive anthocyanins have been reported to be effective as UV screens; acylation reactions have been reported to convert them to better UV-screeners (Agati and Tattini 2010; Woodall and Stewart 1998; da Silva et al. 2012). Accumulation of stilbenes and volatile compounds in the skin of Malbec grapes is also enhanced by the UV received at high altitudes (Berli et al. 2008; Gil et al. 2013).

In *red grape*, the monoglycoside forms of anthocyanins are mostly responsible for cell molecular protection under UV radiation. Further, its esterification with acetyl or coumaryl groups, as well as substitution of hydroxyl or methyl groups (Conde et al. 2007; Springob et al. 2003), increases stabilization and color variation of this pigment (viz. anthocyanins). Such additions could sometimes be essential to allow binding by transporters (Zhao et al. 2011). Pigment accumulation in the skin during berry ripening takes place from *véraison* to harvest, conferring the natural pigmentation to mature fruits (Boss et al. 1996; Famiani et al. 2000).

In *white grape berries*, studies on acclimation and capacity of UV screening simultaneously provide specific traits of commercial value for production of white wine. In white grape berries, hydroxycinnamic acids and flavonols have been identified as the principal phenolic groups conferring UV protection. UV radiation particularly UV-B elicits an increase of the B-ring ortho-dihydroxylated quercetin and the B-ring monohydroxylated Kaempferol in berries that serve as antioxidants.

6.3.3 Arabidopsis

UV acclimated plants under chronic UV-B radiation do not adversely affect photosynthesis or induce expression of stress responsive genes. The process of UV-B acclimation during total plant development has been studied in the short life span of Arabidopsis by measuring radiation effects on morphology, physiology, biochemistry and gene expression. UV-induced morphological changes in acclimated plants include decrease in rosette diameter, decreased inflorescence height, and increased numbers of flowering stems, indicating that chronic UV-B treatment caused redistribution rather than a cessation of growth, thereby affecting traits but not growth of the plants (Hectors et al. 2007). Based on the comparison of gene expression profiles, Hectors et al. (2007) concluded that cell molecular events associated with acclimation to low, chronic dose rates of UV-B is distinct from those associated with acute, stress-inducing high UV-B dose rates. Such studies establish that UV-B-induced morphogenesis is functionally uncoupled from stress responses. *Arabidopsis* mutants of the endonucleases that function in NER, xpf-3, and uvr1-1 show hypersensitivity to UV-B in terms of hypocotyl growth inhibition. Accumulation of UV-B induced photodimers was responsible for the hypocotyl growth phenotype of these NER mutants after UV-B radiation.

Studying EMS generated mutants, viz., uvt1, uvt2, and uvt3, of which uvt1 exhibits remarkable tolerance to UV showing significantly higher UV-B tolerance than the wild types viz. uvs. Studies on an earlier isolated mutant (Lois and Buchanan 1994) and the wild type, Bieza and Lois (2001) demonstrated that uvt1is a single gene dominant mutation that leads to constitutive increase in accumulation of UV-absorbing compounds; this increases the capacity of the leaves to block UV-B radiation and therefore is likely to be responsible for the elevated resistance of this mutant to UV-B radiation. This increase in absorption in the UV region was found to be due to, at least in part, increases in flavonoid and sinapate accumulation.

Leaves from uvt1 plants appear to be significantly darker than those in wild-type or uvs plants, especially in the petioles and midribs of UV-exposed plants, which show a purple coloration (Fig. 6.2). These observations indicated that the increased UV tolerance of uvt1 might be related to increased pigmentation – the purple coloration in leaf tissues being the result of an accumulation of the pigmented flavonoids, anthocyanins (Holton and Cornish 1995). Expression of chalcone synthase (CHS) mRNA was shown to be constitutively elevated in uvt1 plants, suggesting that the increases in UV absorption may that is an index of enhanced flavonoids, is a consequence of changes in gene expression. Expression of CHS in uvt1 was

Fig. 6.2 Effect of different regimes of UV-B radiation on mature wild-type and mutant *Arabidopsis* plants. One-month-old uvt1, wild-type (WT), and uvs plants were exposed to 0.10 W m22 UV-B (mid-UV), 0.18 W m22 UV-B (high UV), or kept under fluorescent lighting (no UV) for 3 days and allowed to recover for 3 days under fluorescent lighting before being photographed (Taken from Lois and Buchanan 1994)

shown to be still inducible by UV, despite being constitutively high; this indicates that the uvt1 lesion may not affect the UV-mediated regulation of CHS gene expression.

6.3.4 Tea

Tea is one of the most important beverage of the world with major tea-growing regions occurring in Southeast Asia and Eastern Africa. UV-upregulated enzymatic splitting of flavonoid glycosides, the end product of the phenylpropanoid pathway, is known to confer aroma quality in tea (Carrao-Panizzi and Bordingnon 2000). Aroma along with antioxidant potential constitutes important traits of commercial importance in tea leaf. Commercially valued Darjeeling tea of high aroma and high antioxidant potential is manufactured from fresh (young) tea leaves obtained in the first/second flush during April–June, when UV radiation is maximum in this region coinciding with summer solstice in this region (i.e Northern hemisphere). Shyam Chowdhury and Sen Mandi (2012) have demonstrated that tea clones growing at the same latitude (27°N) but at different altitude (exhibiting variation in UV irradiance), viz., in Darjeeling (1371 m a.s.l) and in Assam (107 m a.s.l), under high and low (respectively) natural UV irradiance differ in their aroma and antioxidant content. Leaves of high altitude Darjeeling-grown tea clones were found to possess higher UV-absorbing compounds, (subject to varietal difference) viz. flavonoids that in addition to providing UV protection to plants confers high aroma and antioxidant potential (see Chap. 4); this makes Darjeeling tea a commodity of high commercial value. Accumulation of flavonoids in the same (respective) varieties was lower in leaf of the Assam (107 m a.s.l) grown tea clones.

6.4 Molecular Mechanism for UV Adapted Trait Development

Gene X UV (environment) effects are often observed in plants growing in open field at various locations, an in depth understanding of the cell molecular mechanisms causing such variation would help to utilize this natural phenomenon for developing plants with desired traits often with value addition; these genotypes would be better suited for cultivation under the threat of UV-related climate change. Such studies would also help to identify appropriate cultivation sites for naturally available genotypes with an aim toward achieving best qualitative/quantitative traits that could be utilized for human benefit. Reports on extensive cell molecular research towards exploring the unique feature of cellular plasticity operative in plants for developing resilience to ambient environment (viz. UV radiation) are available in the literature. Attempts at exploiting such natural phenomenon for developing desirable (UV adaptive/adapted) traits within limits of genetic variation, should constitute useful endeavor.

Management of plant response to environmental stimuli toward crop improvement, including cultivar selection for developing desirable traits, has often provided interesting end points, as well as value addition in cash crop, viz., tea (Shyam Choudhury and Sen Mandi 2011), biofortification of crops for enhanced nutritional composition (White and Broadley 2009).

Two case studies are presented in the subsequent sections giving details of experiments on (1) agronomic seed quality, viz., vigor–viability in rice maintained through post-harvest storage, and (2) a commercially important leaf quality viz. aroma and antioxidant potential in tea leaf.

Case Study I

6.4.1 UV Acclimation-Related Seed Vigor–Viability Trait in Rice

Several studies indicate that plant response to environmental stimuli could be exploited for crop improvement starting with cultivar selection. A related term is that of "eustress" or positive stress, whereby the elicitation of a stress response provides a beneficial outcome for an organism (Mauch-Mani and Mauch 2005; Ashraf and Foolad 2007). In the study described below, an attempt has been made to explore, using embryonic axis of rice (orthodox seed), cell molecular rearrangement, viz., cellular plasticity (Weinig et al. 2004) related protective mechanism that may confer cell protection from UV damage in embryonic cells of mature seeds during post harvest storage of rice (with UV transparent husk) for enhancing agronomic value (trait) of the seed.

Rice, the staple food of 55 % of the world's population in the tropical and subtropical region where UV stress is high, constitutes an important study material for such a study. It is important to appreciate that despite the use of modern cultivation practice and improved irrigation techniques, high productivity in cultivation will not be achieved unless high vigor-viability seeds are used for sowing. Provided the seeds exhibit uniformly high emergence and that crop stand establishment is accordingly high, crop productivity will also be high. Impacts of improved variety and modern cultivation practices cannot be realized unless high vigor seed is used for cultivation. In most of the tropical and subtropical rice-producing countries where sunrays (and thus UV fluence) is maximum, maintenance of seed vigor–viability for agricultural productivity and for conservation of biodiversity poses a basic problem. From the time of full maturation, a seed (embryonic axis) would lose vigor–viability as a resultant of molecular damage vs countering effect proportionate to the genetically controlled enzymatic processes; dry post harvest seeds however have to take recourse to non enzymatic means viz. accumulation of flavonoids that constitutes a genetically determined cell molecular mechanism. Flavonoids, though constitutively present in plants, may also be upregulated by UV radiation through diversion of primary metabolism to secondary metabolism (Fig. 6.10) during seed maturation-related dehydration stress in open field – the extent of flavonoid production being under genetic control; flavonoids, that due to their structure–activity relationship (SAR) act both as internal UV screen and also as nonenzymatic antioxidants. Being nonenzymatic, these compounds remain stored in 'dry' mature seed serving as internal screening compounds and nonenzymatic antioxidants through "dry"(orthodox) seed storage, thereby serving to retain seed vigor–viability through the storage period.

Seed vigor trait refers to the potential of a hitherto quiescent embryo to initiate germination and growth for establishing crop stand in agricultural field. The trait is defined as germination ability of the seed at fresh harvest; from that time onward, germination ability of the seed is progressively reduced culminating in loss of germination ability, and the seed is then said to have lost viability. This shows that vigor and viability in orthodox seed are interrelated traits. For a comparative study on vigor trait of different variety, it is thus imperative that seeds used for experiments are freshly harvested or stored at 4 °C over $CaCl_2$ just after harvest. Provided that seed vigor is appreciably high, the embryo will establish a successful seedling that grows successfully to maturity under normal field condition and contributes toward crop productivity. However, if left in post harvest dry storage under ambient conditions for long, the embryonic axis of mature seed (at <8 % mc) gradually accumulates (in absence of metabolic repair mechanism in dry embryonic axis) UV-induced cell molecular damages that, under conditions of precluded enzyme (including repair enzymes) activity, gradually lose potential to germinate; such seeds lose vigor – the extent of loss being a function of genetically controlled vigor trait. Remaining quiescent, a mature seed (with metabolic activity suspended) during post-harvest storage exhibits ROS accumulation (Bailly 2004) and associated UV induced cellular damage. At low moisture during seed storage under ambient condition, the UV transparent husk of rice allows UV penetration into the embryonic axis. This causes macromolecular damages in the embryonic axis; the damaging effect being particularly high in the tropical and subtropical region where UV fluence is high. Under such conditions, seeds of several rice varieties are seen to lose viability within a short period – the duration being a reflection of varietal difference in seed vigor–viability trait. Under the circumstance, seeds that accumulate more UV screening compounds, viz., flavonoids, during seed development on mother plant (i.e., before seed maturation-associated dehydration sets in) would exhibit higher cell molecular protection due to higher flavonoids in the seed; such seeds represent seeds of high vigor trait. This allows higher protection of cell molecular entities in the embryo through the period of post harvest storage. Gradual increase in seed deterioration is manifested as loss of seed vigor, the loss occurring faster in stored seeds that accumulate genetically controlled lower content of cell protective compound; the seed finally succumbs to (UV) stress under ambient storage and is rendered nonviable. Low vigor varieties that accumulate low content of protective compounds exhibit lower value of these traits. Crop productivity from aged seeds is progressively reduced with age particularly due to low crop stand establishment of aged seeds in the field. Julliano et al. (1990) have demonstrated that varieties exhibiting low seed vigor trait are also poor storers exhibiting low value of viability trait. Seed vigor–viability are thus important agronomic traits that are interrelated by the unique feature of plasticity in plants. Loss of viability in rice (orthodox) seed leads not only to reduction in crop productivity but also to loss of valuable biodiversity. It is important to appreciate that UV radiation on quiescent mature rice seed in small farmers' ambient storage brings about a reduction in agronomic quality, viz., vigor of seed; seeds with low vigor trait succumb to stressful field condition when set out for cultivation. This brings about massive loss of planting (seed) material resulting in economic loss of the small farmer.

Seed vigor–viability traits in mature orthodox (dry) seed are manifested as a resultant of high dose UV induced cell molecular damage countered by low-dose UV-induced acclimation response. Enzymatic cellular repair being precluded during prolonged dry storage, extent of nonenzymatic antioxidants would determine vigor–viability trait of (orthodox) rice seed in dry storage of tropical/subtropical regions. Low vigor seeds easily succumb to unfavorable field conditions and fail to produce successful plants in the field, thereby losing out on crop productivity.

Experiments presented here explore the extent of adverse effect of UV on cellular macromolecules vis-a-vis genetically determined protective function manifested as internal UV screens, viz., flavonoid (Fig. 5.26) and antioxidative potential (Fig. 5.24). Experimental material was selected from seedlings (sprouting from freshly harvested seeds) grown in pots simulating field conditions (Fig. 6.3) where two distinct groups of contrasting growth characteristics were evident – viz., high vigor trait varieties (varieties A, B, C, D, H, I, J, N, O, P, Q, and R) and low vigor varieties (varieties E, F, G, K, L, M, S, T, U, and V).

From this large number (22) of varieties (Fig. 6.3), six varieties (three of high vigor and three of low vigor) were randomly selected for experiments to determine status of cell molecular events associated with vigor–viability traits. The high vigor varieties (Patnai-23, Pankaj, and Matla) exhibited much higher growth rate than the low vigor varieties (Jaya, Jogen, and Lalat). The selected varieties, viz., Patnai, Pankaj, Matla, Jaya, Jogen, and Lalat, were designated as A, B, C, D, E, and F, respectively. Freshly harvested seeds of each (selected) variety were divided into three groups:

(a) One group of seeds after harvesting was stored at 4 °C over $CaCl_2$ till used for experiments; these seeds were considered as "unaged."
(b) Another group was stored for 1 year under ambient condition; after completion of 1 year, these seeds were stored at 4 °C over $CaCl_2$ until used for experiments. These seeds were considered as "1-year aged" seeds.
(c) The third group was stored for 2 years under ambient condition; after completion of 2 years, these seeds were stored at 4 °C over $CaCl_2$ till used for experiments. These seeds were considered as "2-year aged" seeds.

Seeds are allowed to germinate under suitable germinating condition; the low vigor varieties exhibit low growth rate (compared to the high vigor varieties) manifested as reduced rate of

Fig. 6.3 Representative photographs of freshly harvested (unaged) rice varieties maintained outdoor (at 30–35 °C) on 15 days. **a–v** represent rice varieties. Patnai-23 (**a**), IET 10890 (**b**), Pankaj (**c**), Matla (**d**), IET 13158 (**e**), Badsabhogh (**f**), Tulshimanjuri (**g**), IET 9978 (**h**), Kataribhogh (**i**), Pusabasmati (**j**), Jaya (**k**), Jogen (**l**), Lalat (**m**), IR30 (**n**), Mohan (**o**), Annada (**p**), IR36 (**q**), Tarwari Basmati (**r**), Basmati aman (**s**), Kalojira (**t**), BasmatiT3 (**u**), and NC- 365 (**v**)

seedling growth (Fig. 6.4). Age associated reduction in vigor is seen as reduced speed of germination and concomitant lowering of seedling growth rate (Fig. 6.4). Similar observations have also been reported by Sen and Osborne (1977), Bewley and Black (1994), and Krishnasamy and Seshu (1989). Delay in initiation of germination appears to be due to time required for cell molecular repair of age-associated damage (cf. Osborne 1980); the staggering pattern of germination observed between varieties appear to be due to cell molecular damage incurred to varying extents in seeds of the different varieties. Loss of vigor–viability potential thus appears to be varietally controlled.

Through prolonged aging in tropical/subtropical conditions, vigor (assessed by seedling growth rate) is further reduced resulting in the final catastrophe, i.e., nonviability of the seed (Fig. 6.4a–c).

Figure 6.4a–c) shows germination efficiency of embryonic axis in seeds of differentially aged rice varieties exhibiting different vigor–viability status manifested as variation in root/shoot length in different varieties. Varieties Patnai-23 (A), Pankaj (B), and Matla (C) exhibit high vigor–viability (rapid seedling growth) in early germination, while Jaya (D), Jogen (E), and Lalat (F) exhibit slow seedling growth (low vigor) in aged seed – the extent of

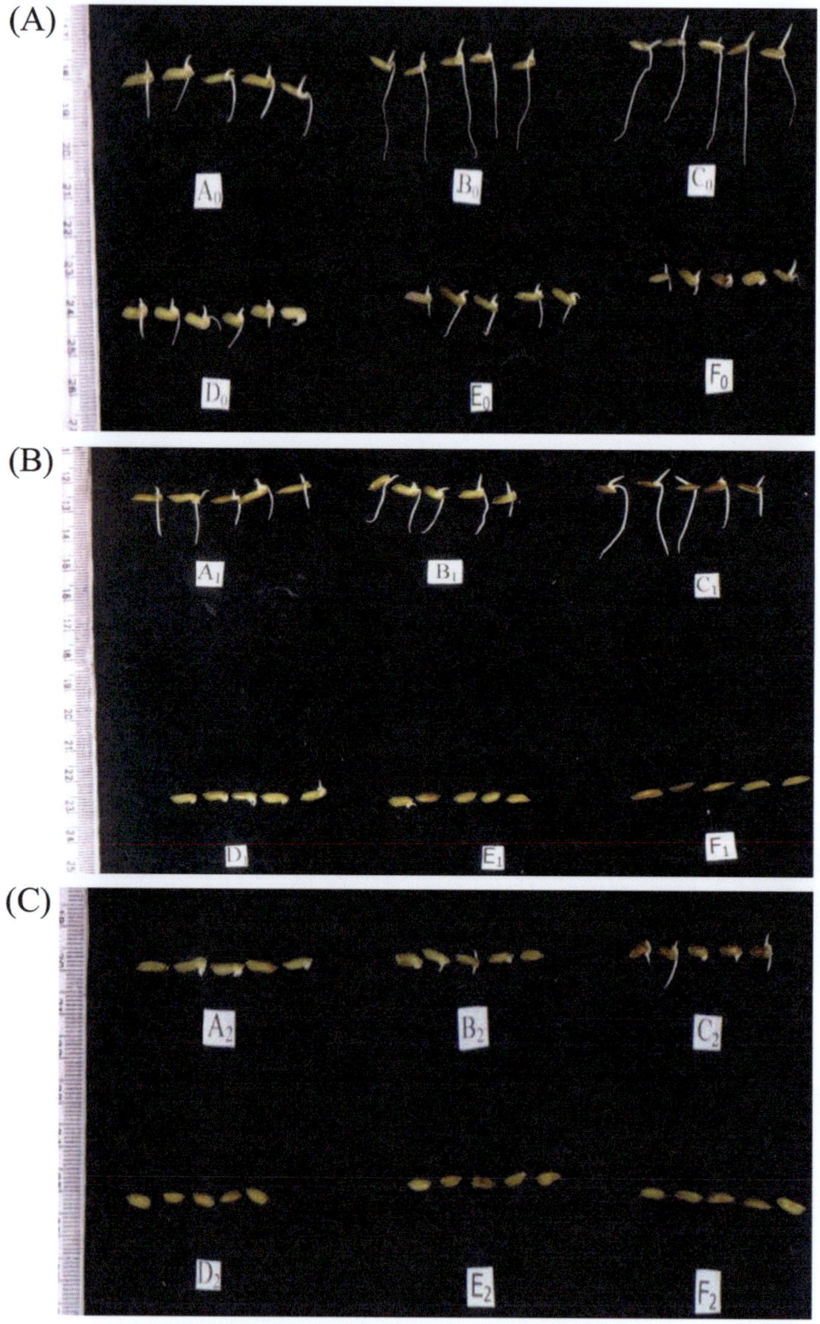

Fig. 6.4 Representative photographs of 48 h old seedlings at 28 ± 2 °C raised on germination plate from (**a**) unaged (designated with 0), (**b**) 1 year-aged (designated with 1), (**c**) 2 year-aged (designated with 2) rice seed stocks. **a** = Patnai-23, **b** = Pankaj, **c** = Matla, **d** = Jaya, **e** = Jogen, **f** = Lalat

reduction in seedling growth of aged seeds is seen to be proportional to varietal difference (Fig. 6.4a–c). Sheng et al. (2005) have described similar findings as reflected in germination index data. Nedeva and Nikolova (1999) with fresh and desiccated wheat seeds have demonstrated that the percent dry matter content and fresh weight increases in fresh high vigor seed appeared higher than in low vigor seed. Ruhland et al. (2007) have demonstrated

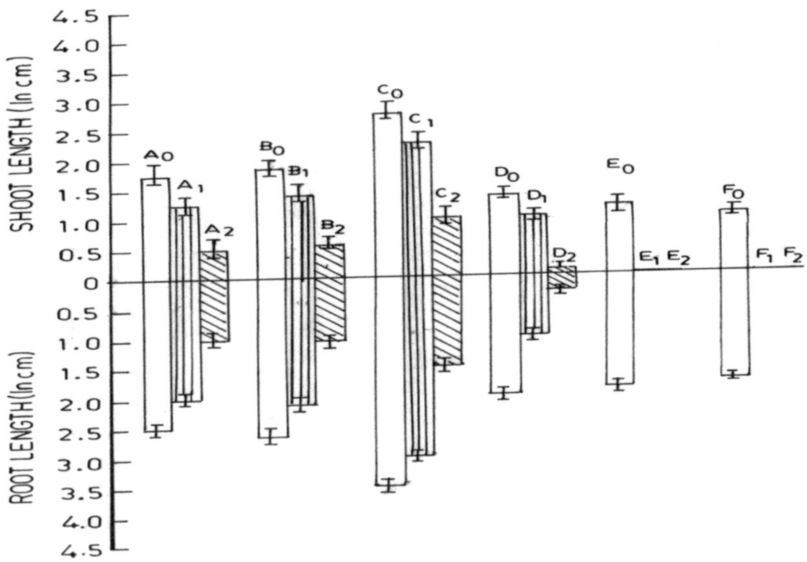

Fig. 6.5 Histogram of root and shoot length of 72 h unaged, 1 and 2 year-aged rice seedlings grown on germination plates in laboratory at 28 ± 2 °C. Unaged seeds designated with 0, 1-year-aged seeds designated with 1- and 2-year-aged seeds designated with 2. A = Patnai-23, B = Pankaj, C = Matla, D = Jaya, E = Jogen, F = Lalat. S.D was calculated for $n = 6$ $P < 0.001$

the reduction in biomass and growth in *Avena sativa* seedlings; the effect in seeds exposed to solar UV radiation being a function of oxidative damage.

A quantitative estimate of seed vigor manifestation of the same varieties showing root/shoot length is given in Fig. 6.5.

In Fig. 6.5, rice varieties A, B, and C show high root/shoot length in comparison to varieties D, E, and F. Difference between varieties is reflected as a function of genetic variation. It may be mentioned here that although varieties studied belong to the different final plant height group (viz., short, medium, and long at maturity), extent of early emergence of shoot above soil (all seeds sown at equal depth under soil) in pots maintained outdoor at 30–35 °C did not reflect such variation between varieties at early stages of germination.

Delayed initiation of germination and reduced seedling growth rate are manifestations of (a) varietal low vigor (Fig. 6.3) and (b) age associated loss of vigor exhibiting low vigor (Fig. 6.4). It is pertinent to mention that age associated low vigor is superimposed on varietal low vigor i.e. varieties that are varietally low vigor lose vigor faster during aging. It may also be noted that at full seed maturity macromolecules exhibit high integrity regardless of varietal vigor studies – variation in integrity becomes evident with progression of aging (see Figs. 4.22 and 4.23). Variation in macromolecular integrity being a function of varietal difference in synthesis of protective compounds viz. flavonoids during seed maturation in open field (Mazza et al. 2000). Reduction in seedling growth rate due to post harvest seed aging reflect accumulation of UV-induced cell molecular damages occurring (as a function of varietal difference) during post-harvest aging in absence of repair enzyme activity in dehydrated embryonic cells; such macromolecular damages are reinstrated with availability of water during early germination (Osborne 1980; SenMandi et al. 1995). Slow-growing seedlings of low vigor stocks that incurs DNA damage during post harvest storage (Figs. 4.22 and 4.23) (Sen Mandi and Bhattacharya 2003) needs to be repaired before onset of growth of the embryonic axis. This study is comparable with the slow-growing hypocotyl of *Arabidopsis thaliana* that take time for cellular repair activities (Biever et al. 2014) before commencement of hypocotyl growth and extension. Provided that cellular damage incurred during post harvest storage is not extensive, the embryonic axis resumes normal growth following completion of requisite repair of DNA and other macromolecules including antioxidative and repair enzymes that confer UV resilience to the emerging seedling. If, however, the extent of damage in the embryonic axis through extensive post-harvest storage is high the embryonic axis succumbs and the seed is found to have become nonviable. The number (%) of germinating seedling in each stock is referred to as percent viability of that stock.

Studies relating to breeding-based introgression of high vigor genes/epigenetic effects through breeding lines where the effect may persist through generations would repay careful investigation (see Sect. 7.4.1 of Chap. 7).

6.4.1.1 Alcohol Dehydrogenase (ADH) Activity Through Progression of Germination

Seeds before sprouting respire by anaerobic means, and thus ADH activity is high at initiation of germination (i.e., at initiation of water imbibitions of seeds). ADH activity is essential for anaerobic respiration (Ho and Scandalios 1975) at preemergence stage of embryo growth/germination. Time taken to achieve resumption of activity of this enzyme thus provides a precise biochemical parameter (marker) for assessment of seed vigor. As sprouting occurs, concentration of ADH decreases as aerobic respiration replaces anaerobic respiration, coincident with emergence of the embryonic axis (at initiation of germination) from within the seed cover – earlier occurrence of this event being a characteristic of high vigor seeds. Thus, time of initiation of ADH activity through progression of "germination" at seed imbibition is thus a reliable parameter for differentiating high and low vigor varieties.

In this study, it is evident that varieties A, B, and C are high vigor varieties since their ADH concentration is almost nil at 48 h, whereas D, E, and F varieties exhibit appreciably high ADH activity at 48 h establishing the need for anaerobic respiration of embryos within seed cover; this correlates with delayed seedling emergence. Such delay in seedling emergence is a reflection of cell molecular repair events in aged seeds that should precede synthesis of cell protective compounds, viz., flavonoids, to provide protection to the emerging embryonic axis. Beyond 48 h, by 72 h ADH activity seems to have started to decline (Fig. 6.6). Comparing the finding with Fig. 6.4, it appears that with the time ADH activity declines in the high vigor varieties coincides with sufficient elongation of the embryonic axis such that the embryos would emerge out of the seed cover. Similar observation has also been observed by Kolloffel (1968) in other seeds, both in monocot and dicot (Ho and Scandalios 1975; De La Fuente and Nicolas 1974).

This time course experiment of ADH activity (taken as marker for embryo elongation and protrusion out of seed cover) in different varieties explains the delayed germination presumably due to requisite cellular repair to be taken up at seed imbibition preceding onset of embryonic growth. This phenomenon may be compared

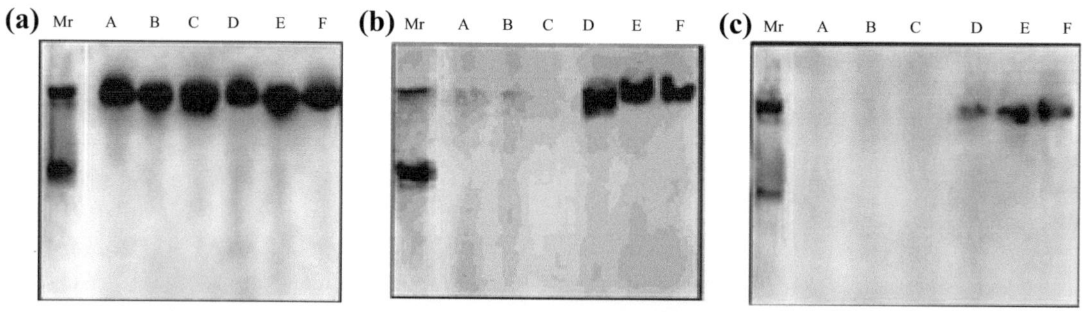

Fig. 6.6 Alcohol dehydrogenase enzyme activity in crude extract obtained from embryonic axis at 0 h (**a**), 48 h (**b**), and 72 h (**c**) imbibition. Lane *Mr* = protein molecular marker (for non-denaturing PA gel) run on the same gel then separated and stained in coomassie *blue*. *A* Patnai-23, *B* Pankaj, *C* Matla, *D* Jaya, *E* Jogen, *F* Lalat

with delayed hypocotyl elongation in dicot seeds due to time required for repairing the damage of DNA caused by UV-B (Biever et al. 2014) and for synthesis of flavonoids which would act as UV screen (Ballare et al. 1996) in the emerging embryonic axis. In these reports, both biosynthesis of flavonoids (Ballare et al. 1996) and DNA repair by photolyase (Biever et al. 2014) may represent cellular plasticity (see Sect. 7.2) specifically operative in plants among higher organisms.

6.4.1.2 Total Lipid Content

Another parameter of seed vigor is rate of lipid biosynthesis associated with cell division and growth of the hitherto quiescent embryonic axis in post harvest stored seed. At initiation of germination, cell membrane lipid is necessary for initiating cell division-associated new cell wall production, while metabolic lipid is required for enhanced metabolism in the growing embryonic axis.

The low total lipid content (reflecting metabolic as well as membrane lipid) in the embryonic axis of varieties D, E, and F (Fig. 6.7) is a feature of varietally low vigor embryos. At fresh harvest condition thus, the high vigor–viable varieties (A,B,C) have higher total lipid content than the low vigor–viable varieties (D,E,F). With onset of post harvest aging (associated metabolic activity remaining precluded) the decline recorded in total lipid content reflects susceptibility of membrane lipid to oxidative stress-related peroxidation (Flood and Sinclair 1981) – the effect being varietally controlled.

6.4.1.3 Molecular Marker Associated with Vigor-Viability Trait in Rice

In the present study, rice seeds of different vigor–viability trait status (matured on mother plant at different locations and during different seasons) were used to explore genome variation.

RAPD analysis of 22 varieties (Fig. 6.3) of rice is shown in Fig. 6.8. In the Gel picture (Fig. 6.8), it is observed that along with several other high vigor varieties, Patnai-23, Pankaj, and Matla (designated as A, C, and D, respectively, in Fig. 6.8) contain a 900 bp size DNA band (marker) which is absent in low vigor varieties (Fig. 6.8) including Jaya, Jogen, and Lalat (designated as K, L, and M, respectively, in Fig. 6.8).

Marker-assisted distinction between varieties of rice with contrasting traits of interest has also been shown through RAPD analysis by Virk et al. (1996), Mishra and Sen Mandi (2004), and Talai and Sen Mandi (2010). Statistical correlative

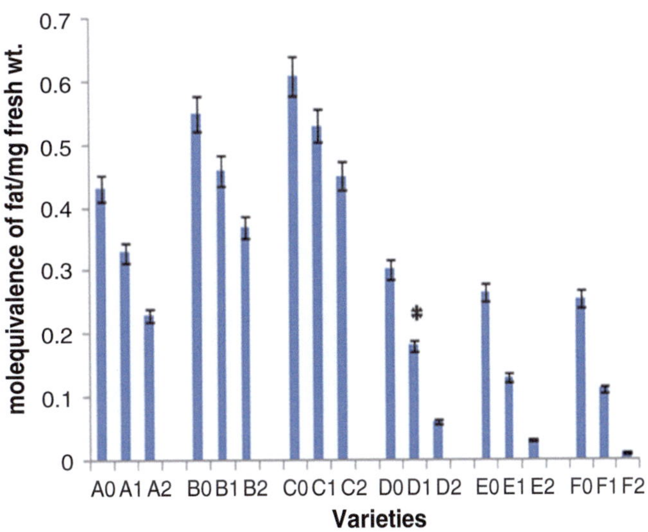

Fig. 6.7 Histogram for spectrophotometric analysis of total lipid (representing metabolic and membrane lipid) content (measured in mol equivalence of fat/mg fresh wt.) of six rice varieties. Unaged seeds designated with 0, 1-year-aged seeds designated with 1 and 2 year-aged seeds designated with 2. *A* Patnai-23, *B* Pankaj, *C* Matla, *D* Jaya, *E* Jogen, *F* Lalat. (S.D calculated for $n = 6, P < 0.001$)

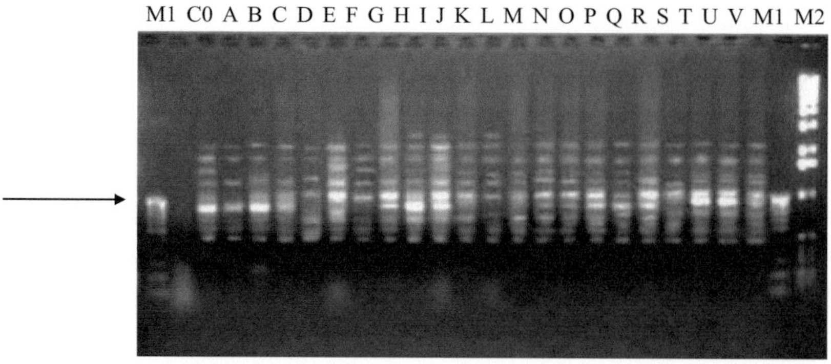

Fig. 6.8 Amplification products from genomic DNA of 22 rice varieties on 1.2 % agarose gel using primer OPAC19. *C0* = negative control without any genomic DNA, *M1* = 100 bp DNA ladder, *M2* = 1 kbp DNA ladder (Promega). Band of interest is marked by *arrow*. **a–v** represent rice varieties: Patnai-23 (A), IET 10890 (B), Pankaj (C), Matla (D), IET 13158 (E), Badsabhogh (F), Tulshimanjuri (G), IET 9978 (H), Kataribhogh (I), Pusa basmati (J), Jaya (K), Jogen (L), Lalat (M), IR30 (N), Mohan (O), Annada (P), IR36 (Q), Tarwari Basmati (R), Basmati aman (S), Kalojira (T), BasmatiT3 (U), and NC-365(V)

analysis involving physiological (growth rate; Fig. 6.4) and biochemical (time course of ADH activity – Fig. 6.6) parameters of seed vigor–viability identified a 900 bp band (marker) generated through use of OPAC 19 primer (Fig. 6.8) which is found to be associated as high vigor–viability varieties of rice (trait-related marker). Varieties (exhibiting the 900 bp band) also exhibit high vigor trait on the basis of ADH activity in time course experiments (Fig. 6.7); varieties that exhibited loss of ADH activity early, represent high vigor embryo (i.e., seed) and were found to be correlated with the 900 bp band in RAPD analysis.

To establish that the 900 bp band present in all high vigor rice varieties are identical (being of same sequence), the fragments (900 bp) were subjected to restriction digestion with three restriction enzymes, viz., Sau96IA, HaeIII, and HindIII (Fig. 6.9). This 900 bp band may be used for selection of rice genotypes possessing the traits of high vigor and viability in wild germplasm screening endeavor.

The gel picture (Fig. 6.8) shows that the 900 bp DNA fragment generated by OPAC19 that appears as marker DNA fragment for high vigor trait in rice appear in all the high vigor varieties studied. Figure 6.9 showed identical restriction fragments, viz., 280 and 620 bp by HaeIII, 320 and 570 bp by HindIII, and 370 and 520 bp by Sau96IA, in selected high vigor varieties (Fig. 6.9). This confirms that the 900 bp bands present in high vigor varieties represent DNA fragments of similar sequence homology.

Multiple regression analysis revealed that the 900 bp band generated by primer OPAC19 is significantly ($P > 0.01$) correlated with high vigor–viability trait – this being absent in low vigor varieties (Fig. 6.8). The 900 bp band after BLAST analysis shows 94 % homology with wheat ACCase. ACCase plays a pivotal role (schematically represented in Fig. 6.10) in total lipid production (primary metabolism) essential for rapid root/shoot growth (manifested as high vigor) that on diversion to the secondary metabolic pathway synthesizes flavonoids (Fig. 6.10). The 900 bp band was used for selection of parent/progeny possessing the contrasting traits of vigor and viability. ACCase gene is known to be upregulated by UV-B radiation, and the regulation occurs through promoter methylation. This appears to be applicable in all high and low vigor varieties collected from a wide range of locations and therefore may be considered as a UV-adaptive trait (see Sect. 7.2 of Chap. 7).

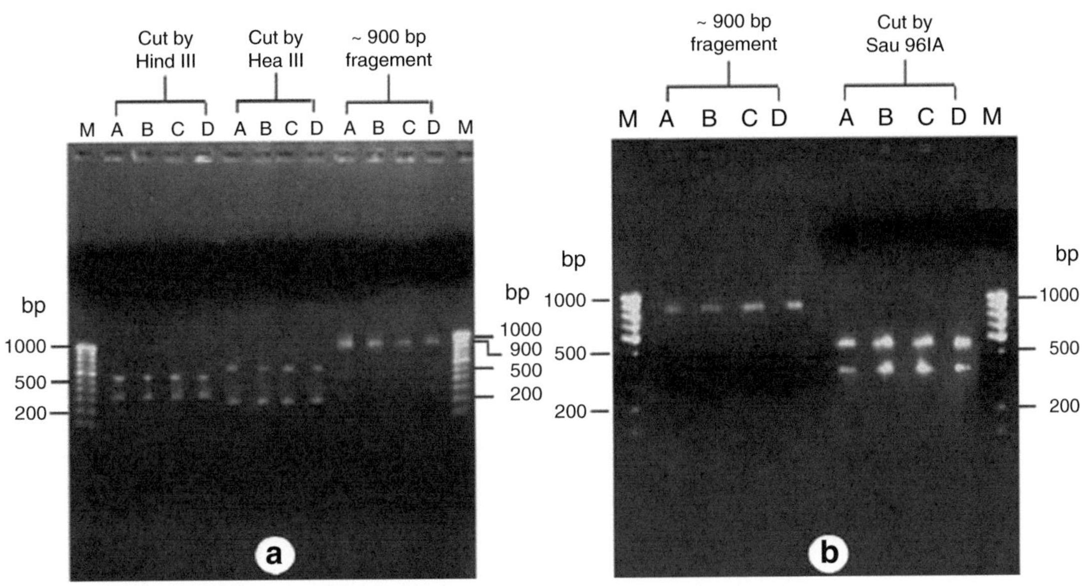

Fig. 6.9 Intact (900 bp DNA fragment) and restriction enzyme-digested products of the 900 bp DNA fragment of high vigor–viability varieties on 1.5 % agarose gel. HindIII and HaeIII (**a**) and Sau96IA (**b**). Lanes $M = 100$ bp DNA ladder (Bangalore Genei). The high vigor varieties shown in this picture are Patnai-23 (A), IET 10890(B), Pankaj (C), Matla (D)

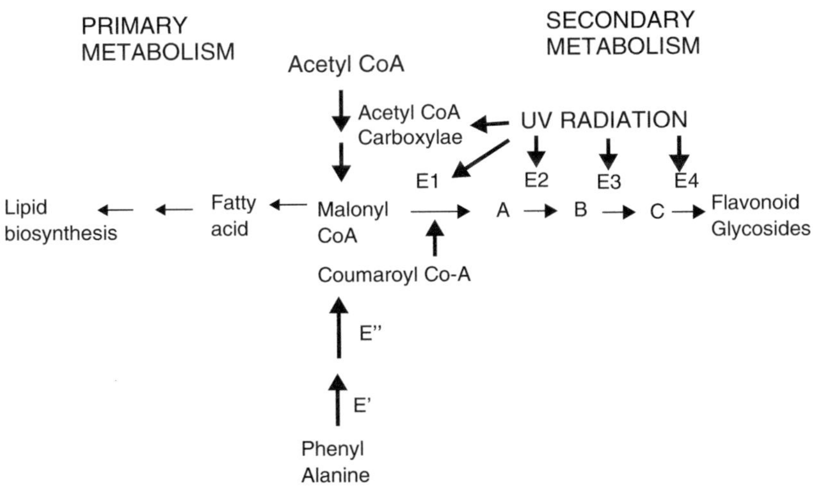

Fig. 6.10 ACCase activity at seed imbibition (for germination) leads to higher production of fatty acid required for onset of cell division at germination of the hitherto quiescent embryo. Dehydration stress at seed maturation causes diversion of ACCase activity toward the onset of secondary metabolism for production of flavonoids that serve as internal UV screen and nonenzymatic antioxidants for protecting seeds under farmers' ambient seed storage. ACCase activity thus plays a pivotal role in mature (dry) rice seed, tilling the balance toward germination (determining vigor trait)/quiescence seed in post-harvest storage that manifests vigor–viability determining viability trait; this trait is manifested on availability of water to hitherto stored seed. [E' – phenyl ammonia lyase (PAL), E''-cinnamate 4′-hydroxylase (C4H), *E1*-chalcone synthase (CHS), *E2*- chalcone isomerase (CHI), *E3*- flavonoid 3′-hydroxylase (F3H), *E4*-dihydroflavonol 4-reductase (DFR), A Chalcone, B flavanone, C dihydroflavonol]

6.4.1.4 Importance of ACCase in Controlling Vigor-Viability Trait in Rice Seed

ACCase activity is the first committed step in lipid biosynthesis (in primary metabolism) that regulates rate of cell division in growing embryo during germination of a hitherto quiescent embryo (in a mature post-harvest seed). Acetyl CoA carboxylase gene is responsible for fatty acid biosynthesis in primary metabolism which provides for faster turnover of lipid membrane development necessary during early period of seed germination (Agarwal et al. 2008; working with soybean seed and rapeseed). As such variation in lipid content among varieties provides an indication of germination rate, i.e., seed vigor, in different varieties. Age-associated loss of lipid content that reflects rate of cell division in growing embryo is correlated with vigor trait in rice seed. Loss of vigor–viability of seeds is thus varietally determined (Fig. 6.7; cf. Fig. 6.4). In dehydrated seed during post harvest storage, enzyme (both metabolic and repair) activity is precluded and seeds remain protected under flavonoids which act both as nonenzymatic antioxidants (effective in stored dry mature seed, Fig. 5.24, and also as internal UV screen, Fig. 5.26). ACCase has a pivotal role in synthesis of flavonoids by diversion to the secondary metabolic pathway (Fig. 6.10) during seed maturation on mother plant in open field. This is a varietally controlled manifestation of cellular plasticity in orthodox (rice) seed.

Diversion of ACCase to the secondary metabolic pathway for biosynthesis of flavonoids in developing seeds before seed maturation leads to accumulation of flavonoids that provides internal UV screen as well as nonenzymatic antioxidants for protecting orthodox seeds through post-harvest ambient storage. A scheme is shown below which depicts primary and secondary metabolic pathway with Acetyl CoA carboxylase at the pivotal position for maintaining seed vigor and viability in seeds subject to genetic variation.

6.4.1.5 Acetyl CoA Carboxylase Assay in Different Rice Varieties

ACCase assay in the contrasting genotypes (in respect to seed vigor–viability) through enzyme kinetics study is shown in Fig. 6.11.

The data (Fig. 6.11) shows higher ACCase enzyme activity in higher seed vigor varieties compared to that in the low seed vigor varieties. This data (Fig. 6.11) together with reports of Baudry et al. (2004), showing high ACCase in high vigor seeds, suggest that seed viability is associated with flavonoids synthesized during seed maturation in field-grown plants. Flavonoids being nonenzymatic antioxidants remain stored in dry mature seed to protect the seed from further damages under UV radiation. This indicates the possibility for ACCase enzyme activity as being a common controlling factor for seed vigor determining growth potential and viability (regulation between growth and death of seeds) traits. ACCase activity being the first committed steps in lipid biosynthesis has an important role in primary metabolism for the initiation of germination; during seed maturation, associated cell dehydration ACCase activity is diverted to the secondary metabolic pathway for synthesis of cell-protective compounds, viz., flavonoids, that, being small phenolic (nonenzymatic) compounds, remain undegraded under UV radiation through ambient seed storage. Flavonoids thus provide UV screen and also antioxidant potential in seeds.

6.4.1.6 Spectrophotometric Assay of ACCase Activity

High ACCase gene expression/enzyme activity induces higher lipid biosynthesis that is necessary for rapid cell division/cell wall material production in rapidly growing embryos manifesting high vigor in high vigor varieties A, B, and C (Fig. 6.3) during early germination. During seed maturation (related dehydration stress) by a diversion from primary metabolism (Logemann et al. 2000), ACCase also feeds (through malonyl CoA) the secondary metabolic pathway, viz., the phenylpropanoid pathway for production of flavonoids during

Fig. 6.11 Enzyme kinetics study of acetyl CoA carboxylase represented as V0 vs [S] where V0 is reduction of absorbance at 340 nm. More production of product NAD+ at per initial time, i.e., at initial 1 min change and [S], is varying substance acetyl CoA concentration in the six different varieties (fresh harvest). **a** = Patnai-23, **b** = Pankaj, **c** = Matla, **d** = Jaya, **e** = Jogen, **f** = Lalat

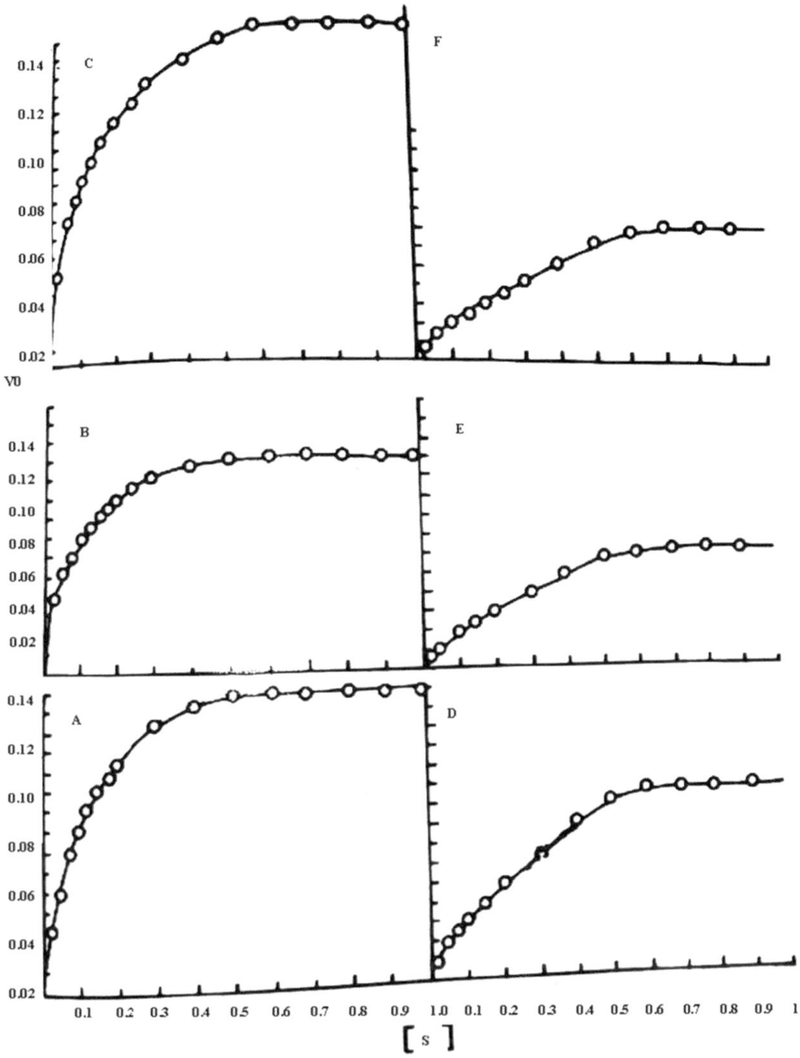

seed development (Bailly 2004). Mature seeds (with accumulated flavonoids) are thus equipped with flavonoids, the protective compounds through post-harvest (dry) seed storage (Li et al. 1993).

By virtue of structure–activity relationship (SAR), flavonoids are capable of serving as antioxidants as well as ultraviolet screens in dry stored seeds (Cockell and Knowland 1999; Kirsch 2001; Amic et al. 2003). Most of the enzymes in this secondary metabolic pathway are upregulated by UV (Kliebenstein et al. 2002) in seeds maturing on plants growing in the field open to UV-B radiation (Mazza et al. 2000). As it is evident from the scheme (Fig. 6.10), acetyl CoA carboxylase is the gene which has role both in vigor and viability and is upregulated by UV.

Spectrophotometric analysis of high and low vigor rice varieties was conducted to demonstrate the content of ACCase in contrasting genotypes (Fig. 6.12).

This data shows that ACCase activity is high in high vigor varieties (A, B, and C) and low in low vigor varieties (D, E, and F). A varietal difference is evident in both. It is pertinent to

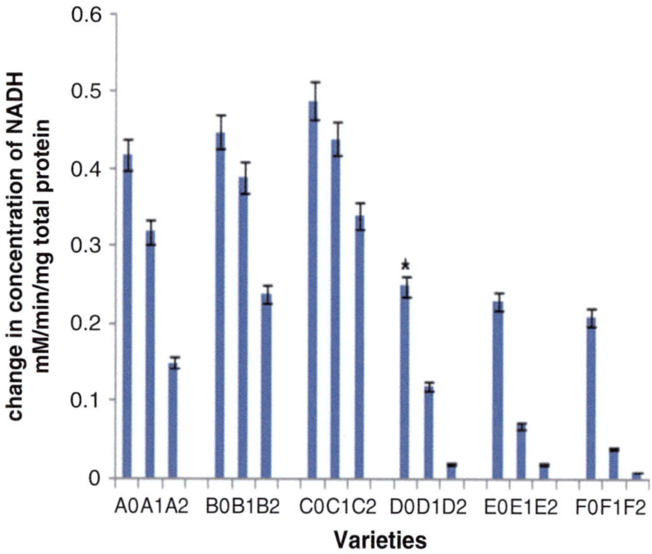

Fig. 6.12 Histogram for spectrophotometric analysis of enzyme acetyl CoA xarboxylase activity (activity expressed in change in NADH concentration/min/mg total protein). Unaged seeds designated with 0, 1-year-aged seeds designated with 1- and 2-year-aged seeds designated with 2. A = Patnai-23, B= Pankaj, C = Matla, D = Jaya, E = Jogen, F = Lalat. (S.D calculated for $n = 6, P < .001$)

mention here that with aging in ambient storage, the amount of ACCase exhibits reduction. While ACCase activity in the unaged seeds reflect only a varietal difference, activity of this enzyme in aged seeds reflects in addition to varietal difference an age-related reduction – the rate of reduction is also varietally controlled. It is also important to note that acetyl CoA carboxylase gene is upregulated by spontaneous UV on maturing plants in the field studies conducted on Italian ryegrass (Martinez-Ghersa et al. 2004), and interestingly the activity of acetyl CoA carboxylase has been reduced under oxidative stress (Ruderman and Saha 2006).

6.4.1.7 Gene Expression Study of ACCase Gene in Different Rice Seed Varieties

Gene expression study of ACCase gene has been performed on 16 rice varieties showing difference in vigor–viability status (Fig. 6.13).

ACCase gene expression varies between these contrasting genotypes as evident by real-time PCR study. Sixteen rice varieties (*Oryza sativa* L. var indica), viz., IET-13158 (R1), Badshabhog (R2), IET-9978 (R3), IET10890 (R4), Tulsi manjari (R5), Kataribhog (R6), Pusha basmati-1 (R7), Joya (R8), Patnai-23 (R9), Jogen (R10), Lalat (R11), Pankaj (R12), Basmati aman (R13), Kalojira R14), Matla (R15), and Mohan (R16), were used in this study. This study on different varieties of rice has shown high ACCase gene expression in high vigor–viability varieties (R3, R4, R6, R7, R9, R12, R15, and R16) and low ACCase expression in low vigor–viability varieties (R1, R2, R5, R8, R10, R11, R13, and R14) as shown in Fig. 6.13. Mpoloka (2008) has also demonstrated direct UV-B radiation results in change of form or function of plants which appears to occur through altered gene activity.

6.4.1.8 Gene Copy Number Determination of ACCase Gene in Different Rice Varieties

Southern blot experiment was carried out to determine ACCase copy number in the different rice varieties of contrasting trait (i.e., vigor–viability).

The Southern blot analysis showed single copy of ACCase gene in all the studied rice varieties irrespective of the vigor–viability status (Fig. 6.14). The data (Figs. 6.13 and 6.14) showed that although there is only one copy of ACCase gene present, gene expression varies between contrasting genotypes.

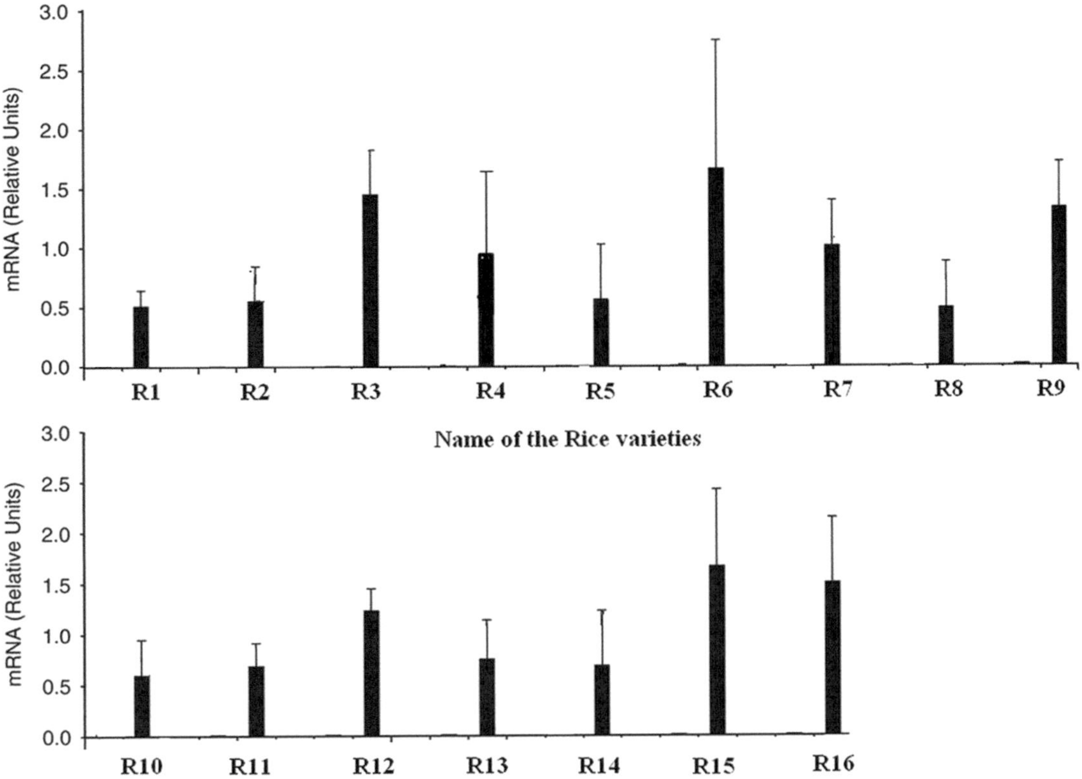

Fig. 6.13 ACCase gene expression study in *Oryza sativa* varieties showing high ACCase gene expression in high vigor–viable varieties (*R3*, *R4*, *R6*, *R7*, *R9*, *R12*, *R15*, *R16*) and low gene expression in low vigor–viable varieties (*R1*, *R2*, *R5*, *R8*, *R10*, *R11*, *R13*, *R14*)

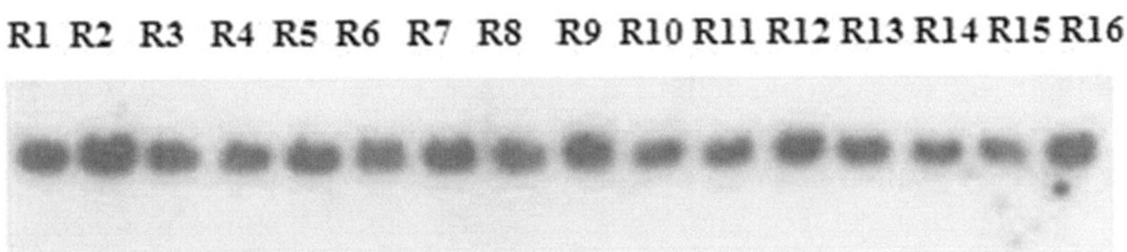

Fig. 6.14 Southern blot analysis showing single copy of ACCase gene in all the studied *Oryza sativa* varieties irrespective of their vigor–viability status (*R1*–*R16*)

6.4.1.9 Epigenetic Mechanism Controlling ACCase Gene in Contrasting Genotypes (hv vs lv)

To determine the possible cause for this differential gene (ACCase) expression (Fig. 6.13) although only one copy of this ACCase gene (Fig. 6.14) is present in both high vigor and low vigor varieties (as shown by Southern Hybridization using ACCase specific probe), Ghosh and Sen Mandi (unpublished data) demonstrated a possible epigenetic mechanism, viz., promoter methylation, could be the cause of differential gene expression in high and low vigor–viability varieties.

Methylation-sensitive restriction enzyme (MSRE) assay revealed methylation in CpG island of ACCase promoter in the phenotypically

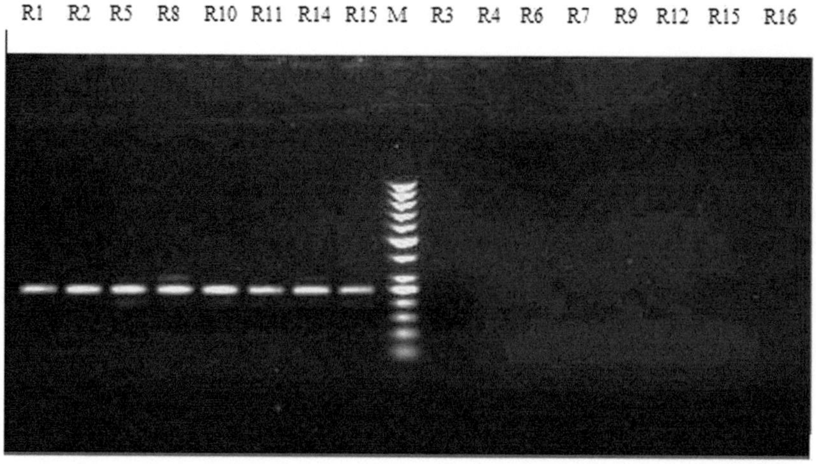

Fig. 6.15 ACCase promoter methylation study showing a 250 bp band in low vigor–viable varieties (*R1*, *R2*, *R5*, *R8*, *R10*, *R11*, *R14*, *R15*) whereas complete absence of that band in high vigor–viable varieties (*R3*, *R4*, *R6*, *R7*, *R9*, *R12*, *R15*, *R16*). M- 50 bp ladder

designated low vigor–viability varieties (Fig. 6.15) confirming that low seed vigor in rice varieties is due to low expression of ACCase gene brought about by its promoter methylation and due to the genomic imprinting of this character this trait may be considered as persisting UV-adapted trait.

DNA methylation of the promoter region of ACCase gene would cause low transcription efficiency and therefore low ACCase enzyme activity resulting in low lipid biosynthesis and thus slow growth exhibiting low vigor (as seen in varieties D, E, and F – see Fig. 6.3).

In this context, a system biology approach to explore the regulatory pathways (between positive feedback loop and cascade regulation) by which acetyl CoA carboxylase controls vigor–viability trait in rice would be a worthwhile endeavor. The finding that delayed embryonic emergence from low vigor seed where DNA damage is found to accumulate during post-harvest dry storage matches with conclusion of Biever et al. (2014) that inhibition of hypocotyl elongation occurs for allowing repairing time of damaged DNA (caused by UV radiation). In the present study, delayed embryo emergence presumably allows time for repair of DNA damage (cf. Thesis of Swati Sen Mandi submitted to Cambridge University in 1975).

Though not common in the literature till date, Katiyar et al. (2012) demonstrated a role of UV-induced DNA methylation in development of photocarcinogenesis. DNA hypermethylation, usually occurring at promoter CpG island, has been shown to be a major epigenetic mechanism in silencing the expression of tumor suppressor genes (Herman and Baylin 2003; Jones and Baylin 2002; Bird 1986). From their studies on epigenetic regulation of adaptive responses in plants, Bräutigam et al. (2013) have suggested that epigenetic variation is likely to contribute to the phenotypic plasticity and adaptive capacity of plant species. The authors report epigenetic mechanisms involved in development processes and responses to environmental cues in plant. Epigenetic processes involved in plant stress response have also been reported by Boyko and Kovalchuk (2008). When the epigenetic mechanisms get imprinted in the genome and are transmitted through breeding lines, they give rise to adapted plants to stress conditions. In our study, as the epigenetically controlled vigor–viability trait is transmitted through breeding lines (see Chap. 7), the vigor–viability trait should be regarded as a UV-adapted trait.

On the basis of rate of germination (Figs. 6.3a–c and 6.5) and ADH time course experiment (Fig. 6.6a–c), out of the six varieties studied, three appeared as low vigor, while the remaining three as high vigor varieties. RAPD analysis of genomes in these contrasting rice genotypes revealed a 900 bp DNA fragment (Fig. 6.8) followed by its molecular similarity in the three high vigor varieties (Fig. 6.9) confirmed this fragment as a molecular marker for high vigor–viability trait in rice (orthodox) seed. BLAST analysis of the sequence of the 900 bp marker indicated ACCase gene as the candidate gene for high vigor–viability. This gene regulates both primary metabolism for conferring vigor trait and secondary metabolism for conferring viability trait (Fig. 6.10). Representative parameters for these two relevant traits, viz., total lipid content for vigor trait (essential for rapid cell division correlated with vigor of embryonic growth (Fig. 6.7) and total flavonoid content (Fig. 5.26), for viability trait providing high UV protection of embryonic cells through dry seed storage showing high value in the same varieties confirm that vigor and viability are interrelated traits. ACCase assay by both enzyme kinetics study (Fig. 6.11) and spectrophotometric analysis (Fig. 6.12) demonstrates high ACCase activity in the high vigor–viability varieties and low ACCase activity in the low vigor–viability varieties. This indicates that ACCase serving as the key enzyme feeds primary/secondary metabolism as per requirement for survival under (UV) environmental condition of the seed during maturation and subsequent storage under post harvest dry seed storage. Although identical copy of ACCase is present in contrasting genotypes (Fig. 6.14), ACCase gene expression was found to vary (Fig. 6.13) between high and low vigor–viability varieties. Data on promoter methylation studies demonstrate an epigenetic regulation (ACCase promoter methylation, Fig. 6.15) controlling vigor–viability trait in rice.

This study establishes that rice varieties of high vigor–viability status develop more cell-protective compounds, viz., flavonoids, during seed maturation in open field (cf. Mazza et al. 2000) than seeds of low vigor–viability status. These compounds (viz., flavonoids) being nonenzymatic remain undegraded and act as internal UV screen and also provide antioxidative protection in (orthodox) rice seed in farmers' ambient storage in open field. It is pertinent to mention here that due to lack of enzymatic (repair) activity in orthodox dry seeds through storage, nonenzymatic protective mechanism is particularly relevant.

Studies presented in this section relate to UV-mediated damaging vs acclimation related trait development brought about by the unique quality of molecular rearrangement-based plasticity in plants. Vigor–viability traits in orthodox (rice) seed thus represent UV-adapted traits that have important implication in global rice productivity.

Case Study II

6.4.2 UV Acclimation-Related Antioxidant and Aroma Quality in Leaf of Tea Clones Growing in Darjeeling/Assam

Tea (*Camellia sinensis* (L.) Kuntze) is one of the most important beverage crops of the world with major tea-growing regions occurring in Southeast Asia and Eastern Africa. This commercially important cash crop is grown across a wide range of altitudes up to 2200 m a.s.l Darjeeling tea, a universally acclaimed beverage based on its "geographic indicator," is universally recognized as a crop growing at altitudes between 100 and 2000 m a.s.l.. Darjeeling tea is ranked high because of its traits of high aroma and medicinal (viz., antioxidant) value. Commercially valued Darjeeling tea of high aroma and high antioxidant potential is manufactured from fresh (young) tea leaves obtained in the first/second flush after a lull period following yearly pruning of the bushes during summer months, viz. April–June, when UV radiation is maximum in this region (Fig. 1.5).

The study presented here was aimed at studying influence of gene X (UV) environment interactions influencing aroma and antioxidant quality in tea leaf. Such studies should provide useful information for developing biotechnological strategies/identifying cultivation sites toward exploiting the beneficial effects of UV radiation naturally fluencing on earth. The present study was conducted on clones of different varieties of tea (representing genetic variation in the experimental material) growing for many years at two different places with variable altitudes (representing variation in UV fluence), viz., high-altitude Darjeeling (1371 m a.s.l.) and low-altitude Assam (197 m a.s.l.); both the sites being in the same latitude (~27°N), thereby ruling out possibility in variation of UV at the two experimental sites due to altered latitude. Average environmental conditions in these two locations are given in Table 4.1. Selected clones of different varieties are routinely cultivated for commercial tea production under appropriate growth condition at these sites. Fresh tea leaves (two leaf and a bud – Fig. 6.16) of the selected varieties were collected from these tea gardens at Darjeeling and Assam during first and second flush. The experiments were conducted for three consecutive years under monthly variation of UV radiation as shown in Fig. 1.5. The objective of the experiment was to determine variation in aroma quality and medicinal compounds in tea leaf at varying altitude, under altitude-related variation of UV fluence effective under variation of seasonal/diurnal UV fluence) condition. Conducting the experiment on a number of varieties also allowed determination of varietal difference on UV-related quality of tea leaf.

It is pertinent to mention that during low-dose UV radiation (due to daily/seasonal variation), plants produce cell-protective compounds, viz., flavonoids, which due to their structure–activity relationship (SAR) acts both as internal UV screen and nonenzymatic antioxidants. UV-upregulated enzymatic splitting of flavonoid glycosides, the end product of the phenylpropanoid pathway (Fig. 6.17), is known to confer aroma quality in tea (Carrao-Panizzi and Bordingnon 2000). Aroma along with antioxidant potential constitutes important traits in tea leaf. The biosynthetic pathway for production of flavonoids/flavonoid glycosides is initiated from phenylalanine ammonia lyase (PAL) caused conversion of phenylalanine via a number of enzymes leading to the production of flavonoid glycosides (Fig. 6.17). The end product of this pathway, flavonoid glycosides, on enzymatic splitting produces an aglycone moiety (Fig. 6.17) that confers aroma in plant parts as found in tea leaf. All enzymes in this pathway are upregulated by UV and therefore are more productive under enhanced UV radiation as

Fig. 6.16 Two leaves and bud used for commercial tea production (Taken from web)

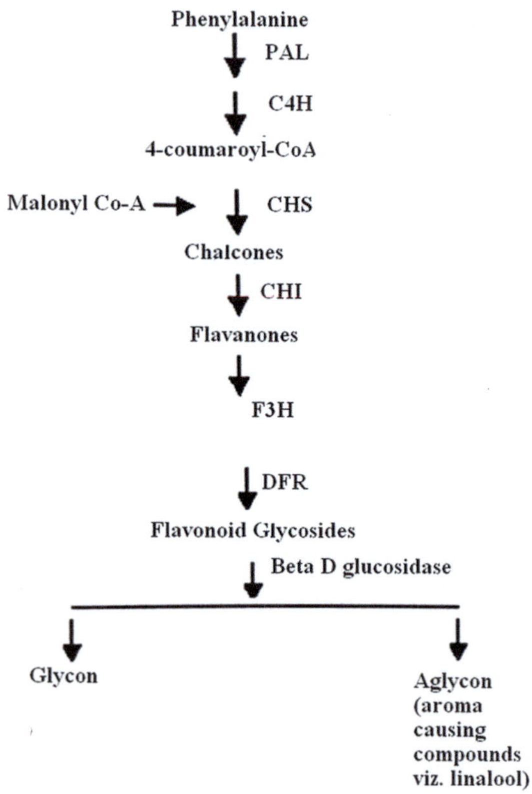

Fig. 6.17 Phenylpropanoid pathway leading to biosynthesis of flavonoid glycosides that on cleavage produces a glycon moiety and an aglycon moiety that in addition to serving as antioxidants also confers aroma in tea leaf (e.g., linalool). (*PAL* phenyl ammonia lyase, *C4H* cinnamate 4′-hydroxylase, *CHS* chalcone synthase, *CHI* chalcone isomerase, *F3H* flavonoid 3′-hydroxylase, *DFR* dihydroflavonol 4-reductase)

would be naturally available at high altitude – the experimental sites being at the same latitude.

To explore the extent of variation in antioxidant and aroma causing compounds under altitude-related difference in UV irradiance (all other factors remaining comparable in these two places – Table 4.1), clones of same genotypes growing in two different altitudes (Darjeeling and Assam) were studied.

6.4.2.1 Assay of Phenylalanine Ammonia Lyase (PAL), Starting Enzyme of Phenylpropanoid Pathway

Phenylalanine ammonia lyase (PAL) is the starting enzyme of the flavonoid biosynthetic pathway. Studies on PAL activity would provide information on the rate of phenylpropanoid pathway operative at different altitudes.

PAL activity observed in tea clones at high altitude (viz., Darjeeling) was higher than in the clones growing at low altitude (viz., Assam). This was true for all the varieties studied at the two locations. Further, a varietal difference in PAL activity was recorded in different varieties growing in each altitude. From similar studies on pea, Pluskota et al. (2005) have demonstrated UV stress-induced enhancement in flavonoid biosynthesis through signal transduced transcription (upregulation) of genes for phenylalanine ammonia lyase. Casati and Walbot (2005) working with different varieties of maize have also demonstrated the higher activity of phenylalanine ammonia lyase at higher altitude under higher UV irradiance.

6.4.2.2 Flavonoid Glycoside: Aroma Causing Substrate Compound

Flavonoids which occur in plants in the form of flavonoid glycoside are subject to variation due to altitude-related alteration in UV fluence. Thin-layer chromatography was used to determine the qualitative difference in flavonoid glycoside content in tea clones growing at different altitudes.

The TLC picture shows qualitative difference in flavonoid glycoside content in the same clones growing at different altitudes (Fig. 6.19). Tea clones grown in Darjeeling contain more number of flavonoid glycosides whereas the same clones grown in Assam contain less number of flavonoid glycosides. Carro-Panazzi and Bordingnon (2000) have also reported varietal difference in (iso)flavone glycosides as altered by the environment. All the Darjeeling grown varieties exhibit much higher number of flavonoids compared to the number evident in the Assam-grown clones.

For quantitative analysis, rutin (a representative flavonoid glycoside formed by the condensation of the flavonol quercetin and the disaccharide rutinose) (Lucci and Mazzafera 2009) content was analyzed spectrophotometrically in the same tea clones growing at contrasting altitudes (i.e., Assam and Darjeeling).

The data represents the same trend as shown in the TLC experiments. Rutin content was found to be higher in the varieties in the high UV fluencing region, viz., Darjeeling than in the Assam-growing clones – the enhancement exhibiting varietal proportionality in each altitude.

Flavonoid glycosides have also been reported to accumulate in leaves of other higher plants to screen out harmful UV radiation (Tevini and Teramura 1989; Stapleton 1992; Mazza et al. 2000) and also to serve as antioxidants. Varietal difference in flavonoid glycoside content evident from this study (Figs. 6.19 and 6.20) is similar with the observation of Hofmann et al. (2000) who have reported that in the leaf of *Trifolium repens* quercetin (flavonoid) glycoside, i.e., rutin, shows varietal expression, varying in amount in different cultivars; the authors also reported pronounced UV-B induced increase in the flavonol glycosides. Studies on maize (Stapleton and Walbot 1994) and on parsley (Logemann et al. 2000) have also shown higher flavonoid glycoside accumulation in plants growing under high (artificial) UV radiation than in the untreated control plants.

The flavonoid glycosides found in tea leaf of higher altitude tea plants are additionally useful for providing the typically desirable aroma in the beverage tea. Leaves of the same tea clones growing in Assam at low altitude not upregulating high UV protective compounds do not possess this distinctive aroma. These studies may be related with the previously reported studies of Krizek et al. (1998), who found that exclusion of UV-A and UV-B significantly reduced accumulation of flavonoids and anthocyanins in lettuce compared to plants grown in the presence of UV-A and UV-B radiation.

6.4.2.3 Aroma Releasing Enzyme β-D Glucosidase

The enzyme β-D glucosidase splits flavonoid glycosides to aglycone (viz., linalool) and glycon moieties in tea plants.

The data of β-D glucosidase enzyme activity reveals varietal difference (Fig. 6.21). The tea clones growing in Darjeeling contain high amount of β-D glucosidase in comparison to the same tea clones growing in Assam. The enhancement in β-D glucosidase values of the varieties at high altitude is presumably due to altitude-related UV enhancement. Studying two tea cultivars, Jang et al. (2010) demonstrated higher activity of β-D glucosidase in cultivars under enhanced artificial UV radiation – the higher enzyme activity being a function of intensity of UV irradiance on the plants.

6.4.2.4 Aroma Causing Aglycone Linalool Compounds

The aglycone moiety, linalool, liberated from flavonoid glycosides by β-D glucosidase activity on flavonoid glycosides is responsible for tea aroma (Moon et al. 1996; Mizutani et al. 2002; Shyam Choudhury and Sen Mandi 2011).

Linalool content was found to be enhanced in Darjeeling tea clones (belonging to the same genetic stock) indicating higher value addition in Darjeeling tea clones. This provides a desirable quality trait in Darjeeling tea. It is clear from the data of linalool content (Fig. 6.22) that high accumulation of linalool in tea leaf is a function of high-altitude environment over and above the varietal difference in values. In comparison with data in Fig. 6.21, it appears that high-altitude related UV irradiance caused enhanced activity of the enzyme β-D glucosidase resulting in higher production of this aglycone moiety, viz., linalool. Linalool has been reported to be high in black tea, oolong tea, and green tea (Sakata et al. 1995). Varietal difference in linalool content in different altitude (under different dose of UV irradiance)-grown *Ocimum basilicum* was reported by Ioannidis et al. (2002). It appears that the high aroma known for Darjeeling orthodox tea (Lachman et al. 2003) is due to the high level of aglycone moiety, viz., linalool, occurring in fresh leaf of Darjeeling grown tea clones.

Lower value of lipid peroxidation (Fig. 5.27), higher antioxidant activity (Table. 4.2), high accumulation of flavonoid glycosides (Figs. 6.19 and 6.20), as well as higher activity of β-D glucosidase (Fig. 6.21) that splits flavonoid glycosides causing higher production of aroma causing compounds, viz., linalool, (Fig. 6.22) in high-altitude grown tea clones compared to a lower value in the same clones grown at lower altitude (i.e., Assam) reflect an interaction between gene X and UV environment in providing cellular protection in high altitude-grown tea plants.

Higher PAL activity (Fig. 6.18) in tea clones at high altitude compared to those at low altitude establishes UV enhancing effect on phenylpropanoid pathway for synthesis of flavonoids, the extent of enhancement in activity however exhibiting genetic control. The end product of this pathway, viz., flavonoid glycosides in general and rutin (a specific flavonoid glycoside), were also found to be higher at high altitude evidently providing higher UV protection under high UV fluence at high altitude than at low altitude (Figs. 6.19 and 6.20). The aroma-releasing enzyme, β-D glucosidase (Fig. 6.21) and linalool, the aroma conferring compound (Fig. 6.22), were found to be higher at high altitude – their value varying under genetic control at both altitudes; such effects were recorded through three consecutive growing seasons. This is analogous to effects of UV-B radiation over several growing seasons in long-lived woody plants (Caldwell et al. 1998). These values provide evidence that tea plants (leaf) at high altitude are better protected from high UV fluence at high altitude. The high antioxidant value providing protection and high aroma content providing value addition also constitutes the basic value addition quality in Darjeeling tea leaf; this has been exploited for commercialization of Darjeeling-grown tea

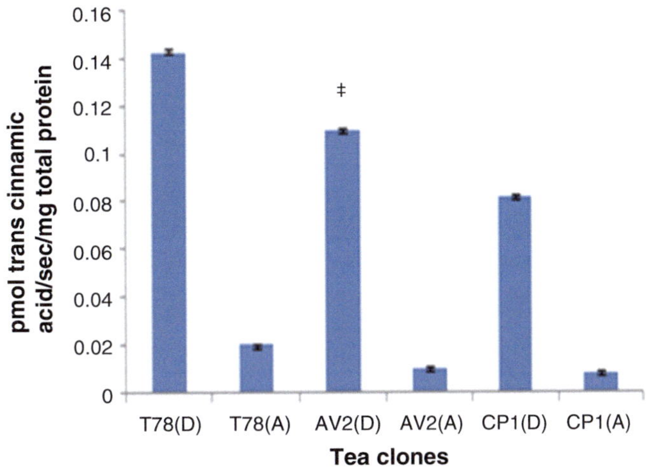

Fig. 6.18 Histogram for spectrophotometric assay of enzyme phenylalanine ammonia lyase activity (pmol trans-Cinnamic acid/s/mg of total protein) of tea clones, viz., T78, AV2, CP1. *D* Darjeeling, *A* Assam. (S.D calculated for $n = 6, P < 0.001$)

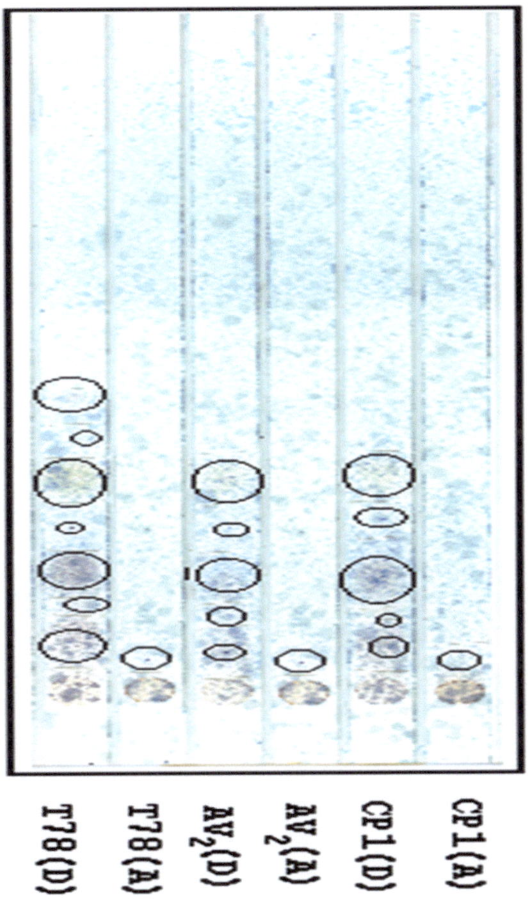

Fig. 6.19 Representative picture of thin-layer chromatographic separation on silica plate for flavonoid glycoside analysis of tea clones growing in Darjeeling and Assam, viz., T78, AV2, CP1. *D* Darjeeling, *A* Assam. Spots are marked to specify the color produced for flavonoid glycosides specifically

which is considered as a beverage of universal acclaim. It is pertinent to mention here that Darjeeling represents not only a high altitude location but also a place at 27^0N latitude that enjoys high UV fluence due to solar solstice in the Northern Hemisphere during the annual first flush tea leaf-plucking season (May–June) (Fig. 1.5)

Studies of Pontin et al. (2010) relating to high-altitude-related UV effect on grapevine (*Vitis vinifera* cv. Malbec) leaves through transcriptome analysis identified functional categories commonly modulated under UV-B treatments as well as transcripts specifically regulated in an UV-B intensity-dependent way. The authors also provide an extensive list of genes regulating multiple metabolic pathways involved in the response of grapevine to UV-B that can be used for future research. Teramura and Sullivan (1991) have reported that plants collected along a 3000 m a.s.l elevational gradient in Hawaii showed differences in sensitivity which were correlated with elevation. Most plants native to low elevations were sensitive to UV-B, but plants from the higher elevations, where UV-B fluence is greater, were very tolerant to UV-B radiation. The data presented demonstrate that plant cell plasticity based molecular rearrangements in plants at high-altitude-related high UV radiation induced enhanced production of secondary metabolites not only to acclimatize plants (with flavonoid – Fig. 5.29) but is also associated with cellular mechanism for value addition, viz., aroma by flavonoid glycoside

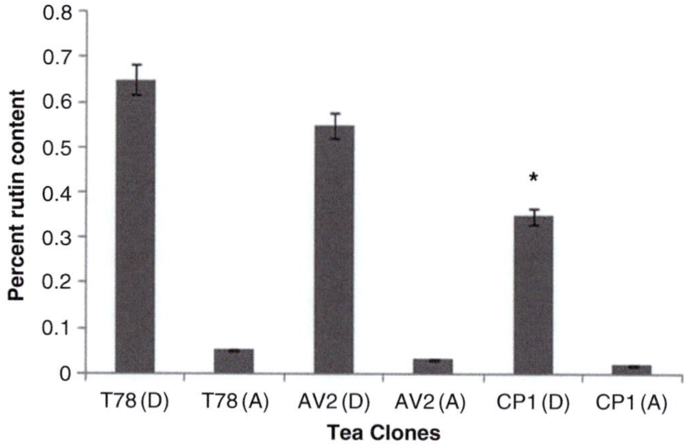

Fig. 6.20 Histogram of percent rutin content in different tea clones, viz., T78, AV2, CP1. *D* Darjeeling, *A* Assam. ($P < .001$, for S. D $n = 6$)

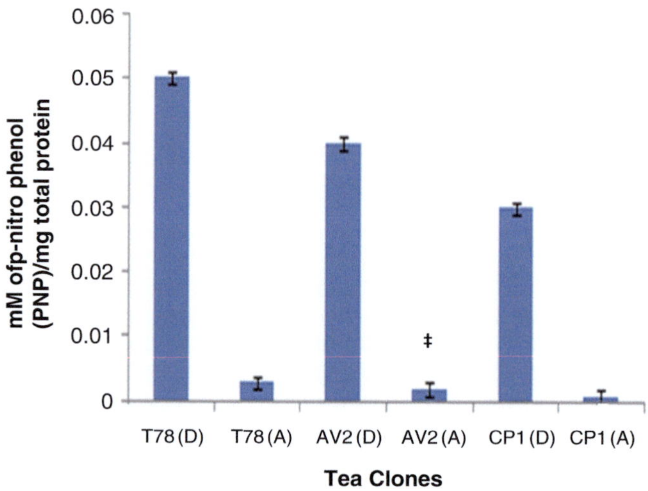

Fig. 6.21 Histogram for spectrophotometric assay of β -D glucosidase enzyme activity (mM of p-Nitrophenol (PNP)/mg of total protein) of tea clones, viz., T78, AV2, CP1. *D* Darjeeling, *A* Assam (S.D calculated for $n = 6$, $P < 0.001$)

(Fig. 6.19)-derived linalool (Fig. 6.22) – such effects exhibiting variation at the genetic/varietal level. The value of these compounds being higher in high altitude Darjeeling tea clones matches with the universal concept that Darjeeling tea is recognized as the highest value tea of commerce.

In several studies, Sullivan and Teramura (1988) have demonstrated that newly emergent loblolly pine seedlings are deleteriously affected by increased levels (high dose) of UV-B radiation. At low dose of UV radiation, an acclimatization effect (also see Sect. 4.4 of Chap. 4) was observed in the saplings that responded to UV-B radiation in a dose-specific manner by increasing needle flavonoid concentrations. Such increases in flavonoids appeared concomitant with reduction in photosynthetic capacity and growth presumably as a seasonal manifestation evidently by rearrangement between primary metabolic and secondary metabolic pathways representing adaptive adjustments (made feasible by the unique feature of plasticity in plants) as a means of survival strategy. This study represents variation in UV fluence as well as in different developmental stages of the plants reflecting constitutive phenolics, viz., hydroxycinnamic acid/anthocyanin and upregulated flavonoids. Incidence of UV-B radiation on the earth's surface in increasing altitude and decreasing latitude provides the opportunity of studying plants under natural

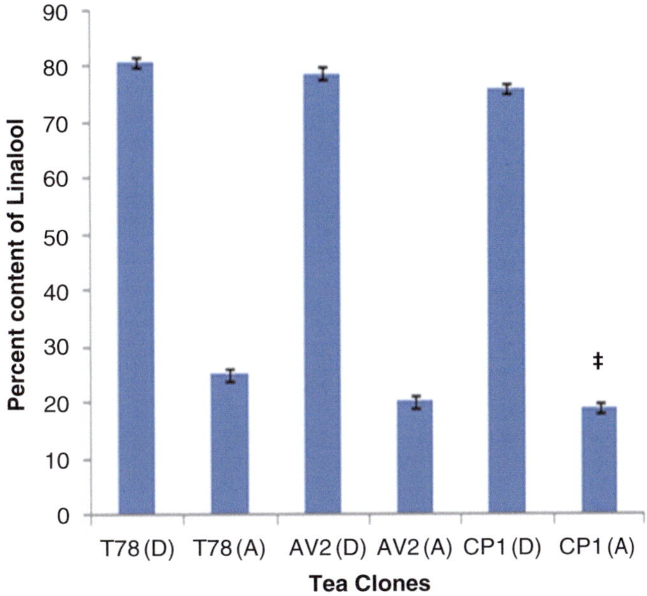

Fig. 6.22 Histogram for percent content of linalool (conferring antioxidant and aroma) in different tea clones, viz., T78, AV2, CP1. *D* Darjeeling, *A* Assam (S.D calculated for $n = 6$, $P < 0.001$)

variation of UV radiation. It is pertinent to mention here that ozone depletion over mid and higher latitude causes increased UV fluence in these latitudes, thereby affecting traits of interest. These studies along elevational gradients have been used to explore plant responses to natural gradients of solar UV-B radiation (Sullivan et al. 1992; Ziska et al. 1992; Rozema et al. 1997). Kozlowska et al. (2007) have reported difference in flavonoid glycoside content varies among three species of conifers growing in the same latitude, thereby supporting the fact that flavonoid glycoside content differ at genotype level. Such studies on accumulated effects of UV-B radiation over several growing seasons in long-lived woody plants as well as the study presented here on tea leaf demonstrate effect of UV radiation through several growing seasons, thereby exemplifying long term UV-adaptive traits (Chap. 7).

In another instance, cell molecular mechanism that correlates with high flavonoid content in vegetatively propagated plants (viz., ginger) originally growing at high altitude was also found to be correlated with a single nucleotide polymorphism in the CHS gene (Ghosh and Sen Mandi 2015) that constitutes the rate-limiting enzyme in flavonoid biosynthetic pathway. Developing in-depth understanding from such studies should help to develop strategy for best exploitation of gene X UV environment effect on desirable traits including value addition in plants of economic importance. Due to their high plasticity, plants respond with characteristic phenotypic acclimation processes to UV-B such as reduced growth and/or an increased accumulation of phenolic compounds, which act in epidermal cells as a sunscreen. UV induced upregulation of sunscreen/antioxidant compounds, viz., flavonoids that constitutes chemicals of medicinal/nutritional value in plants, emphasizes the need for developing in depth understanding into cell molecular events associated with this upregulation event in plants. Attempts at exploiting the benefits conferred by such UV adapted traits would repay careful investigation.

Bibliography

Agarwal GK, Hajduch M, Graham K, Thelen JJ (2008) In-depth investigation of the soybean seed-filling proteome and comparison with a parallel study of rapeseed. Plant Physiol 148:504–518

Agati G, Tattini M (2010) Multiple functional roles of flavonoids in photoprotection. New Phytol 186:786–793

Ajavon AN, Albritton DL, Watson RT (2007). Scientific assessment of ozone depletion: 2006, Rep. 50, Global Ozone Res. and Monit. Project, World Meteorol. Organ., Geneva

Allen DJ, Nogués S, Morison JIL, Greenslade PD, McLeod AR, Baker NR (1999) A 30% increase in UV-B has no impact on photosynthesis in well-watered and droughted pea plants in the field. Glob Chang Biol 5(2):235–244

Ambler JE, Krizek DT, Semeniuk P (1975) Influence of UV-B radiation on early seedling growth and translocation of eSZinc from cotyledons in cotton. Physiol Plant 34(3):177–181

Amic D, Davidovic-Amic D, Beslo D, Trinajstic N (2003) Structure-radical scavenging activity relationships of flavonoids. Croat Chim Acta 76:55–61

Antonovics J (1975) Predicting evolutionary response of natural populations to increased UV radiation. In: Nachtwey DS, Caldwell MM, Biggs RH (eds) Climatic Impact Assessment Program (CIAP), Monograph 5. U. S. Dept. Transport, Report No. DOT-TST-75-55, National Techn. Infor. Serv., Springfield, pp 8-7–8-27

Ashraf M, Foolad MR (2007) Roles of glycine betaine and proline in improving plant abiotic stress resistance. Environ Exp Bot 59:206–216

Atkinson RJ, Mattewhs WA, Newman PA, Plumb RA (1989) Nature 340:290–294

Bailly C (2004) Active oxygen species and antioxidants in seed biology. Seed Sci Res 14:93–107

Ballare CL, Scopel AL, Stapelton AE, Yanovsky MJ (1996) Solar ultraviolet-B radiation affects seedling emergence dna integrity plant morphology growth rate and attractiveness to herbivore insects in *Datura ferox*. Plant Physiol 112:161–170

Ballaré CL, Caldwell MM, Flint SD, Robinson SA, Bornman JF (2011) Effects of solar ultraviolet radiation on terrestrial ecosystems. Patterns, mechanisms, and interactions with climate change. Photochem Photobiol Sci 10:226–241

Barnes PW, Shinkle JR, Flint SD, Ryel RJ (2005) UV-B radiation, photomorphogenesis and plant-plant interactions. Prog Bot 66:313–340

Baroniya SS, Kataria S, Pandey GP, Guruprasad KN (2011) Intraspecific variation in sensitivity to ambient ultraviolet-B radiation in growth and yield characteristics of eight soybean cultivars grown under field conditions. Braz J Plant Physiol 23 (3):197–202

Baudry A, Heim MA, Dubreucq B, Caboche M, Weisshaar B, Lepiniec L (2004) TT2, TT8, and TTG1 synergistically specify the expression of BANYULS and proanthocyanidin biosynthesis in *Arabidopsis thaliana*. Plant J 39:366–380

Bennett JH (1981) Photosynthesis and gas diffusion in leaves of selected crop plants exposed to ultraviolet-B radiation. J Environ Qual 10(3):271–275

Berli F, D'Angelo J, Cavagnaro B, Bottini R, Wuilloud R, Silva MF (2008) Phenolic composition in grape (*Vitis vinifera* L. Cv. Malbec) ripened with different solar UV-B radiation levels by capillary zone electrophoresis. J Agric Food Chem 56:2892–2898

Berli FJ, Fanzone M, Piccoli P, Bottini R (2011) Solar UV-B and ABA are involved in phenol metabolism of Vitis vinifera L. increasing biosynthesis of berry skin polyphenols. J Agric Food Chem 59 (9):4874–4884

Bewley JD, Black M (1994) Seeds: physiology of development and germination, 2nd edn. Plenum Press, New York, p 445

Biever JJ, Brinkman D, Gardner G (2014) UV-B inhibition of hypocotyl growth in etiolated *Arabidopsis thaliana* seedlings is a consequence of cell cycle arrest initiated by photodimer accumulation. J Exp Bot 65:2949–2961

Bieza K, Lois R (2001) An *Arabidopsis* mutant tolerant to lethal ultraviolet-B levels shows constitutively elevated accumulation of flavonoids and other phenolics. Plant Physiol 126:1105–1115

Biggs RH, Kossuth SV (1978) Impact of solar UV-B radiation on crop productivity–effects of ultraviolet-B radiation enhancements on eighty-two different agricultural species. In: UV-B Biological and Climatic Effects Research (BA CER), FY 77-78 Research report on the impacts of ultraviolet b radiation on biological systems: a study related to stratospheric ozone depletion. Final Report, Vol. II, SIRA File No. 142.23, EPA-IAG-D6-0168, USDA-EPA, Stratospheric Impact Research and Assessment Program (SIRA), US Environmental Protection Agency, Washington, DC, 79 pp

Biggs RH, Kossuth SV, Teramura AH (1981) Response of 19 cultivars of soybeans to ultraviolet-B irradiance. Physiol Plant 53(1):19–26

Bird AP (1986) CpG-rich islands and the function of DNA methylation. Nature 321(6067):209–213

Björn LO, Callaghan TV, Johnsen I (1997) The effects of UV-B radiation on European heathland species. Plant Ecol 128:252–264

Blumthaler M, Ambach W (1990) Indication of increasing solar ultraviolet-B radiation flux in alpine regions. Science 248:206–208

Bornman JF, Teramura AH (1993) Effects of ultraviolet-B radiation on terrestrial plants. In: Young AR, Bjo¨rn LO, Moan J, Nultsch W (eds) Environmental UV photobiology. Plenum Press, New York, pp 427–472

Boss PK, Davies C, Robinson SP (1996) Analysis of the expression of anthocyanin pathway genes in developing *Vitis vinifera* L. cv shiraz grape berries and the implications for pathway regulation. Plant Physiol 111:1059–1066

Boyko A, Kovalchuk I (2008) Epigenetic control of plant stress response. Environ Mol Mutagen 49:61–72

Bräutigam K, Vining KJ, Lafon-Placette C, Fossdal CG, Mirouze M, Guitierrez MJ et al (2013) Epigenetic regulation of adaptive responses of forest tree species to the environment. Ecol Evol 3:399–415

Brosche M, Strid A (2003) Molecular events following perception of ultraviolet-B radiation by plants. Physiol Plant 117:1–10

Brown BA, Cloix C, Jiang GH, Kaiserli E, Herzyk P, Kliebenstein DJ et al (2005) A UV-B-specific signaling component orchestrates plant UV protection. Proc Natl Acad Sci U S A 102:18225–18230

Caldwell MM, Robberecht R, Nowak RS, Billings WD (1982) Differential photosynthetic inhibition by ultraviolet radiation in species from the arctic-alpine life zone. Arct Alp Res 14:195–202

Caldwell MM, Flint SD, Searles PS (1994) Spectral balance and UV-B sensitivity of soybean: a field experiment. Plant Cell Environ 17:267–276

Caldwell MM, Lo B, Bornman JF, Flint SD, Kulandaivelu G, Teramura AH, Tevini M (1998) Effects of increased solar ultraviolet radiation on terrestrial ecosystems. J Photochem Photobiol B46:40–52

Carrão-Panizzi MC, Bordingnon JR (2000) Activity of beta-glucosidase and levels of isoflavone glycosides in soybean cultivars affected by environment. Pesqui Agrop Brasileira 35(5):873–878

Casati P, Walbot V (2005) Differential accumulation of maysin and rhamnosylisoorientin in leaves of high-altitude landraces of maize after UV-B exposure. Plant Cell Environ 28:788–799

Chen L, Niu K, Wu Y, Geng Y, Mi Z, Flynn DFB, He J-S (2013) UV radiation is the primary factor driving the variation in leaf phenolics across Chinese grasslands. Ecol Evol 3(14):4696–4710

Cockell CS, Knowland J (1999) Ultraviolet radiation screening compounds. Biol Rev 74:311–345

Conde C, Silva P, Fontes N, Dias ACP, Tavares RM, Sousa MJ, Agasse A, Delrot S, Geros H (2007) Biochemical changes throughout grape berry development and fruit and wine quality. Food 1:1–22

Corlett JE, Stephen J, Jones HG, Woodfin R, Mepsted R, Paul ND (1997) Assessing the impact of UV-B radiation on the growth and yield of field crops. In: Lumsden PJ (ed) Plants and UV-B: responses to environmental change, Society for Experimental Biology, Seminar Series 64. Cambridge University Press, Cambridge

Correia CM, Areal ELV, Torres-Pereira MS, Torres-Pereira JMG (1998) Intraspecific variation in sensitivity to ultraviolet-B radiation in maize grown under field conditions. I. Growth and morphological aspects. Field Crops Res 59:81–89

da Silva PF, Paulo L, Barbafina A, Eisei F, Quina FH, Macanita AL (2012) Photoprotection and the photophysics of acylated anthocyanins. Chemistry 18:3736–3744

De La Fuente B, Nicolas G (1974) Respiratory activity during germination of seeds of *Cicer arietinum* L. glycolysis and fermentation. Plant Sci Lett 3:143–148

Dumpert K, Boscher J (1982) Response of different crop and vegetable cultivars to UV-B irradiance: preliminary results. In: Bauer H, Caidwell MM, Tevini M, Worrest RC (eds) Biological effects of UV-B radiation, Bereich Projekttr/igerschaften, Gesellschaft fiir Strahlen- und Umweltforschung mbH, Munich, West Germany, pp 102–107

Famiani F, Walker RP, Tecsi L, Chen ZH, Proietti P, Leegood RC (2000) An immunohistochemical study of the compartmentation of metabolism during the development of grape (*Vitis vinifera* L.) berries. J Exp Bot 51:675–683

Filella I, Peñuelas J (1999) Altitudinal differences in UV absorbance, UV reflectance and related morphological traits of Quercus ilex and Rhododendron ferrugineum in the Mediterranean region. Plant Ecol 145:157–165

Filippos VF, Emmanouil T, Carl D, Guenter V, Georg K, Nickolas P (2007) Biotechnology of flavonoids and other phenylpropanoid-derived natural products. Part I: chemical diversity, impacts on plant biology and human health. Biotechnol J 2(10):1214–1234

Fiscus EL, Booker FL (1995) Is increased UV-B a threat to crop photosynthesis and productivity. Photosynth Res 43:81–92

Fisher RA (1930) The genetical theory of natural selection. Dover Publications, New York

Flood RG, Sinclair A (1981) Fatty acid analysis of aged permeable and impermeable seeds of Trifolium subterraneum (subterranean clover). Seed Sci Technol 9:475–477

Frohnmeyer H, Staiger D (2003) Ultraviolet-B radiation-mediated responses in plants. Balancing damage and protection. Plant Physiol 133(4):1420–1428

Gardner G, Lin C, Tobin EM, Loehrer H, Brinkman D (2009) Photobiological properties of the inhibition of etiolated Arabidopsis seedling growth by ultraviolet-B radiation. Plant Cell Environ 32:1573–1583

Garrard LA, Van TK, West SH (1976) Plant response to middle ultraviolet (UV-B) radiation: carbohydrate levels and chloroplast reactions. Soil Crop Sci Soc Fla Proc 36:184–188

Ghosh S, Mandi SS (2015) SNP in Chalcone synthase gene is associated with variation of 6-gingerol content in contrasting landraces of *Zingiber officinale*. Roscoe, Gene (2015), http://dx.doi.org/10.1016/j.gene.2015.04.042

Gil M, Bottini R, Berli FJ, Pontin M, Silva MF, Piccoli P (2013) Volatile organic compounds characterized from grapevine (*Vitis vinifera* L. cv. Malbec) berries increase at pre-harvest and in response to UV-B radiation. Phytochemistry 96:148–157

Gonzales R, Paul ND, Percy K, Ambrose M, McLaughlin CK, Barnes JD, Areses M, Wellburn AR (1996) Responses to ultraviolet-B radiation (280–315 nm) of pea (*Pisum sativum*) lines differing in leaf surface wax. Physiol Plant 98:852–860

Hectors K, Prinsen E, De Coen W, Jansen MA, Guisez Y (2007) *Arabidopsis thaliana* plants acclimated to low dose rates of ultraviolet B radiation show specific

changes in morphology and gene expression in the absence of stress symptoms. New Phytol 175 (2):255–270
Herman JG, Baylin SB (2003) Gene silencing in cancer in association with promoter hypermethylation. N Engl J Med 349:2042–2054
Ho DT, Scandalios JG (1975) Regulation of alcohol dehydrogenase in maize scutellum during germination. Plant Physiol 56:56–59
Hofmann RW, Swinny EE, Bloor SJ, Markham KR et al (2000) Responses of nine *Trifolium repens* L. populations to ultraviolet-B radiation: differential flavonol glycoside accumulation and biomass production. Ann Bot 86:527–537
Holton TA, Cornish EC (1995) Genetics and biochemistry of anthocyanin biosynthesis. Plant Cell 7 (7):1071–1083
Hutchinson F (1987) A review of some topics concerning mutagenesis by ultraviolet light. Photochem Photobiol 45:897–903
Ioannidis D, Bonner L, Johnson C (2002) UV-B is required for normal development of oil glands in *Ocimum basilicum* L. (sweet basil). Ann Bot 90:453–460
Jaakola L, Hohtola A (2010) Effect of latitude on flavonoid biosynthesis in plants. Plant Cell Environ 33:1239–1247
Jang J, Yang YC, Zhang GH, Chen H, Lu JL, Du YY, Ye JH, Ye D, Borthakur D, Zheng XQ, Liang YR (2010) Effect of ultra-violet B on release of volatiles in tea leaf. Intl J Food Prop 13(3):608–617
Jansen MAK, Gaba V, Greenberg BM (1998) Higher plants and UV-B radiation: balancing damage, repair and acclimation. Trends Plant Sci 3:131–135
Jones PA, Baylin SB (2002) The fundamental role of epigenetic events in cancer. Nat Rev Genet 3:415–428
Jordan BR (1996) The effects of ultraviolet-B radiation on plants: a molecular perspective. Adv Bot Res 22:97–162
Juliano BO, Perez CM, Kaosaard M (1990) Grain quality characteristics of export rices in selected markets. Cereal Chem 67:192–197
Kakani VG, Reddy KR, Zhao D, Sailaja K (2003) Field crop responses to ultraviolet-B radiation: a review. Agric For Meteorol 120:191–218
Kalbina I, Strid A (2006) Supplementary ultraviolet-B radiation reveals differences in stress responses between Arabidopsis thaliana ecotypes. Plant Cell Environ 29:754–763
Kataoka I, Sugiyama A, Taira T (2003) Role of ultraviolet radiation in accumulation of anthocyanin in berry section of "Gros Colman" grapes (*Vitis vinifera* L.). J Jpn Soc Hortic Sci 72:1–6
Katiyar SK, Singh T, Prasad R, Sun Q, Vaid M (2012) Epigenetic alterations in ultraviolet radiation-induced skin carcinogenesis: interaction of bioactive dietary components on epigenetic targets. Photochem Photobiol 88(5):1066–1074
Kirsch JD (2001) Flavonoids: the good, the bad and the ugly. Free Rad Biol Med 77:222, Spring 2001
Kliebenstein DJ, Lim JE, Landry LG, Last RL (2002) *Arabidopsis* UVR8 regulates ultraviolet-B signal transduction and tolerance and contains sequence similarity to human regulator of chromatin condensation. Plant Physiol 130:234–243
Kolb CA, Kpecký J, Riederer M, Pfündel EE (2003) UV screening by phenolics in berries of grapevine (Vitis vinifera). Funct Plant Biol 30:1177–1186
Kolloffel C (1968) Activity of alcohol dehydrogenase in the cotyledons of peas germinated under different environmental conditions. Acta Bot Neerlandica 17:70–77
Kozłowska M, Brzezińska E, Stobiecki M (2007) Sensitivity differences and accumulation of screening compounds in three conifer plants under enhanced UV-B radiation. Pol J Environ Stud 16(6):823–830
Krishnasamy V, Seshu DV (1989) Seed germination rate and associated characters in rice. Crop Sci 29:904–908
Krizek DT, Britz SJ, Mirecki RM (1998) Inhibitory effects of ambient levels of solar UV-A and UV-B radiation on growth of cv. New Red Fire lettuce. Physiol Plant 103:1–7
Kubasek WL, Shirley BW, Mckillop A, Goodman HM, Briggs W, Ausubel FM (1992) Regulation of flavonoid biosynthetic genes in germinating Arabidopsis seedlings. Plant Cell 4(10):1229–1236
Kumagai T, Sato T (1992) Inhibitory effects of increase in near-UV radiation on the growth of Japanese rice cultivars (*Oryza sativa* L.) in a phytotron and recovery by exposure to visible radiation. Jpn J Breed 42:545–552
Lachman J, Orsak M, Pivec V, Dudjak J, Krym O (2003) Polyphenol content in green black and oolong tea (Camellia sinensis (L.)Kuntze) infusions in different times of tea maceration. Sci Agric Bohem, (Czech Republic) 34:22–28
Li J, Ou-Lee T, Raba R, Amundson RG, Last RL (1993) Arabidopsis flavonoid mutants are hypersensitive to UV-B radiation. Plant Cell 5:171–179
Lidon FC (2012) UV-B radiation in rice: interaction between micronutrients accumulation and the photosynthetic performance. J Plant Inter 7(1):19–28
Logemann E, Tavernaro A, Schulz W, Somssich IE, Hahlbrock K (2000) UV Light selectively induces supply pathways from primary metabolism and flavonoid secondary product formation in parsley. Proc Natl Acad Sci 97:1903–1907
Lois R, Buchanan BB (1994) Severe sensitivity to ultraviolet radiation in an *Arabidopsis* mutant deficient in flavonoid accumulation. II. Mechanisms of UV-resistance in Arabidopsis. Planta 194:504–509
Lucci N, Mazzafera P (2009) Distribution of rutin in fava d'anta (Dimorphandra mollis) seedlings under stress. J Plant Interact 4:203–208
Martin JT, Juniper BE (1970) The cuticles of plants. Edward Arnold, London
Martínez-Ghersa MA, Vila-Aiub MM, Ghersa CM, Gundel PE, Satorre EH (2004) Herbicide selection of

Italian ryegrass under different levels of UV-B radiation. J Environ Qual 33:1376–1386

Martínez-Ghersa MA, Olszyk D, Radosevich SR (2008) Growth and yield responses of Italian ryegrass (Lolium multiflorum) to diclofop-methyl and ozone. Weed Res 48:68–77

Mauch-Mani B, Mauch F (2005) The role of abscisic acid in plant–pathogen interactions. Curr Opin Plant Biol 8:409–414

Mazza CA, Boccalandro HE, Giordano CV, Battista D, Scopel AL, Ballaré CL (2000) Functional significance and induction by solar radiation of ultraviolet-absorbing sunscreens in field-grown soybean crops. Plant Physiol 122:117–125

Mishra RK, Sen Mandi S (2004) Molecular profiling and development of DNA marker associated with drought tolerance in Darjeeling. Curr Sci 87:60–66

Mizutani M, Nakanishi H, Ema J-I, Ma S-J, Noguchi E, Inohara-Ochiai M, Fukuchi-Mizutani M, Nakao M, Sakata K (2002) Cloning of β-Primeverosidase from tea leaves, a key enzyme in tea aroma formation. Plant Physiol 130:2164–2176

Moon JH, Watanabe N, Ijima Y, Yagi A, Ina K, Sakata K (1996) cis- and trans-Linalool 3,7-oxides and methyl salicylate glycosides and (Z)-3- hexenyl β-d-glucopyranoside as aroma precursors from tea leaves for oolong tea. Biosci Biotechnol Biochem 60:1815–1819

Mpoloka SW (2008) Effects of prolonged UV-B exposure in plants. Afr J Biotechnol 7(25):4874–4883

Nachtwey DS, Rundel RD (1975) Ozone change: biological effects. In: Bower FA, Ward RB (eds) Stratospheric ozone and man. CRC Press, Boca Raton, pp 81–121

Nedeva D, Nikolova A (1999) Fresh and dry weight changes and germination capacity of natural or premature desiccated developing wheat seeds. Bulg J Plant Physiol 25(1–2):3–15

Osborne DJ (1980) Senescence in seeds. In: Thimann KV (ed) Senescence in plants. CRC Press, Boca Raton, pp 13–37

Pluskota WE, Michalczyk DJ, Gorecki RJ (2005) Control of phenylalanine ammonia-lyase gene promoters from pea by UV radiation. ACTA Physiol Plant 27:229–236

Pontin MA, Piccoli PN, Francisco R, Bottini R, Martinez-Zapater JM, Diego L (2010) Transcriptome changes in grapevine (Vitis vinifera L.) cv. Malbec leaves induced by ultraviolet-B radiation. BMC Plant Biol 10:224

Potters G, Pasternak TP, Guisez Y, Jansen MAK (2009) Different stresses, similar morphogenic responses: integrating a plethora of pathways. Plant Cell Environ 32:158–169

Ren J, Dai W, Xuan Z, Yao Y, Korpelainen H, Li C (2007) The effect of drought and enhanced UV-B radiation on the growth and physiological traits of two contrasting poplar species. Forest Ecol Manag 239:112–119

Robberecht R, Caldwell MM (1978) Leaf epidermal transmittance of ultraviolet radiation and its implications for plant sensitivity to ultraviolet-radiation induced injury. Oecologia 32:277–287

Rozema J, Lenssen GM, Van de Staaij JWM, Tosserrams M, Visser AJ, Broekman RA (1997) Effects of UV-B radiation on terrestrial plants and ecosystems: interaction with CO_2 enrichment. Plant Ecol 128:183–191

Ruderman NB, Saha AK (2006) Metabolic syndrome: adenosine monophosphate-activated protein kinase and malonyl coenzyme A. Obesity 14:25–33

Ruhland C, Fogal M, Buyarski C, Krna M (2007) Solar ultraviolet-B increases phenolic content and ferric reducing antioxidant power in *Avena sativa*. Molecules 12:1220–1232

Sakata K, Guo WF, Moon JH (1995). Molecular basis of alcoholic aroma formation in Oolong tea. In Proceedings of 95 International Tea-quality – human health symposium in Shanghai. Hangzhou Tea Society of China 1995, pp 175–187

Sato T, Kumagai T (1993) Cultivar differences in resistance to the inhibitory effects of near-UV radiation among Asian ecotype and Japanese lowland and upland cultivars of rice (*Oryza sativa* L.). Jpn J Breed 43:61–68

Sen S. (1975) PhD thesis title: germination and viability of rye embryos: nucleic acid and Protein synthesis during early hours of germination. Submitted to University of Cambridge

Sen S, Osborne DJ (1977) Decline in ribonucleic acid and protein synthesis with loss of viability during the early hours of imbibition of rye *Secale cereale* L. embryos. Biochem J 166:33–38

SenMandi S, Bhattacharya S (2003). Varietal difference in cellular damage associated with ageing in dry stored seed. Indian J Plant Physiol 8(1):210–216

SenMandi, S., Nandi, S., Das, G. Chakraborti, S. 1995. Bio-molecular and physiological studies on hydration-dehydration seed invigoration pretreatment: an attempt to overcome constraints of wet seeding in rice. International workshop on "Constraints, opportunities and innovations of wet-seeded rice". IRRI. Discussion Paper, Series (Edited by K. Moody) No. 10: 199–218

Sheng M, Tang M, Chen H, Yang BW, Zhang FF, Huang YH (2005) Influence of arbuscular mycorrhizae on the root system of maize plants under salt stress. Can J Microbiol 55:879–886

Shyam Choudhury S, Sen Mandi S (2011) Natural ultra violet irradiance related variation in antioxidant and aroma compounds in tea (*Camelia sinensis* L. Kuntze) plants grown in two different altitudes. Int J Environ Biol 2(1):1–6

Shyam Choudhury S, Sen Mandi S (2012) Natural ultra violet radialtion on field grown rice (*Oryza sativa* L.) plants confer protection against oxidative stress in seed during storage under subtropical ambience. Environ Pollut 1:2

Springob K, Nakajima J, Yamazaki M, Saito K (2003) Recent advances in the biosynthesis and accumulation of anthocyanins. Nat Prod Rep 20:288–303

Stapleton AE (1992) Ultraviolet radiation and plants: burning questions. Plant Cell 4:1353–1358

Stapleton AE, Walbot V (1994) Flavonoids can protect maize DNA from the induction of Ultraviolet radiation damage. Plant Physiol 105(3):881–889

Steinmuller D, Tevini M (1985) Action of ultraviolet radiation (UV-B) upon cuticular waxes in some crop plants. Planta 164:557–564

Sullivan JH, Teramura AH (1988) Effects of ultraviolet radiation on seedling growth in the pinaceae. Am J Bot 75:225–230

Sullivan JH, Teramura AH, Ziska LH (1992) Variation in UV-B sensitivity in plants from a 3,000-m elevational gradient in Hawaii. Am J Bot 79:737–743

Talai S, Sen-Mandi S (2010) Seed vigour-related DNA marker in rice shows homology with Acetyl CoA Carboxylase gene. Acta Physiol Plant 32:153–167

Tarrant AM, Reitzel AM, Kwok CK, Jenny MJ (2014) Activation of the cnidarian oxidative stress response by ultraviolet radiation, polycyclic aromatic 3 hydrocarbons and crude oil. J Exp Biol 217:1444–1453

Tello J, Cordero-Bueso G, Aporta I, Cabellos JM, Arroyo T (2015) Genetic diversity in commercial wineries: effects of the farming system and vinification management on wine yeasts. J Appl Microbiol 112:302–315

Teramura AH, Sullivan JH (1991) Potential impacts of increased solar UV-B on global plant productivity. In: Riklis E (ed) Photobiology. Plenum Press, New York, pp 625–634

Teramura AH, Ziska LH, Sztein AE (1991) Changes in growth and photosynthetic capacity of rice with increased UV-B radiation. Physiol Plant 83:373–380

Tevini M, Teramura AH (1989) UV-B effects on terrestrial plants. Photochem Photobiol 50:479–487

Usmanov PD, Mednik IG, Lipkind BI, Giller Y (1988) genotypic characteristics of the response of plants to medium-wave ultraviolet radiation. Sov Plant Physiol 34:578–586

Van TK, Garrard LA (1976) Effect of UV-B radiation on net photosynthesis of some C3 and C4 crop plants. Soil Crop Sci Soedin Fla Proc 35:I-3

Virk PS, Ford-Lloyd JBV, Pooni MT, Clemeno TP, Newbury HJ (1996) Predicting quantitative variation within rice germplasm using molecular markers. Heredity 76:296–304

Vu CV, Allen LH Jr, Garrard LA (1978) Effects of supplemental ultraviolet radiation (UV-B) on growth of some agronomic crop plants. Soil Crop Sci Soc Fla Proc 38:59–63

Wang Q-W, Hidema J, Hikosaka K (2014) Is UV-induced DNA damage greater at higher elevation? Am J Bot 101(5):796–802

Weinig C, Gravuer KA, Kane NC, Schmitt J (2004) Testing adaptive plasticity to UV: costs and benefits of stem elongation and light-induced phenolics. Evolution 58:2645–2656

White PJ, Broadley MR (2009) Biofortification of crops with seven mineral elements often lacking in human diets-iron, zinc, copper, calcium, magnesium, selenium and iodine. New Phytol 182:49–84

Woodall GS, Stewart GR (1998) Do anthocyanins play a role in UV protection of the red juvenile leaves of Syzygium? J Exp Bot 49:1447–1450

Zhang ZZ, Li XX, Chu YN, Zhang MX, Wen YQ, Duan CQ, Pan QH (2012) Three types of ultraviolet radiation differentially promote expression of shikimate pathway genes and production of anthocyanins in grape berries. Plant Physiol Biochem 57:74–83

Zhao J, Huhman D, Shadle G, He XZ, Sumner LW, Tang YH, Dixon RA (2011) MATE2 mediates vacuolar sequestration of flavonoid glycosides and glycoside malonates in *Medicago truncatula*. Plant Cell 23:1536–1555

Ziska LH, Teramura AH, Sullivan JH (1992) Physiological sensitivity of plants along an elevational gradient to UV-B radiation. Am J Bot 79:863–871

Plant Survival Under Natural UV Radiation on Earth: UV Adaptive/UV-Adapted Traits 7

7.1 Introduction

Early life forms had remained under oceanic water for protection from UV radiation (a component of solar radiation) fluencing on earth. Their migration from marine to terrestrial habitat occurred during the Ordovician era by when a (partial) UV restricting stratospheric ozone layer had already been formed – the layer varying in thickness as a function of latitudinal variation resulting from ozone drift after formation at the equator. The latitudinal variation in ozone accumulation/thickness of stratospheric ozone layer allows proportional fluence of UV on earth's surface (Lucas et al. 2006). UV fluence on earth may further be affected by altitudinal variations as a factor of rarification of air proportional to height above sea level; as per records at the Light Climatic Observatory in Arosa, Switzerland – Blumthaler and Ambach (1990) reported that UV radiation in earth's atmosphere, reflecting altitude variation, shows an annual total increase of UV fluence: 19 %/1000 m (UV-B), 11 %/1000 m (UV-A), and 9 %/1000 m (global radiation, referring to total low wavelength radiation). The ratio of daily totals of UV-B in terms of global radiation shows a significant seasonal variation with the maximum UV radiation fluencing in summer. The ratio of daily totals of UV-A/global radiation was found to show no significant seasonal variation.

The threat of enhancement of UV in earth's atmosphere calls for studies on cell molecular mechanisms operative in plants for developing resilience/acclimation under enhancing UV radiation in terms of (a) countering UV-induced macromolecular damage (Hutchinson 1987; Dubest et al. 2002) and (b) developing UV acclimation processes by upregulation of phenolic cell protective compounds, viz., flavonoids (Alvero-Bascos and Ungson 2012) for developing internal UV screen as well as antioxidative potential in plant cells. Such cell protective molecular mechanism would confer fitness advantage manifested as UV adaptive traits in plants evident at the genetic level (Dixon et al. 2001). Molecular events related to resilience against UV-B radiation brought about by upregulation of phenolic/flavonoid/wax synthesis may often be manifested at the morphological level exhibiting photomorphogenic traits (Ballare et al. 1995; jansen et al. 2010). Several studies on growth and development under variation in UV environment highlight varietal difference (Kumagai and Sato 1992; Hidema et al. 1996).

Studies on evaluation of flavonoid and anthocyanins content in Iranian borage (Echium amoenum) accessions growing in different habitats in north of Iran indicate a gene X (UV) environment effect in plants (Sajirani et al. 2014; Shyam Choudhury and Sen Mandi 2012). Hofmann and Jahufer (2011) from their studies on white clover and other species suggested merit for future genotype X environment stress. While

S. Sen Mandi, *Natural UV Radiation in Enhancing Survival Value and Quality of Plants*,
DOI 10.1007/978-81-322-2767-0_7

high-intensity UV radiation induces cell molecular damages, low dose UV (resultant from attenuation due to passage through cloud cover) epigenetically induces signal transduction mediated upregulation of genes associated with cell protection processes viz. (a) DNA repair enzymes (Brosche and Strid 2003) and (b) enzymes of the Phenyl Propanoid pathway for biosynthesis of phenolics viz. flavonoids (Brown and Jenkins 2008; Frohnmeyer et al. 1999; Ulm et al. 2004; Kliebenstein et al. 2002) that provide internal UV screening as well as antioxidative potential in plant cells. Cell molecular processes associated with need based upregulation of genes related to UV protection involve macromolecular rearrangement enabled by the unique feature of adaptive (cellular) plasticity, particularly operative in plants; this allows UV acclimation-related appropriate trait development that exhibit resilience to UV (Kumagai and Sato 1992) without causing stress symptoms (Hectors et al. 2007). Functioning of enzymes of the phenylpropanoid (secondary metabolism) pathway is affected through molecular rearrangements involving temporal diversion of primary metabolism to secondary metabolism (Caretto et al. 2015) for synthesis of UV protective compounds providing internal UV screens and nonenzymatic antioxidative activity. Accumulation of flavonoids in cells confers major aspects of UV acclimation in plants, exposure resulting from brief periods of high UV exposure and associated low dose UV exposure related flavonoid synthesis; the UV tolerance thus developed is manifested as UV adaptive traits that, exhibiting rapid interchange of UV adaptive status under variation of ambient UV environment, confers evolutionary significance. Such cell molecular events occurring under variation of exposure to UV-B fluence constitute regulatory components orchestrating typical phenotypic outcomes (manifested as UV adaptive traits). Such plants, exhibiting rapid interchange of UV adaptive status under variation of ambient UV environment evidently exhibit ecological significance.

In this way plants may modify their UV induced phenotype for matching their environment. Such developmental plasticity is commonly viewed as a mechanism enabling organisms to adaptively grow and develop, matching their phenotype to local UV conditions thereby overcoming environmental stress. Studies of Sullivan et al. (1992) on UV adaptivity in plants collected along a 3000 m.a.s.l. elevational gradient in Hawaii showed differences in sensitivity that exhibited correlation with elevation (exemplifying UV adaptive traits). Survival of plants along the altitudinal gradient demonstrates plant's capacity to adjust to elevation-related variation in UV radiation by developing cellular plasticity-enabled trait variation. Such UV-related traits manifesting both at the genetic and also at the phenotypic level, enables plants to survive and flourish appropriately under variation of UV fluence. Exemplifying UV adaptive trait, many organisms continually modify their UV-related phenotype in response to local conditions with the induced character state rapidly disappearing in noninductive UV environment (Weinig et al. 2004). Acclimation strategies in plants (exhibited as trait variation) enable plants to establish and survive in the open field (Caldwell et al. 1994; Björn et al. 1997; Allen et al. 1999) with associated variation in UV fluence occurring during plant migration to areas of latitudinal/longitudinal difference. Such flexibility of trait variation rapidly across variable (UV) environment evidently represents "UV adaptive trait" (Niklas 1992; Casal et al. 1994) that relates to acclimation associated adaptive plasticity for optimizing growth and survival of plants (Bradshaw 1965; Levins 1968; Lively 1986, 1998) for developing resilience under different UV locations (such as evidenced in plants growing under altitudinal/latitudinal variation). Such traits confer evolutionary advantage in plants (Schmitt and Wulff 1993; Dudley and Schmitt 1996).

Reports on temporal/persistent trait development relating to location-based UV variation for plant establishment in open field indicate UV as a trait-selective agent. Similar conclusion has been drawn by Chen et al. (2013) from their studies demonstrating that leaf phenolics and other UV-absorbing compounds are significantly higher in plants on the Tibetan Plateau than in those on the inner Mongolian grasslands, independent of phylogenetic relationships between the species. The trait variation reported in this study evidently relates to UV adaptive trait. Regression analyses showed that the variation in leaf phenolics was strongly affected by climatic factors, particularly UV radiation.

As different from the temporal molecular readjustment-based trait development, plants under prolonged existence in same UV condition, the epigenetic (UV adaptive) trait related to temporal upregulation of UV protective compounds, through generations may become imprinted on the genomes (Surani 1998; Martienssen and Richards 1995), thereby giving rise to "fixed" traits (with respect to synthesis of phenolics) that eventually perform as constitutive traits (Agrawal 1999; Agrawal et al. 1999; Van Tienderen 1991) in UV adapted genotypes. This would result in developing sustainable tolerance to UV-B radiation regardless of current status of UV under variations of location. This evidently reflects "UV adaptive fixity"/"fixed traits" (Alpert and Simms 2002) simulating UV-adapted/constitutive trait. Plants from the high elevations, where UV-B is greatest, are thus found to be tolerant to UV-B radiation under a wide range of location – exemplifying "fixed/constitutive" traits (Teramura and Sullivan 1992). However plants native to low elevations (i,e the plants that have not experienced high UV radiation and thus have not been preadapted to high UV condition) exhibit sensitivity to UV-B when shifted to location of high UV fluence. Such UV related variation in trait development indicates UV as being a trait selective agent.

A ranking of cultivars according to the UV adaptation/UV tolerance ability would provide useful information on plant-specific cellular plasticity based development of UV protective traits in plants (Biggs et al. 1981). According to Alpert and Simms (2002), cellular adaptive plasticity-enabled molecular rearrangements leading to genetic differentiation evident as (UV) adaptive trait and adaptive fixity representing UV adapted traits are likely to be the three major alternative evolutionary paths along which plants may evolve in response to natural selection. UV radiation-induced variation in traits represents natural adaptation involving UV-related changes in anatomical, morphological, and biochemical traits in plants (Teramura and Sullivan 1991). The flexibility of UV adaptive plants for rapidly modifying their UV adaptive status to cope with environmental UV variation and adjust with altered UV environment make them ecologically significant and evolutionarily potential. This provides a process of major evolutionary significance. This establishes that natural selection through evolutionary changes in environment will favor (environmental need based temporal) adaptive traits Information on location based UV adapted traits in plant would also be useful for planning suitable cultivation sites in the face of enhancing UV fluence on earth. During prolonged preexposures under high intensity UV involving several spans of intermittent low dose fluence, epigenetic changes may become imprinted in genomes (Martienssen and Richards 1995). Such plants constitutively produce high amount of UV protective compounds (representing UV adapted plants) that exhibit genetic stability with respect to trait of interest irrespective of environmental variation. Such plants represent UV adapted genotypes exhibiting a major evolutionary process resulting from cellular adaptive plasticuiity. In the alternative situation temporal UV protection traits developed under short term variation in UV environment confer UV adaptive traits resulting from cellular adaptive plasticity. Genotypes with such traits would appear to be of ecological significance. UV adapted plants are better suited for exploitation in production of useful products through Marker Assisted Breeding/tissue culture/hairy root culture practices as well as modern biotechnological means.

Traits of interest in the context of plant survival under varying UV would include:

(a) short term inhibition of growth to allow time for repair of UV induced macromolecular damages that occur during post harvest storage of 'dry' orthodox seed – manifested as photomorphogenic effect evident in delayed hypocotyl extension (Ballare et al. 1995). Such (temporal) inhibition in growth rate on the basis of time required to repair damage incurred in embryonic axis is also evident in germination related delay in embryo emergence (Ganguli et al. 1992). This represents UV adaptive trait variation observed in different genotypes. In addition to such morphological characteristics accumulation of epicuticular waxes also represents photomorphogenic traits (Gonzalez et al. 1996; Kakani et al. 2003)

(b) long term plant protection conferred by accumulation of protective compounds that confer UV acclimation in plants through variation of UV fluence (Brown et al. 2005).

7.2 UV Adaptive Trait

Although high-intensity UV in sunlit sky induces cell molecular damage (Niklas 1992; Casal et al. 1994), the same radiation (particularly UV-B) at low dose (caused due to transmission of the radiation through temporal variation of weather-related aerosol/cloud cover) in turn epigenetically upregulates enzymes for synthesis of UV protective compounds. This confers acclimation potential to plants in open field (Mazza et al. 2000; Dixon et al. 2001) through variation in ambient (UV) environment. Such rapid change in cellular metabolism for upregulation of protective compounds, provides: (a) internal UV screening compounds viz. flavonoids for restricting UV entry into cells and (b) non-enzymatic antioxidants (by virtue of dual role of flavonoids) for cell molecular protection from UV induced oxidative damage and also (c) upregulation of DNA repair enzymes viz. photolyase. This is enabled by adaptive plasticity, a feature specifically operative in plants, reflecting a cellular retaliation in plants to restrain/counter UV related adverse effects. Such processes confer UV acclimation, providing fitness advantage (Thompson 1991; Bradshaw 1965; Levins 1968; Dudley and Schmitt 1996) to plants manifested as UV adaptive traits relating to survival potential in life forms. UV radiation may thus be seen as an UV survival trait-selecting agent. UV adaptive traits are difficult to select particularly in microorganisms; detection of such traits are complicated by the fact that (a) many organisms continually modify their phenotype in response to rapidly changing local UV environmental conditions, and (b) an UV-induced character state may rapidly disappear in noninductive UV environments.

The flexibility of UV adaptive plants to cope with environmental (UV) variation through interconversion (enabled by cellular plasticity) between primary and secondary metabolism determines plants' ability to adjust in altered UV environment. UV adaptive plants are thus of evolutionary significance. UV adaptive traits thus participate in useful plant developmental processes of ecological and evolutionary significance (Schmitt and Wulff 1993; Dudley and Schmitt 1996). Various levels of UV-absorbing compounds, viz., phenolics, in different ecotypes are among the factors that may contribute to the observed ecotypic variation in UV-B tolerance (Hidema et al. 1996; Sato et al. 1997). Exhibiting epigenetic effects that are not imprinted in genomes, UV adaptive genotypes are at liberty to undertake need-based modification for survival under variations of environment; these genotypes (with UV adaptive trait) thus exhibit higher evolutionary potential. This would help in selection of appropriate need-based cultivars (Teranishi et al. 2004) for cultivation at different locations.

Transgenic and mutant plants are considered powerful tools for testing adaptive significance of phenotypic plasticity (Van Kleunen and Fischer 2005); for understanding the limits of phenotypic plasticity, it is critical to understand the mechanism of plasticity-based variation in plants (Via and Lande 1985; Smith 1990, Van Tienderen and Van der Toorn 1992). At the genetic level cellular plasticity-related regulatory control can reflect allelic sensitivity (Schlichting and Pigliucci 1993); this entails environmental variation-related epigenetic upregulation of genes in the pathway for synthesis of (UV) protective compounds, viz., flavonoids, thereby developing UV adaptive traits. Such regulatory control modulating environment-related transcript involves epistatic modulation of development by one or more regulatory genes whose function exhibits sensitivity to specific environmental (including UV radiation) cues, thereby generating environmentally suitable phenotype (Smith 2000; Rollo 1994; Arnone and Davidson 1997; Purugganan 2000).

A commonly studied UV adaptive trait is manifestation of reduced stem elongation/ inhibited hypocotyl (during the process of germination of seeds) that is associated with temporary diversion of primary metabolism (exemplifying UV adaptive trait) to secondary metabolism for timely enhancement in synthesis of UV protective compounds, viz., flavonoids, in plants (Barnes et al. 1990; Sullivan et al. 1992; Ballare' et al. 1995; Dixon et al. 2001). This will be of greater significance for survival under high intensity UV radiation at locations of sparse plant population where UV radiation freely fluences over the plant population.

Need based oscillation between metabolic pathways involving macromolecular rearrangements in cells represents manifestation of the unique feature of 'developmental plasticity' particularly operative in plant cells – the feature involving a temporal trade off between primary and secondary metabolism (Caretto et al. 2015). Developmental plasticity in cells during temporal enhancement in UV fluence is related to enhanced synthesis of phenolics through epigenetic upregulation of the constitutive trait for flavonoid biosynthesis towards need based (UV adaptive) trait development for restraining UV penetration into cells (Perkins et al. 2005). Variation in UV adaptive trait has been reported in plants of different species growing under variations of natural UV-B radiation (Tevini and Teramura 1989; Caldwell et al. 1989; Day et al. 1992). In studies along an altitudinal gradient, Filella and Peñuelas (1999) specifically demonstrated variation in trait as a manifestation of gene X (UV) environment: the concentration of UV-B radiation absorbing pigments was 15% higher in sun leaves of *Quercus ilex* at 1200 m a.s.l than in those from 200 m a.s.l. altitude. Sajirani et al. (2014) have reported development of enhanced flavonoid based internal UV screen in cells for restriction of UV entry thereby conferring higher UV tolerance in plants at high altitude. A multivariate analysis by Hofmann and Jahufer (2011) demonstrates explicitly that a trade-off between flavonoid accumulation and plant growth/productivity attributes constitutes an appropriate mechanism manifesting adaptive plasticity. It is also important to appreciate that among dense population of plants, not requiring enhanced internal protection (because of plant density related low (solar) UV radiation fluencing at the region) the status of metabolism would retain primary metabolism that is associated with plant productivity (Fig. 7.1). This phenomenon exemplifies a situation where enhancement of secondary metabolism for enhancing flavonoid content may render the plant maladaptive and undesirable with respect to energy cost. From studies on a large number of species involving different plant types, Poorter and Villar (1997) have reported that synthesis of phenolic uses a lot of energy (on average 2.6 g glucose/g phenolic). Under the circumstance a large (unnecessary) investment in phenolic compounds would adversely affect the plant's growth and vitality, since biosynthesis of flavonoids (in plants) occurs at the cost of primary metabolism. The authors also concluded that interspecific variation in cost can be considerable if reports on individual species are compared.

7.3 UV-Adapted Traits

During prolonged persistence under high UV intensity, the epigenetic change (induced by temporal low dose UV fluencing through cloud cover) in genome (allowing upregulation of genes related to UV acclimation manifested as UV adaptive traits) in the course of evolution would likely become imprinted in the genome. This confers UV adapted/constitutive trait development in plants, thereby manifesting "fixed" traits for survival under variable ambient UV environment (Schmitt et al. 1995; Kingsolver

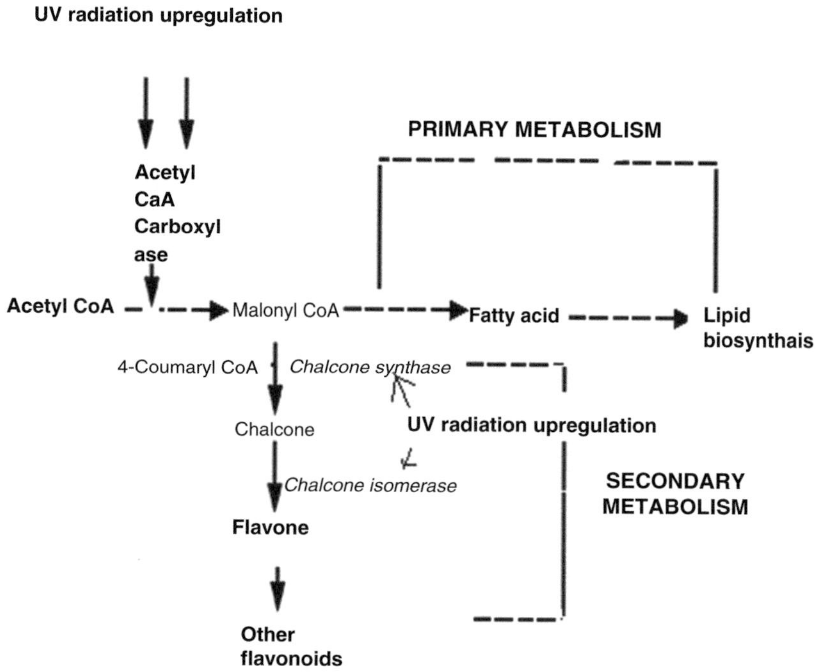

Fig. 7.1 Interrelation between primary and secondary metabolic pathways in field grown plants

1995; Dudley and Schmitt 1996; Agrawal et al. 1999). Such "ecologically specialized" genotypes with UV-adapted traits exhibit fixed (or canalized) trait expression, remaining neutral across (UV) environmental changes (Weinig et al. 2004). Data from extensive and intensive field experiments under tropical field conditions conducted during an international assessment of the impacts of UV-B radiation on rice productivity indicate that rice yields likely will not be affected by increases in UV-B radiation (Dai et al. 1997) indicating presence of UV-adapted traits. Correia et al. (1999), Torabinejad and Caldwell (2000), Kalbina and Strid (2006), and Talai and Sen Mandi (2010) have reported variation in UV adapted traits among ecotypes within a given species of plants. Cell molecular mechanism leading to genetic imprints evidently occurring in UV adapted genotypes and exhibiting canalized (fixed) gene expression, as well as plastic/UV adaptive genotypes exhibiting epigenetic mechanism would be equally expected to coexist in the same field (Turck and Coupland 2014). Jansen et al. (2010) have reported that at high altitude, both highly tolerant and moderately tolerant *Arabidopsis thaliana* populations coexist. Variation in UV-B tolerance among the different ecotypes of *Arabidopsis* and the apparent control of at least some tolerance traits by additive gene effects should contribute to survival of this species in natural populations exposed to substantially elevated UV-B radiation.

Development of in depth understanding on conditions leading to UV-adapted/constitutive traits and associated gene expression would be a worthwhile endeavor. A recent short-term study by Jansen et al. (2010) assessed the natural variation in constitutive (inherent) and induced protection from UV induced damage of photosynthetic function using 224 lines (varieties) of *Arabidopsis thaliana* The authors found a continuum of constitutive UV protection levels, with some accessions being UV sensitive and others UV tolerant. Correlation between acclimative/adaptive response to UV-B radiation and habitat origin has also been reported for other plants – (Yao et al. 2007 in buckwheat); the authors by principal component analysis of intraspecific responses of tartary buckwheat to UV-B radiation under field conditions reported that the sensitivity of plants to UV-B radiation is not only associated with the ambient UV-B level in natural habitats but also with inherent growth rate reflecting rate of primary metabolism vis-a-vis secondary metabolism and other factors. Sullivan et al. (1992) have studied 33 different species growing in the same field (i.e., under same UV radiation); the data obtained with respect to UV effect showed significant reduction in plant height in 14 species and biomass reduction in 8 species; biomass was also noted to increase in four species. Such data indicate variation in UV radiation at location of prolonged of the species. Such data establish that UV induced oscillation between primary and secondary metabolism in plants under UV fluence through clear/cloud covered sky as affected by UV fluence at the site of origin would be effective in capacity building in plants for survival during prolonged UV persistence under enhanced UV fluence. In terms of natural selection, production and accumulation of flavonoid under intermittent spells of low dose UV under high intensity UV radiation is important: under low UV intensity natural selection would favor genotypes with flavonoids accumulated to an extent that would not involve high energy cost for production of secondary metabolites.

Thus, most plants native to low elevations (i.e., under low UV fluence) exhibit sensitivity to UV-B, while plants from higher elevations (where UV-B is high) generally manifest enhanced UV-B tolerance (Teramura and Sullivan 1994) regardless of UV ambience under locational variation. Gene expression in such plants reflect genomic imprinting of UV adaptive traits thereby conferring UV tolerance in plants under widely variable location, thus representing UV-adapted plants; these plants would not enter into upregulation of flavonoids in the changed location since this would be unnecessary and would entail unnecessary energy loss.

UV-adapted genotypes are better suited (compared to genotypes with UV adaptive genotypes) for exploitation of desirable UV-adapted beneficial traits such as antioxidants, naturally produced in plants. UV adapted plants may be exploited for incorporation of their (desirable) UV adapted/UV-fixed traits (i.e., epigenetic trait imprinted in genome) into otherwise desirable character bearing plants through tissue culture technology (Alvero-Bascos and Ungson 2012)/planned breeding (Caldwell et al. 1995)/ biotechnological intervention including transgenic approach (Ramachandra Rao and Ravishankar 2002).

7.4 Breeding for Exploiting Benefits of UV-Adapted Traits

Marker assisted breeding for incorporation of UV adapted traits manifesting genomic imprints of UV adaptive epigenetic traits (Surani 1998; Martienssen and Richards 1995) would be a worthwhile endeavor. Observing significant genotypic variation in flavonoid glycoside quercetin among white clover populations Hofmann and Jahufer (2011) has drawn attention to the potential for using this compound in breeding programmes targeted at improving stress resistance.

Torabinejad and Caldwell (2000) have reported inheritance of UV-B tolerance in seven ecotypes of *Arabidopsis thaliana*. Their studies demonstrated that UV-B-tolerant *Arabidopsis* pure lines can be developed for UV tolerance trait. Findings of the authors suggest that such breeding efforts may also be useful for developing UV tolerant/UV-adapted crop plants.

Studies of Hidema and Kumagai (2006) on breeding for variation in CPD photolyase activity (a crucial factor identifying difference in UV-B sensitivity) demonstrate that rice cultivars with one- or two base substitution in the CPD photolyase gene exhibit altered activity of the enzyme and associated resistance of the plant to UV-B radiation. This indicates the possibility of increasing resistance of rice plants to UV-B radiation by selective breeding between plants of contrasting CPD photolyase activity. Teramura et al. (1991) from their experiments on rice cultivars suggest that selective breeding might be successfully used to develop UV-B-tolerant rice cultivars (also see Sect. 7.5 – a case study).

Several rice ecotypes of Bengal (located in subtropical region) growing in different regions are noted to produce seed of dissimilar vigour–viability trait (Talai and SenMandi 2010). Such data assayed as seed vigour status manifested as a resultant of high UV-dose-induced macromolecular damage and damage controlling cellular mechanism (under low-dose UV radiation) establish that seed vigour–viability traits in orthodox (rice) seed represent UV-adapted trait, the extent of adaptivity varying with the genotype (Figs. 6.3 and 6.4); seed vigour–viability in orthodox seed represents a trait that confers survival value of post-harvest orthodox seed (8 % mc) under ambient (UV) storage; this determines survival potential of the genotype. Incorporation of high vigour trait into varieties with otherwise desirable traits through the process of breeding (preferably marker assisted breeding) would thus constitute useful endeavor under enhancing UV radiation worldwide.

Transcriptome analysis related genetic screening (Jenkins 2014) would be useful in selecting parents in breeding programs for development of trait variation in the progeny. A heterologous expression approach would be useful for a high-throughput characterization of gene functions in agricultural plant species. In this context Sakurai et al. (2011) have developed a system named FOX hunting (full-length cDNA overexpressor gene hunting) using full-length cDNAs (fl-cDNAs). The RiceFOX has useful search functions, and it is possible to refer to the introduced rice full-length cDNA (fl-cDNA) in addition to the morphological features of a broad range of expressive traits and also non-expressive traits.

As different from genetic traits, epigenetic traits were for long thought to cause temporary changes occurring under location-based environmental variation; these traits (unless imprinted in genomes) were thought to be effective in one generation where the change had occurred. More recently, however, traits associated with DNA methylation related epigenetic effects have been found to be heritable and persisting through breeding lines (Tsaftaris and Polidoros 2000; Talai and Sen Mandi 2010). Ghosh and Sen Mandi (unpublished data) have demonstrated that seed vigour–viability trait associated ACCase gene, if methylated at the promoter region, confers low vigour trait (Fig. 6.14) in rice seed. Whether this feature could be utilized for long-term plant adaptation for varietally controlled survival under UV environmental challenges would constitute interesting study. Bearing in mind that varietal low vigour trait (as different from age related low vigour manifestation during seedling growth) though exhibiting slow germination rate exhibit a 'make up' effect during later stages of growth thereby exhibiting productivity comparable to plants from high vigour seeds (Ganguli et al. 1992). It is suggested that the low germination rate of low vigour varieties (Fig. 6.4) represents survival strategy, representing UV adaptive trait under challenges of high UV ambience whereby seedlings before emergence would undertake flavonoid biosynthesis (cf Ballare et al. 1995).

Use of modern technology, in transcriptome analysis and marker assisted breeding, for developing understanding in the area of UV radiation-caused damage and protection through experiment-based discussion in the context of a paradigm shift in photobiology together with linkages between fundamental plant biology related plant quality outcome (viz., UV adaptation/UV-induced value addition in traits) through exploiting natural phenomenon would repay careful investigation. The challenge now is to develop understanding on cell molecular mechanisms for manifestation and persistence of desirable traits in plants. This would also help to select genotypes with beneficial genes and incorporate beneficial traits (e.g., of seed vigour–viability-related crop stand establishment and thus crop productivity) into otherwise elite varieties through marker assisted breeding; use of biotechnology undertaking a transgenic approach would also be a worthwhile endeavor. It is important to note that despite bearing traits of desirable quality, cultivated crops tend to be more sensitive to UV-B than wild plant species (Weinig et al. 2004); this is presumably because during breeding, UV-tolerant traits are often lost in "trade-off" between commercial value and biodiversity survival trait (regardless of UV tolerance trait in crop plants). Thus, being able to introgress high UV adaptive vigour traits (that would ensure crop stand establishment) from wild type plants into cultivated crop species for conferring survival of elite varieties developed through breeding would constitute useful endeavor. Against a backdrop of knowledge development on UV effect in plants and associated modification of trait to cope with the predicament of UV enhancement in the atmosphere, there is a need for harnessing the beneficial traits (developed through gene X UV environment) by incorporating the beneficial traits into otherwise desirable varieties through marker assisted breeding.

In the present age of progressively increasing UV radiation, attempts at incorporating UV-adapted trait into otherwise desirable plants would constitute worthwhile endeavor for ensuring better survival under enhanced UV radiation in "post-ozone hole era." That adaptive plasticity may lead to persisting UV adapted traits through breeding involving genomic imprinted characteristics is evidenced when long term genetic/epigenetic traits may pass through breeding lines (Tsoftraris and Polidorus 2000). Such mechanisms have greatly influenced the evolution of terrestrial plants (Rozema et al. 1997). In the study presented below, for exploring the possibility of incorporating the UV-adapted trait of high seed vigour (manifested in seed of some varieties) into otherwise desirable varieties, cross was conducted between a high and low seed vigour varieties. Planned breeding programs involving landraces/varieties of rice (*Oryza sativa*) exhibiting variation in sensitivity to UV in expression of vigour–viability would constitute a useful means of incorporating high seed vigour trait into otherwise desirable variety.

Case Study

7.4.1 Breeding Rice Varieties Toward Introgression of UV-Adapted Seed Vigour Trait Through Marker-Assisted Breeding

From studies on 22 cultivars of rice from diverse origins, Barnes et al. (1993) concluded that in rice, as in other grasses, shoot morphology may be more responsive to solar UV-B change than plant productivity indicating that plant productivity more than shoot morphology (both being outcome of primary metabolism) represents UV adaptive trait. Seed vigour reflected as shoot growth at early times of germination that determines crop stand establishment in the field is directly correlated with crop productivity. This important trait, central to conferring plant survival/seed viability (as a function of post-harvest seed survival through post-harvest dry storage) and crop productivity, was found to be species specific. Seeds produced through long-term cultivation schedule under different locations and exhibiting appreciably similar seed vigour status at fresh harvest indicate a UV-adapted feature of this trait. Studying the effect of vigour–viability of sowing material (i.e., stored orthodox rice seed) where under ambient condition in farmers' seed storage rice seed with UV transparent husk faces the predicament of fluencing UV radiation, more so in the tropical region where this experiment was conducted bearing in mind that loss of seed vigour–viability quality in farmers' ambient storage is of particular concern; this experiment was conducted with an aim toward translational biology. Provided that crop stand establishment is successful, crop productivity will not be affected. Ganguli et al. (1992) demonstrated that provided a seedling is established in the field, successful crop stand and thus crop productivity will not exhibit appreciable reduction. Bearing in mind that adverse effect (of UV-B) on rice production could have profound impact on the world's staple food supply, exploring whether seed vigour, a UV adaptive trait, could be manipulated through breeding programs would provide useful information.

A genetic analysis on the UV-B sensitivity of rice, using progeny from a cross between UV resistant (Sasanishiki) and UV sensitive (Norin-1) strains, showed that UV-B sensitivity is controlled by more than two major recessive genes (Sato et al. 1994). While information on QTL contribution to trait development provides information of interest, this is not useful for planning precise (marker-assisted) breeding strategies. For such information to be useful, it is important to identify the major gene tightly linked with the trait of interest. Identification of gene function and associated trait is important not only for basic research but also for translational biology, especially with regard to improvement of traits associated with plant survival/crop productivity in farmer's field. In this context, identification of ACCase as the major gene tightly linked with high seed vigour–viability trait (Talai and Sen Mandi 2010) provides useful information.

This section presents preliminary studies on (a) introgression of ACCase associated DNA marker (associated with high vigour character in seeds) through hybridization between high vigour and low vigour parents and (b) germination performance of the F1 seeds resulting from the cross. The experimental material used in this experiment consisted of contrasting varieties exhibiting high and low seed vigour–viability trait. Criteria for selection of parents for the breeding experiment were (a) growth characteristics (Fig. 6.3) and (b) RAPD analysis studies demonstrating the presence of a DNA marker associated with high seed vigour trait and its absence in low seed vigour varieties (Figs. 6.8 and 6.9). Variety IET10890 was selected as high vigour parent (marked variety "B" in Fig. 6.3). This variety also exhibited the high vigour-associated marker (900 bp) in RAPD analysis of genome (Figs. 6.8 and 6.9). IET13158 was selected as low vigour parent due to its low vigour-related morphology and growth features (variety "E" in Fig. 6.3); this variety did not contain high vigour-associated marker (900 bp) in RAPD analysis (Figs. 6.8 and 6.9).

Direct and reciprocal crosses were carried out between the selected high vigour (IET 10890) and low vigour (IET13158) parents. RAPD analysis of both parents and hybrid progeny of each cross is shown in Fig. 7.2a.

The gel pattern exhibits a 900 bp DNA band (marker) in high vigour parent IET10890 (lane

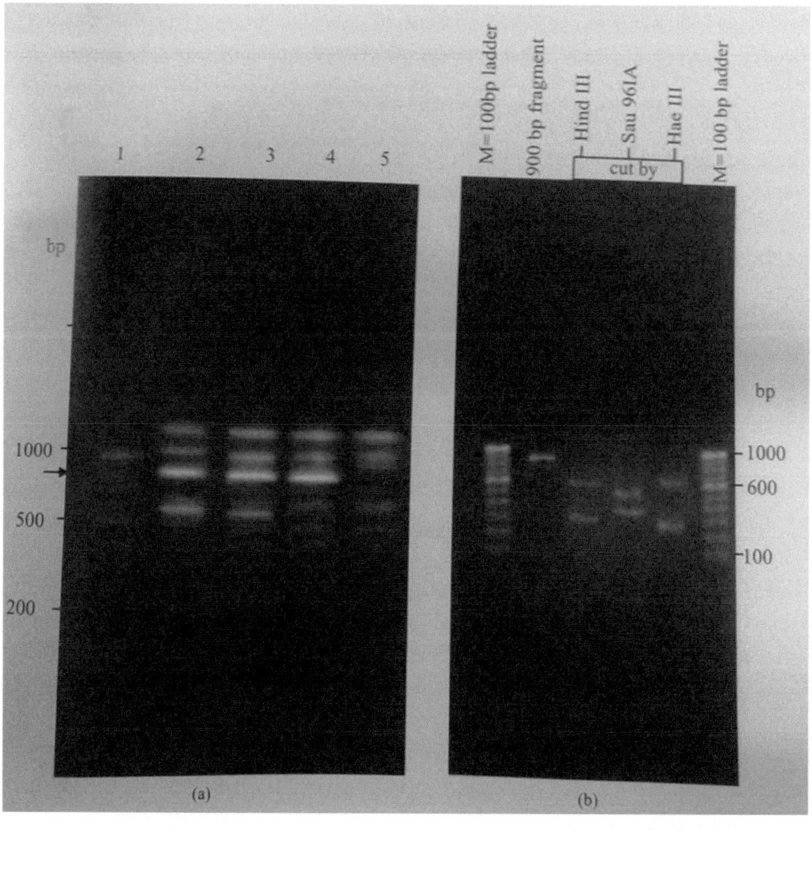

Fig. 7.2 (**a**) Genome analysis by RAPD of parents (IET 10890 and IET 13158) and F1 hybrids obtained by PCR amplification of genomic DNA using primer OPAC 19 in 1.2 % agarose gel. Lane *1*, 100 bp ladder; lane *2*, high vigour parent (IET 10890); lane *3*, F1 seed hybrid: (♂ IET 13158 X ♀ IET 10890); lane *4*, F1 seed of reciprocal cross hybrid: (♂ IET 10890 X ♀ IET 13158); lane *5*, low vigour parent (IET 13158). (**b**) Intact and restriction enzyme (Hind III, Sau 961 A, and Hae III) digested fragments of the 900 bp DNA exhibiting authenticity of the introgressed DNA fragment observed in hybrid genomes, run on 1.5 % agarose gel

2 of Fig. 7.2a) and also in the hybrids obtained through direct and reciprocal crosses (lane 3 and 4 of Fig. 7.2a). The low vigour parent (IET13158) lacks this 900 bp marker (lane 5 of Fig. 7.2a). RE cut pattern of this 900 bp marker in F1 hybrid using Hind III and Hae III demonstrated that this 900 bp band (marker) was the same as that in high vigour parent variety (Fig. 7.2b cf. Fig. 6.9). The molecular authenticity of the 900 bp marker present in high vigour parent and hybrid varieties established by RE analysis of the 900 bp band are shown in Fig. 7.2b. Thus, the experiments depicted in Fig. 7.2a, b demonstrate introgression of the high vigour-associated DNA marker into the hybrid varieties. Germination performance of the F1 hybrid seed is given in Table. 7.1.

Germination performance (expressed as percent germination) of parents and F1 hybrid seeds both in direct and reciprocal crosses is shown in Table 7.1. The hybrid seeds, derived from both direct and reciprocal crosses, manifest the - UV-adapted (high vigour) trait, thereby establishing that inheritance is genetic and that the trait does not exhibit cytoplasmic inheritance. It is important to note that the vigour trait is particularly relevant for early germination times (up to 48 h) within which the seed emerges from within seed cover and field soil cover. Once the seed has emerged into open air, the embryonic axis is no more constrained for growth and thus plant productivity (cf. Ganguli et al. 1992). Data on germination performance of the hybrid seeds establishes that the high vigour trait has been introgressed into the low vigour genetic background. Data on Table 7.1 confirms introgression of the 900 bp marker associated with vigour trait (feature of the high vigour parent) into plants grown from F1 hybrid seed.

Successful introgression of the 900 bp (i.e., high vigour–viability associated marker) fragment into the hybrids establishes the feasibility of incorporating high vigour–viability trait into F1 hybrids. This data suggests the possibility of crop improvement by incorporating high vigour trait (for improved crop stand establishment) into otherwise desirable varieties with basic low seed vigour (and thus low crop stand establishment) trait.

This molecular mechanism should serve to ensure persistence of the seed vigour and thus plant survival under unfavorable environmental conditions. The introgression of rice vigour–viability trait, controlled through epigenetic mechanism, through breeding lines suggests the transmission of genomic imprinting such as promoter methylation as observed in this study (see Chap. 6). Similar finding is also reported in *E. coli* where DNA methylation-induced gene turn on and off expression is inherited for many generations (Lim and Oudenaarden 2007). The fact that epigenetic traits in addition to genetic traits may constitute a relevant entity in the process of breeding (Tsaftaris and Polidoros 2000) suggests that exploring UV-adapted/UV-benefited traits for use in plant improvement programs would repay careful investigation. From their studies on plant improvement through breeding focusing on tolerance to UV radiation, Torabinejad and Caldwell (2000) suggest that both pure lines and hybrids of *Arabidopsis* can be developed for greater tolerance of UV-B radiation. Studies of Caldwell et al. (1995) reporting on intraspecific variation in responsiveness to

Table 7.1 Germination percentage of high and low vigour (parents) rice varieties and their F1 progeny for 120 h at 28 ± 2 °C. Means of three replicates with standard errors are given

	Growth percentage (G%) at different hour of germination				
Genotypes	24 h	48 h	72 h	96 h	120 h
1	39.5 ± 0.96	97.88 ± 1.12	100	100	100
2	10 ± 1.28	65.42 ± 1.69	96.52 ± 1.67	99.24 ± 0.018	99.24 ± 0.018
3	44.86 ± 1.01	97.01 ± 1.62	100	100	100
4	36.12 ± 1.04	98.45 ± 0.84	100	100	100

Row 1 = IET 10890, high vigour parent; row 2 = IET 13158, low vigour parent; row 3 = ♂ IET 10890 X ♀ IET 13158; row 4 = ♂ IET 13158 X ♀ IET 10890

UV-B suggest that breeding for UV-B tolerance should be feasible in terrestrial plants.

Developing desirable UV tolerance traits through breeding/biotechnological involvements for incorporation of UV-adapted/UV-fixed epigenetic trait would provide wide scope for utilizing cell molecular variation. Enabled by genotypic plasticity, different varieties are utilized for plant improvement. That use of biochemical and morphological features for plant survival could provide basis for developing appropriate strategies for plant improvement under natural UV radiation has also been reported by Barthod et al. (2007). Genetic modification of cultivars to suit possible enhancement in natural UV resulting from depletion of ozone layer would help to provide desirable cultivars for sustenance in future.

7.5 Biotechnological Means for Production of UV-Tolerated Plant

Modern biotechnology provides a powerful tool in achieving plant improvement through (a) marker assisted breeding of plants (Se'venier et al. 2002) and (b) development of UV tolerant plant through transgenic approach. For maintaining steady production or even enhanced production of desirable compounds in genotypes producing low amount of the same, biotechnological approach would constitute a useful proposition. In the past couple of decades, a lot of genes associated with production of enzymes responsible for secondary metabolite production have been transformed for overexpression either in the original plant or in other plant species for achieving enhanced production of the compound of interest.

Biotechnological tools are useful for the multiplication and genetic enhancement of medicinal potential in plants (Abdin and Kamaluddin 2006). Secondary metabolites are known to play a major role in the adaptation of plants to their environment (Chap. 4) while also providing an important source of pharmaceuticals (Ramachandra Rao and Ravishankar 2002). Of these, flavonoids and terpenoids constitute the most important natural products for use as antioxidant-based nutraceuticals and pharmaceuticals for humans. Development of transgenic plants overexpressing UV-upregulated secondary metabolites (viz. flavonoid and terpenoid) and thereby utilizing the positive effects of UV radiation (Muir et al. 2001; Mahmoud and Croteau 2001; Deavours and Dixon 2005) constitutes a useful endeavor toward exploiting UV-adapted beneficial traits.

Some examples of transgenic plant overexpressing secondary metabolites such as flavonoids and terpenoids of human benefit are described in the following sections:

7.5.1 Alfalfa

Deavours and Dixon (2005) developed transgenic alfalfa (*Medicago sativa*) which expresses isoflavone synthase (IFS – isoflavone synthase catalyzes the first committed step of isoflavone biosynthesis, a branch of the phenylpropanoid pathway) producing genistein glucoside as end product. Genistein glucoside was thus constitutively produced by genetic engineering approach in alfalfa (*Medicago sativa*) leaves by inserting overexpressed isoflavone synthase gene (MtIFS1) from *Medicago truncatula*. Although MtIFS1 was highly expressed in all organs, genistein was found to be accumulated in leaves. MtIFS1-expressing lines accumulated several additional isoflavones, including formononetin and daidzein, in response to UV-B.

7.5.2 Tomato

Many transgenic lines of tomato that produces high quantities of flavonoids/terpenoids that provide valuable natural products for use of humans have been generated on tomato. Chalcone isomerase (CHI), an early enzyme of the phenylpropanoid pathway, is the key enzyme for enhanced production of flavonol (Muir et al. 2001). Overexpression of the *Petunia CHI* gene causes 78-fold enhancement of flavonoid production in the tomato peel. After generating such tomatoes, a 21-fold increase of flavonols in tomato paste was reported in comparison to non-transgenic lines (Verpoorte and Memelink 2002). In tomatoes, overexpression of an *S*-linalool synthase (gene responsible for linalool synthesis through terpenoid biosynthetic pathway) causes many fold enhancement in the production of the monoterpenoid flavor compound *S*-linalool in comparison to control plants, although there is no change in the levels of other terpenoids (Wang et al. 2001).

Further, β-carotene (an antioxidant that provides protection from antioxidative damage) content was increased up to threefold by overexpression of a bacterial phytoene desaturase in tomato plastids. In this process, however, the total carotenoid content, including the direct product of the enzyme lycopene, was decreased (Romer et al. 2000). Thus, while several of the carotenoid enzymes were upregulated, the decrease of total carotenoid content was probably due to feedback inhibition somewhere in the carotenoid biosynthesis pathway. Another biotechnological approach for enhancement of carotenoid biosynthesis in tomato involved the incorporation of a bacterial phytoene synthase in tomato fruits; this causes twofold to fourfold increase in the amount of total carotenoids. No concomitant reduction in levels of other plastidial isoprenoids as well as activities of other enzymes in the pathway occurred (Fraser et al. 2002).

7.5.3 Rice

As rice plants do not contain provitamin A in their endosperm, recombinant technologies are the only way for production of provitamin A (β-carotene). The introduction of β-carotene biosynthesis pathway through recombinant technology into this major staple food, i.e., rice, by overexpression of phytoene synthase, phytoene desaturase, and lycopene β-cyclase (these genes are required for production of carotenoids through carotenoid biosynthetic pathway) is an important achievement (Ye et al. 2000) as vitamin A deficiency causes symptoms ranging from night blindness to those of xerophthalmia and keratomalacia, leading to total blindness, more prevalent in Southeast Asia. Such achievement of introducing vitamin into rice that constitutes staple food for the large population of the low-income group people is indeed commendable.

Isoflavonoids are derived from a flavonone intermediate, naringenin, that is ubiquitously present in plants and play a critical role in plant development and defense response. Rice like other cereals do not produce isoflavones in natural condition. Synthesis of isoflavones causes enhancement in its nutritional value and also facilitates entering into a symbiotic relationship with rhizobia. In rice, the key enzyme responsible for conversion of flavonoids to isoflavonoids is the isoflavone synthase (IFS). With an aim to develop variety having the ability to induce nodulation (nod) genes in rhizobia, the isoflavone synthetase (IFS) gene from soybean was incorporated into rice (*Oryza sativa* L. cv. Murasaki R86).

7.5.4 Artemisia

Farnesyl diphosphate synthase produces precursors for the biosynthesis of essential isoprenoids like carotenoids, withanolides, ubiquinones, dolichols, and sterols and also helps in farnesylation and geranylation of proteins. Overexpression of a chimeric farnesyl diphosphate synthase gene in *Artemisia annua* was reported to increase the flux in the sesquiterpenoid biosynthetic pathway resulting in a twofold to threefold increase. This enhanced production is used for the synthesis of antimalaria drug (Lewinshon et al. 2001) and also acts as molecular target of bisphosphonates for drug development against tumors as well as human pathogens (Dhar et al. 2013).

7.5.5 Mint

Overexpression of the gene-encoding deoxyxylulose phosphate reductoisomerase (*DXR* – which is the product of plastidial methylerythritol 4-phosphate (MEP) pathway) in mint resulted in twofold to fourfold higher DXR activities (required for monoterpenoid production) in mint (Mahmoud and Croteau 2001). The transgenic plants produced had a normal phenotype with almost 50 % increase in essential oil (monoterpenoid) content. Increasing menthol levels in mint oil is reported by blocking one of the competitive branches in the monoterpenoid metabolic pathway which produces menthofuran (Mahmoud and Croteau 2001).

Bibliography

Abdin MZ, Kamaluddin (2006) Improving quality of medicinal herbs through physiochemical and molecular approaches. In: Abdin MZ, Abrol YP (eds) Traditional systems of medicine. Narosa Publishing House Pvt. Ltd, New Delhi, pp 30–33

Agrawal AA (1999) Induced responses to herbivory in wild radish: effects on several herbivores and plant fitness. Ecology 53:1713–1723

Agrawal AA, Strauss SY, Stout MJ (1999) Costs of induced responses and tolerance to herbivory in male and female fitness components of wild radish. Evolution 53:1093–1104

Allen DJ, Nogue's S, Morison JIL, Greenslade PD, McLeod AR, Baker NR (1999) A 30% increase in UV-B has no impact on photosynthesis in well-watered and droughted pea plants in the field. Glob Chang Biol 5(2):235–244

Alpert P, Simms EL (2002) The relative advantages of plasticity and fixity in different environments: when is it good for a plant to adjust? Evol Ecol 16:285–297

Alvero-Bascos EM, Ungson LB (2012) Ultraviolet-B (UV-B) radiation as an elicitor of flavonoid production in callus cultures of Jatropha (Jatropha curcas L.). Philipp Agric Sci 95(4):335–343

Arnone ML, Davidson EH (1997) The hard-wiring of development: organization and function of genomic regulatory systems. Development 124:1851–1864

Ballare' CL, Barnes PW, Flint SD, Price S (1995) Inhibition of hypocotyls elongation by ultraviolet-B radiation in de-etiolating tomato seedlings. II. Time-course, comparison with flavonoid responses and adaptive significance. Physiol Plant 83:593–601

Barnes PW, Flint SD, Caldwell MM (1990) Morphological responses of crop and weed species of different growth forms to ultra-violet-B radiation. Am J Bot 77:1354–1360

Barnes PW, Maggard S, Holman SR, Vergara B (1993) Intraspecific variation in sensitivity to UV-B radiation in rice. Crop Sci 33:1041–1046

Barthod S, Cerovic Z, Epron D (2007) Can dual chlorophyll fluorescence excitation be used to assess the variation in the content of UV-absorbing phenolic compounds in leaves of temperate tree species along a light gradient? J Exp Bot 58:1753–1760

Bieza K, Lois R (2001) An Arabidopsis mutant tolerant to lethal ultra-violet B levels shows constitutively elevated accumulation of flavonoids and other phenolics. Plant Physiol 126:1105–1115

Biggs RH, Kossuth SV, Teramura AH (1981) Response of 19 cultivars of soybeans to ultraviolet-B irradiance Physiol. Plant 53:19–26

Björn LO, Callaghan TV, Johnsen I (1997) The effects of UV-B radiation on European heathland species. Plant Ecol 128:252–264

Blumthaler M, Amback W (1990) Indication of increasing solar UV-B radiation flux in alpine regions. Science 248:206–208

Bradshaw AD (1965) Evolutionary significance of phenotypic plasticity. Genetics 13:115–155

Brosche M, Strid A (2003) Molecular events following perception of ultraviolet-B radiation by plants. Physiol Plant 117:1–10

Brown BA, Jenkins GI (2008) UV-B signaling pathways with different fluence-rate response profiles are distinguished in mature *Arabidopsis* leaf tissue by requirement for UVR8, HY5, and HYH. Plant Physiol 146:576–588

Brown BA, Cloix C, Jiang GH, Kaiserli E, Herzyk P, Kliebenstein DJ et al (2005) A UV-B-specific signaling component orchestrates plant UV protection. Proc Natl Acad Sci U S A 102:18225–18230

Caldwell MM, Teramura AH, Tevini M (1989) The changing solar ultraviolet climate and the ecological consequences for higher plants. Trends Ecol Evol 4:363–367

Caldwell MM, Bjorn LO, Bomman JF, Flint SD, Kulandaivelu G, Teramura M, Tevini M (1994) Effect of increased solar ultraviolet radiation on terrestrial ecosystem. J Photochem Photobiol 46 (5):40–52

Caldwell MM, Teramura AH, Tevini M, Bornman JF, Bjorn LO, Kulandaivellu G (1995) Effects of increased solar ultraviolet radiation on terrestrial plants. Ambio 24:166–173

Caretto S, Linsalata V, Colella G, Mita G, Lattanzio V (2015) Carbon fluxes between primary metabolism and phenolic pathway in plant tissues under stress. Int J Mol Sci 16(11):26378–26394

Casal JJ, Ballare CL, Tourn M, Sanchez RA (1994) Anatomy, growth and survival of a long-hypocotyl mutant of Culcuinis satihus deficient in phytochrome B. Ann Bot 73:569–575

Chen L, Niu K, Wu Y, Geng Y, Mi Z, Flynn DFB, He J-S (2013) UV radiation is the primary factor driving the variation in leaf phenolics across Chinese grasslands. Ecol Evol 3(14):4696–4710

Correia CM, Areal ELV, Torres-Pereira MS, Torres-Pereira JMG (1999) Intraspecific variation in sensitivity to ultraviolet-B radiation in maize grown under field conditions. I. Growth and morphological aspects. Field Crop Res 59:81–89

Dai Q, Peng S, Chavez AQ, Miranda MML, Vergara BS, Olszyk DM (1997) Supplemental Ultraviolet-B

radiation does not reduce growth or grain yield in rice. Results from a 7-season field study. Agro J 89:793–799
Day TA, Vogelmann TC, DeLucia EH (1992) Are some plant life forms more effective than others in screening out ultraviolet-B radiation? Oecologia 92:513–519
Deavours BE, Dixon RA (2005) Metabolic engineering of isoflavonoid biosynthesis in Alfalfa. Plant Physiol 138:2245–2259
Dhar MK, Koul A, Kaul S (2013) Farnesyl pyrophosphate synthase: a key enzyme in isoprenoid biosynthetic pathway and potential molecular target for drug development. New Biotechnol 30(2):114–123
Dixon P, Weinig C, Schmitt J (2001) Susceptibility to UV damage in impatiens capensis (Balsaminaceae): testing for opportunity costs to shade-avoidance and population differentiation. Am J Bot 88:1401–1408
Dubest S, Gallego ME, White CI (2002) Role of the AtRad1p endonuclease in homologous recombination in plants. EMBO Rep 3:1049–1054
Dudley SA, Schmitt J (1996) Testing the adaptive plasticity hypothesis: density dependent selection on manipulated stem length in Impatiens capensis. Am Nat 147:445–465
Filella I, Peñuelas J (1999) Altitudinal differences in UV absorbance, UV reflectance and related morphological traits of *Quercus ilex* and *Rhododendron ferrugineum* in the Mediterranean region. Plant Ecol 145:157–165
Fraser PD, Ro¨mer S, Shipton CA, Mills PB, Kiano JW, Misawa N, Drake RG, Schuch W, Bramley PM (2002) Evaluation of transgenic tomato plants expressing an additional phytoene synthase in a fruit-specific manner. Proc Natl Acad Sci U S A 99:1092–1097
Frohnmeyer H, Loyall L, Blatt MR, Grabov A (1999) Millisecond UV-B irradiation evokes prolonged elevation of cytosolic-free Ca^{2+} and stimulates gene expression in transgenic parsley cell cultures. Plant J 20:109–117
Ganguli S, Das G, Ganguli S, Das G, Sen-Mandi (1992) Plant emergence and productivity of different vigour of wheat (Triticum aestivum) seeds. Indian J Agr Sci 62(3):224–227
Ghosh S, Sen Mandi S. Methylation status of ACCase promoter affects seed vigour-viability trait in Oryza sativa L. varieties. Unpublished data
Gonzalez R, Paul ND, Percy K, Ambrose M, McLaughlin CK, Barnes JD, Areses M, Wellburn AR (1996) Responses to ultraviolet-B radiation (280–315 nm) of pea (Pisum sativum) lines differing in leaf surface wax. Physiol Plant 98:852–860
Hectors K, Prinsen E, De Coen W, Jansen MA, Guisez Y (2007) Arabidopsis thaliana plants acclimated to low dose rates of ultraviolet B radiation show specific changes in morphology and gene expression in the absence of stress symptoms. New Phytol 175:255–270
Hidema J, Kumagai T (2006) Sensitivity of rice to ultraviolet-B radiation. Ann Bot 97:933–942
Hidema J, Kang H-S, Kumagai T (1996) Differences in the sensitivity to UV-B radiation of two cultivars of rice (Oryza sativa L.). Plant Cell Physiol 37:742–747
Hofmann RW, Jahufer MZZ (2011) Tradeoff between biomass and flavonoid accumulation in white clover reflects contrasting plant strategies. PLoS One 6(4): e18949. doi:10.1371/journal.pone.0018949
Hutchinson F (1987) A review of some topics concerning mutagenesis by ultraviolet light. Photochem Photobiol 45:897–903
Jansen MAK, Coffey AM, Prinsen E (2010) UV-B induced morphogenesis. Plant Signal Behav 7 (9):1185–1187
Jenkins GI (2014) Structure and function of the UV-B photoreceptor UVR8. Curr Opin Struct Biol 29:52–57
Kakani VG, Reddy KR, Zhao D, Mohammed AR (2003) Effects of ultraviolet-B radiation on cotton (*Gossypium hirsutum* L.) morphology and anatomy. Ann Bot 91:817–826
Kalbina I, Strid A (2006) Supplementary ultraviolet-B radiation reveals differences in stress responses between Arabidopsis thaliana ecotypes. Plant Cell Environ 29:754–763
Kingsolver JG (1995) Viability selection on seasonally polyphenic traits: wing melanin pattern in western white butterflies. Evolution 49:932–941
Kliebenstein DJ, Lim JE, Landry LG, Robert L (2002) Arabidopsis UVR8 regulates ultraviolet-B signal transduction and tolerance and contains sequence similarity to human regulator of chromatin condensation I. Plant Physiol 130(1):234–243
Kumagai T, Sato T (1992) Inhibitory effects of increase in near-UV radiation on the growth of Japanese rice cultivars (*Oryza sativa* L.) in a phytotron and recovery by exposure to visible radiation. Jpn J Breed 42:545–552
Levins R (1968) Evolution in changing environments. Princeton University Press, Princeton
Lewinsohn E, Schalechet F, Wilkinson J, Matsui K, Tadmor Y, Kyoung-Hee Nam K-H et al (2001) Enhanced levels of the aroma and flavor compound S-Linalool by metabolic engineering of the terpenoid pathway in tomato fruits. Plant Physiol 127:1256–1265
Lim H, Oudenaarden AV (2007) A multistep epigenetic switch enables the stable inheritance of DNA methylation states. Nat Genet 39(2):260–275
Lively CM (1986) Canalization verses developmental conversion in a spatially variable environment. Am Nat 128:561–572
Lively CM (1998) Developmental strategies in spatially variable environments: barnacle shell dimorphism and strategic models of selection. In: Tollrian J, Harvell CD (eds) The evolution of inducible defenses. Princeton University Press, Princeton, pp 245–258
Lucas R, McMichael T, Smith W, Armstrong B (2006) Solar ultraviolet radiation. Global burden of disease from solar ultraviolet radiation. World Health Organization, Environmental Burden of Disease Series, 13
Mahmoud SS, Croteau RB (2001) Metabolic engineering of essential oil yield and composition in mint by altering expression of deoxyxylulose phosphate reductoisomerase and menthofuran synthase. Proc Natl Acad Sci U S A 98:8915–8920

Martienssen RA, Richards EJ (1995) DNA methylation in eukaryotes. Curr Opin Genet Dev 5:234–242

Mazza CA, Boccalandro HE, Giordano CV, Battista D, Scopel AL, Ballare CL (2000) Functional significance and induction by solar radiation of ultraviolet- absorbing sunscreens in field-grown soybean crops. Plant Physiol 122:117–125

Muir SR, Collins GJ, Robinson S, Hughes S, Bovy S, De Vos CH, van Tunen AJ, Verhoeyen ME (2001) Overexpression of petunia chalcone isomerase in tomato results in fruit containing increased levels of flavonols. Nat Biotechnol 19:470–474

Niklas KJ (1992) Plant biomechanics-an engineering approach to plant form and function. University of Chicago Press, Chicago

Perkins MC, Roberts CJ, Briggs D, Davies MC, Friedmann A, Hart CA, Bell GA (2005) Surface morphology and chemistry of Prunus laurocerasus L. leaves: a study using X-ray photoelectron spectroscopy, time-of-flight secondary-ion mass spectrometry, atomic-force microscopy and scanning-electron microscopy. Planta 221:123–134

Poorter H, Villar R (1997) The fate of acquired carbon in plants: chemical composition and construction costs. In: Bazzaz FA, Grace J (eds) Plant resource allocation. Academic, San Diego, pp 39–72

Purugganan MD (2000) The molecular population genetics of regulatory genes. Mol Ecol 9:1451–1461

Ramachandra Rao S, Ravishankar GA (2002) Plant cell cultures: chemical factories of secondary metabolites. Biotechnol Adv 20:101–153

Rollo CD (1994) Phenotypes: their epigenetics, ecology and evolution. Chapman and Hall, New York

Romer S, Fraser PD, Kiano JW, Shipton CA, Misawa N, Schuch W, Bramley PM (2000) Elevation of the provitamin A content of transgenic tomato plants. Nat Biotechnol 18:666–669

Rozema J, Staaij J, Bjorn LO, Caldwell M (1997) UV-B as an environmental factor in plant life: stress and regulation. Trends Ecol Evol 12:22–28

Sajirani EB, Hadian J, Abdossi V, Larijani K (2014) Evaluation content of flavonoids and anthocyanins in Iranian borage (*Echium amoenum* fisch & mey) subjected in eshkevari accessions affected by different habitats in North of Iran. J Biodivers Environ Sci (JBES) 4(2):364–368

Sakurai T, Kondou Y, Akiyama K, Kurotani A, Higuchi M (2011) RiceFOX: a database of Arabidopsis mutant lines overexpressing rice full-length cDNA that contains a wide range of trait information to facilitate analysis of gene function. Plant Cell Physiol 52 (2):265–273

Sato T, Kang H-S, Kumagai T (1994) Genetic study of resistance to inhibitory effects of UV radiation in rice (*Oryza sativa* L.). Physiol Plant 91:234–238

Sato T, Ueda T, Fukuta Y, Kumagai T, Yano M (1997) Mapping of quantitative trait loci associated with ultraviolet-B resistance in rice (*Oryza sativa* L.). Theor Appl Genet 107:1003–1008

Schlichting CD, Pigliucci M (1993) Evolution of phenotypic plasticity via regulatory genes. Am Nat 142:3660370

Schmitt J, Wulff RD (1993) Light spectral quality, phytochrome and plant competition. Trends Ecol Evol 8:47–50

Schmitt J, McCormac AC, Smith H (1995) A test of the adaptive plasticity hypothesis using transgenic and mutant plants disabled in phytochrome-mediated elongation responses to neighbors. Am Nat 146:937–953

Sévenier R, van der Meer IM, Bino R, Koops AJ (2002) Increased production of nutriments by genetically engineered crops. J Am Coll Nutr 21 (3 Suppl):199S–204S

Shyam Choudhury S, Sen Mandi S (2012) Natural ultra violet radialtion on field grown rice (*Oryza sativa* L.) plants confer protection against oxidative stress in seed during storage under subtropical ambience. Environ Pollut 1(2):32–39

Smith H (1990) Signal perception, differential expression within multigene families and the molecular basis of phenotypic plasticity. Plant Cell Environ 13:585–594

Smith H (2000) Phytochromes and light signal perception by plants – an emerging synthesis. Nature 407:585–591

Sullivan JH, Teramura AH, Ziska LH (1992) Variation in UV-B sensitivity in plants from a 3000m elevational gradient in Hawaii. Am J Bot 79:737–743

Surani MA (1998) Imprinting and the initiation of gene silencing in the germ line. Cell 93:309–312

Talai S (2010) Development of bio-molecular markers for vigour in rice seeds, Ph.D thesis, submitted to University of Calcutta

Talai S, Sen-Mandi S (2010) Seed vigour-related DNA marker in rice shows homology with Acetyl CoA Carboxylase gene. Acta Physiol Plant 32:153–167

Teramura AH, Sullivan JH (1991) Effects of UV-B radiation on photosynthesis and growth of terrestrial plants. Photosynth Res 39:463–473

Teramura AH, Sullivan JH (1992) Potential impacts of increased solar UV-B on global plant productivity. In: Riklis E (ed) Photobiology. Plenum Press, New York, pp 625–634

Teramura AH, Sullivan JH (1994) Effects of UV-B radiation on photosynthesis and growth of terrestrial plants. Photosynth Res 39:463–473

Teramura AH, Ziska LH, Sztein AE (1991) Changes in growth and photosynthetic capacity of rice with increased UV-B radiation. Physiol Plant 83:373–380

Teranishi M, Iwamatsu Y, Hidema J, Kumagai T (2004) Ultraviolet-B sensitivities in Japanese lowland rice cultivars: cyclobutane pyrimidine dimer photolyase activity and gene mutation. Plant Cell Physiol 45:1845–1856

Tevini M, Teramura AH (1989) UV-B effects on terrestrial plants. Photochem Photobiol 50:479–487

Thompson JD (1991) Phenotypic plasticity as a component of evolutionary change. Trends Ecol Evol 6:246–249

Torabinejad J, Caldwell MM (2000) Inheritance of UV-B tolerance in seven ecotypes of *Arabidopsis thaliana* L. Heynh. and their F_1 hybrids. J Hered 91:228–233

Tsaftaris AS, Polidoros A (2000) DNA methylation and plant breeding. In: Janick J (ed) Plant breeding reviews, vol 18. Wiley, New York, pp 18: 87–176. ISBN 0-471-35567-4

Turck F, Coupland G (2014) Natural variation in epigenetic gene regulation and its effects on plant developmental traits. Evolution 68(3):620–631

Ulm R, Baumann A, Oravecz A, Mate Z, Adam E, Oakeley J, Schafer E, Nagy F (2004) Genome-wide analysis of gene expression reveals HY5 function in the UV-B response of *Arabidopsis*. Proc Natl Acad Sci U S A 101:1397–1402

Van Kleunen M, Fischer M (2005) Constraints on the evolution of adaptive phenotypic plasticity in plants. New Phytol 166:49–60

Van Tienderen P (1991) Evolution of generalists and specialists in spatially heterogeneous environments. Evolution 45:1317–1331

Van Tienderen PH, Van der Toorn J (1992) Genetic differentiation between populations of Plantago lanceolata. I. Local adaptation in three contrasting habitats. J Ecol 79:27–42

Verpoorte R, Memelink J (2002) Engineering secondary metabolite production in plants. Curr Opin Biotechnol 13(2):181–187

Via S, Lande R (1985) Genotype-environment interaction and the evolution of phenotypic plasticity. Evolution 39:5055522

Wang E, Wang R, DeParasis J, Loughrin JH, Gan S, Wagner GJ (2001) Suppression of a P450 hydroxylase gene in plant trichome glands enhances natural-product-based aphid resistance. Nat Biotechnol 19:371–374

Weinig C, Gravuer KA, Kane NC, Schmitt J (2004) Testing adaptive plasticity to UV: costs and benefits of stem elongation and light-induced phenolics. Evolution 58(12):2645–2656

Yao Y, Xuan Z, He Y, Lutts S, Korpelainen H, Li C (2007) Principal component analysis of intraspecific responses of tartary buckwheat to UV-B radiation under field conditions. Environ Exp Bot 61:237–245

Ye X, Al-Babili S, Klöti A, Zhang J, Lucca P, Beyer P, Potrykus I (2000) Engineering the Provitamin A (-β-carotene) biosynthetic pathway into (Carotenoid-Free) rice endosperm. Science 287:303–305

Conclusion

UV radiation of low wavelength (viz UV-B and UV-A) and consequently high energy is generally known to inflict abiotic stress in living forms; plants with stationary habit are particularly subject to this predicament. For survival under the brunt of UV radiation, plants have obligatorily developed an alternative system of self-protection involving molecular mechanism-based trait variation thereby avoiding stress symptoms. The UV fraction (comprising 0.5% of solar UV radiation) fluencing on earth through the UV restraining (stratospheric) ozone layer consists of UV-B (5–10%) and UV-A (90%); the UV-C fraction of solar radiation remains totally restrained by the stratospheric ozone layer. Before the development of a protective ozone layer in the stratospheric region of earth's atmosphere, the then small population of simple life forms remained under oceanic water protection in aqueous habitat. With the gradual accumulation of oxygen given out by living organisms pioneered by Cyanobacteria in Archaean era and subsequently carried out by green plants a protective layer of ozone was formed within the stratosphere by reaction between molecular oxygen and atomic oxygen at the equatorial region. After formation at the equatorial region, ozone drifts with wind flow toward the poles, forming a protective ozone layer over the entire earth's surface being thickest at the poles and subsequently accumulated in reduced thickness at mid- and lower latitude. Formation of the UV protective ozone layer enabled the migration of organisms from their UV protective aqueous habitat to terrestrial surface (in Ordovician era) making a ground breaking event in earth's evolutionary history. Since then an equilibrium was established between fluencing UV radiation through stratospheric ozone layer and acclimation strategies in plants developed through macromolecular rearrangements as a manifestation of cellular plasticity uniquely operative in plants. Early research for developing rapid understanding on UV effects reports on UV induced damage in plants come from studies conducted in green house/growth chamber where light conditions do not simulate natural light that consists of interaction between UV and other wave bands including the Photosynthetically Active Radiation (PAR); consequently, effect of UV on plants in closed chamber often produces exaggerated effects compared to the plants growing in open field under solar radiation. With gradual thinning of the stratospheric ozone hole over the Antarctic and later over the Arctic and subsequently in the mid- and lower latitudes caused ozone thinning based enhancement in UV fluence over these regions through the breach in ozone layer. This was caused by enhanced accumulation of "greenhouse gases" (as an outcome of adverse anthropogenic activity) causing global warming-associated stratosphere cooling in the Antarctic (the coolest continent) resulting in disruption of the earlier established equilibrium between UV radiation and persistence of life forms on earth.

S. Sen Mandi, *Natural UV Radiation in Enhancing Survival Value and Quality of Plants*,
DOI 10.1007/978-81-322-2767-0

This stimulated urgent research for (i) understanding cellular processes associated with UV caused molecular damage and acclimation strategies enabling plants to survive under variations (locational / temporal) of UV fluence on earth and (ii) setting up of regulatory bodies such as Kyoto Protocol and Montreal Protocol for restricting emission of harmful gases into the atmosphere including greenhouse gases that originated through adverse anthropogenic activity into the atmosphere.

Every ecosystem such as terrestrial and aquatic (marine and freshwater) is affected by the deleterious effects of ultraviolet radiation, viz., UV-B and UV-A, as a resultant of gene X UV environment mediated cell functioning. Realization of the threat from enhancing UV radiation studies on damaging and cell protective mechanisms brought research on UV effects to high priority.

Studies reveal that UV-B radiation adversely affects cell molecular integrity either directly or indirectly by ROS: UV-B by direct impinge causes damage to macromolecules such as DNA, proteins, and lipids; indirectly, UV-B through various species of ROS adversely affects cell molecular mechanisms. As different from UV-B, UV-A radiation causes direct damage only to DNA; UV-A-induced damage to other macromolecules is mediated by reactive oxygen species, particularly singlet oxygen, through photosensitizers. Persistence of life on earth under the (commonly believed) UV damaging potential is enabled by evolutionary development of UV acclimatory cell molecular processes. This in turn is epigenetically upregulated via signal transduction for enhanced synthesis of UV protective compounds, viz., flavonoids. Flavonoids by virtue of their unique structural specialization perform a dual role: (a) forming internal (in plant cells) UV screens and (b) functioning as antioxidative compounds naturally in plant cells and through the dietary process in animals. This is a unique example of genetically influenced all-pervading protection of life forms on earth under UV radiation.

UV-induced upregulation of protective compounds in plants that temper the high-intensity UV radiation-induced damaging effects and thereby facilitate plants to survive in their stationary lifestyle in open field is brought about by low-dose UV radiation (resulting from transmission through aerosol/cloud cover). Short-term location-based oscillatory variation in intensity of UV fluence induces molecular rearrangements between primary and secondary metabolic pathways that would determine extent of synthesis of protective compounds. This is enabled by the unique feature of cellular plasticity, specifically operative in plants. This phenomenon is manifested as a trade-off between growth/biomass development involving primary metabolism and survival trait involving enhanced secondary biosynthesis of secondary metabolites functioning as protective compounds, viz., flavonoids. Studies over the years on different crop plants/cash crops and also other wild/cultivated plants have revealed that plants, by exploiting their unique feature of plasticity, may lead to trait variation to ensure survival and associated value-added quality development.

During prolonged existence under high UV radiation as in high altitude, UV adaptive traits often become imprinted on genomes and manifest as UV-adapted traits. Such traits confer long-term UV tolerance to plants under variation of ambient UV fluence. UV adaptive traits exhibit flexibility in trait variation for survival under variation of ambient fluence and thus exhibit evolutionary advantage. UV-adapted traits exhibit stable traits making them amenable to exploitation through breeding/biotechnology-based transgenic approach for developing desirable traits in otherwise elite varieties.

MIX
Papier aus verantwortungsvollen Quellen
Paper from responsible sources
FSC® C105338

If you have any concerns about our products,
you can contact us on
ProductSafety@springernature.com

In case Publisher is established outside the EU,
the EU authorized representative is:
Springer Nature Customer Service Center GmbH
Europaplatz 3, 69115 Heidelberg, Germany

Printed by Libri Plureos GmbH
in Hamburg, Germany